PROJECT MANAGEMENT

FOR

ENGINEERING AND TECHNOLOGY

David L. Goetsch

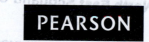

Boston   Columbus   Indianapolis   New York   San Francisco   Upper Saddle River
Amsterdam   Cape Town   Dubai   London   Madrid   Milan   Munich   Paris   Montreal   Toronto
Delhi   Mexico City   São Paulo   Sydney   Hong Kong   Seoul   Singapore   Taipei   Tokyo

Editorial Director: Vernon R. Anthony
Senior Acquisitions Editor: Lindsey Gill
Editorial Assistant: Nancy Kesterson
Director of Marketing: David Gesell
Senior Marketing Coordinator: Alicia Wozniak
Marketing Assistant: Les Roberts
Program Manager: Maren L. Beckman
Project Manager: Janet Portisch
Procurement Specialist: Deidra M. Skahill
Art Director: Jayne Conte
Cover Designer: Karen Salzbach
Cover Image: © Dusit/Shutterstock
Manager, Rights and Permissions: Mike Lackey
Media Director: Leslie Brado
Lead Media Project Manager: April Cleland
Full-Service Project Management: Jogender Taneja, Aptara®, Inc.
Composition: Aptara®, Inc.
Printer/Binder: Courier/Westford
Cover Printer: Lehigh/Phoenix Color Hagerstown
Text Font: 10/12 ITC Garamond Std

Credits and acknowledgments borrowed from other sources and reproduced, with permission, in this textbook appear on the appropriate page within text.

Library of Congress Cataloging-in-Publication Data

Goetsch, David L.
   Project management for engineering & technology/by David L. Goetsch.
      pages cm
   ISBN-13: 978-0-13-281640-3
   ISBN-10: 0-13-281640-7
   1. Project management.   I. Title. II. Title: Project management for engineering and technology.
   T56.8.G63 2014
   658.4'04—dc23   2013042178

10 9 8 7 6 5 4 3 2 1

PEARSON

ISBN 10:   0-13-281640-7
ISBN 13: 978-0-13-281640-3

BRIEF CONTENTS

CONTENTS

PART THREE   Project Management—People Functions

PREFACE

BACKGROUND

In the fields of engineering and technology, efficient, effective project management is critical. All engineering and technology projects—design, manufacturing, quality improvement, process development—share common goals: The projects are to be completed on time, within budget, and according to specifications. These goals cannot be achieved without effective project management. Engineering and technology projects range from the design and manufacture of the largest jetliner to the smallest circuit board, and projects must be well managed if they are to be completed successfully.

This unrelenting demand to complete projects on time, within budget, and according to specifications has created a pressing need for specialized education and training for those who manage engineering and technology projects. Project managers must know how to manage processes and lead people. The process aspects of project management include: cost estimation, planning/scheduling, procurement, risk management, monitoring, and closeout. The people aspects of project management include: leadership, motivation, communication, and efficient/effective management of time, change, diversity, and adversity. Project management has become a specialized field within the broad fields of engineering and technology, a specialized field requiring specialized instructions in both the process and people aspects of the job.

WHY IS THIS BOOK WRITTEN AND FOR WHOM?

This book is written to fulfill the need for a comprehensive, up-to-date, practical teaching resource that focuses on helping engineering and technology students become effective project managers. This book is developed in accordance with specifications contained in *A Guide to the Project Management Body of Knowledge (PMBOK Guide)* maintained by the Project Management Institute (PMI), Pennsylvania. It provides comprehensive coverage of both aspects of project management—process management and leading people—specifically from the perspective of engineering and technology projects. Educators and students in engineering and technology disciplines will benefit from the material presented herein. The direct, straightforward presentation of material focuses on making the principles of project management practical, understandable, and useful for students. Up-to-date research has been integrated throughout the text along with real-world activities and cases.

ORGANIZATION OF THE BOOK

The text contains 18 chapters organized in three parts. Part Two covers all of the process skills needed by project managers. Part Three covers all of the people skills needed by project managers. The chapters are presented in an order that is compatible with the typical organization of a course in project management, and a standard chapter format is maintained throughout the book. In addition to text, photos, and illustrations, each chapter contains a list of chapter topics, summary, key terms and concepts, review questions, and practical application activities. Every chapter contains a case study of a challenging engineering project that illustrates for students how complex the projects they might work on can be and why effective project management is so critical.

DOWNLOAD INSTRUCTOR RESOURCES FROM THE INSTRUCTOR RESOURCE CENTER

Supplementary teaching and learning materials are provided online. These materials include a PowerPoint presentation covering all chapters in the book, a comprehensive test bank, and an Instructor's Manual. To access supplementary materials online, instructors need to request an instructor access code. Go to www.pearsonhighered.com/irc to register for an instructor access code. Within 48 hours of registering, you will receive a confirming e-mail including an instructor access code. Once you have received your code, locate your text in the online catalog and click on the Instructor Resources button on the left side of the catalog product page. Select a supplement, and a login page will appear. Once you have logged in, you can access instructor material for all Pearson textbooks. If you have any difficulties accessing the site or downloading a supplement, please contact Customer Service at http://247pearsoned.custhelp.com/.

HOW THIS BOOK DIFFERS FROM OTHERS

The approach taken in this book is the result of more than 100 interviews with project managers, students, and professors in engineering and technology disciplines. Through these interviews the author learned that most textbooks on project management take a generic approach in an attempt to reach the broadest possible market. Consequently, this text focuses solely on project management as it relates to engineering and technology so that all text, illustrations, cases, and activities can be specific to engineering and technology and so that concepts can be treated in greater depth than is possible in a generic text.

ABOUT THE AUTHOR

David L. Goetsch is Emeritus Vice President and Professor at Northwest Florida State College. Prior to entering higher education full time, Dr. Goetsch had a career in the private sector that included project management positions in engineering and manufacturing settings. He served as a project manager in an engineering and manufacturing firm that designed, manufactured, and assembled the components for nuclear reactors. He now serves on the board of directors of Fort Walton Machining, Inc. in Fort Walton Beach, Florida. Dr. Goetsch has been selected as Professor of the Year at Northwest Florida State College and the University of West Florida, Florida's Outstanding Technical Instructor of the Year, and was also the recipient of the U.S. Secretary of Education Award for having the Outstanding Technical Program in the United States in 1984 (Region 10).

CHAPTER 1

Overview of Project Management

Much of the work of engineering and technology firms consists of projects completed by teams of individuals using specific resources and processes. Assume that an engineering and technology firm wins a contract to design and develop a communication system for a new aircraft, a special seat for a luxury automobile, or a software package for a new application. All of these contracts would become the basis for projects. Assume that a firm is asked to develop a prototype for a new landing gear system for a jet aircraft or set up a new process for manufacturing printed circuit boards for the next generation of computers. Once again, these assignments would become projects.

Projects in engineering and technology firms are completed by project teams. Project teams are led by project managers who are responsible for ensuring that their projects are completed on time, within budget, and according to specifications. Consequently, engineering and technology professionals should be prepared to manage projects and lead project teams. Becoming an effective project manager requires the development of specific process and people skills. Preparing engineering and technology students and professionals to be effective project managers is the purpose of this book. This preparation begins with a definition of project management as a concept.

PROJECT DEFINED

Groups of college students are sometimes required to work together on class projects. When this happens, the new group usually calls a meeting to: (1) select a group leader to coordinate the activities of individual members and to ensure that all of the work is completed properly and on time and (2) divide the work and assign it to different group members. In this example, the assignment from the professor is a project and the person selected to lead the group is the project manager.

The class project consists of multiple individual assignments that require people, resources, and processes to complete them. There is a definite start and end date for the project and specific grading criteria (success criteria). For example, assume that a professor divides his class into five groups and gives each group the assignment of solving a well-defined problem. To complete their assignments, the students in each group are to use such resources as computers, books, the Internet, calculators, paper, and, of course, human knowledge and skills. In developing their presentation, the students apply such processes as research, word processing, and public speaking. There is a definite date on which the assignment is made and a definite date on which the final presentation must be made. During the presentation, the professor applies specific success criteria in arriving at a grade.

All projects—whether in college classes or in engineering and technology firms—have these same characteristics. Hence, a project can be defined as follows:

> A project is a fully coordinated group of interdependent tasks that are completed by people using resources and processes. Projects have definite starting and ending dates and success criteria.

There are several important concepts in this definition of a project including the following: fully-coordinated interdependent tasks, people, resources, processes, starting date, ending date, and success criteria (see Figure 1.1). Project managers should understand all of these concepts and their significance.

Fully Coordinated Interdependent Tasks

Assume that you and several friends are driving in a car and have a flat tire. The group is in a hurry so it is important to get the tire changed as quickly as possible. To expedite the process, each individual in the group agrees to complete a different task. One individual might get the jack out of the trunk while another retrieves the spare tire. One might loosen the lug nuts while another stands by to take the flat tire off and put the spare tire on. The individual who took the lug nuts off is prepared to put them back on once the spare tire is in place. The individual who jacked the car up is prepared to let it down and put the jack and flat tire back in the trunk of the car.

All of the individual tasks that must be performed in order to change the flat tire are interdependent. This means that one task depends on another for its successful completion, and all of the individual tasks must be done in the proper order for the project to be successfully completed. For example, the flat tire cannot be taken off until the car has been jacked up and the lug nuts have been removed. Then, the car cannot be let down until the new tire has been put on and the lug nuts tightened. Projects in engineering and technology firms are like this example in that they involve a lot of different tasks to be performed—some

---

Project-Related Concepts

- Fully coordinated interdependent tasks
- People, processes, and resources
- Starting and ending dates
- Success criteria

FIGURE 1.1 Important project-related concepts.

simultaneously and some in a specific order. Coordinating all of these interdependent tasks and making sure they are performed in the right way and in the proper order is the job of the project manager.

People, Processes, and Resources

Returning to the example of changing the flat tire, the project—like all projects—required people, processes, and resources. People do the work required to complete the project. In doing the work of the project, people use processes and resources. In the flat tire example, the people riding in the car used such processes as jacking up the car, loosening the lug nuts, removing the flat tire, putting on the spare tire, and tightening the lug nuts. In applying these processes they used resources including a jack, a lug-nut wrench, a spare tire, and time. In addition, the people who did the work of changing the tire were, collectively, a resource. Resources are simply assets that are needed to complete a project. The human resource is an important resource in any project.

Starting and Ending Dates

Projects typically start once a contract has been awarded or shortly thereafter. They also have a definite ending date—a deadline by which all work on the project must be finalized. Ensuring that projects are completed on time is one of the most important responsibilities of project managers. Some contracts received by engineering and technology companies contain penalty clauses that are activated if the project is not completed on time and according to specifications. Specifications are the next project component. They specify the quality criteria.

Success Criteria

When engineering and technology firms receive a contract, it is accompanied by specifications. Specifications explain in detail how the project is supposed to turn out—what a successful project will look like. For example, the publisher that received the contract to produce this book needed to know certain things before it could proceed. It needed to know what kind of page layout was desired, the type and size of font for text and headings, if the book would be produced in color or black and white, page size, hardback or soft cover, and cover design to name just a few areas of concern. This information was provided in the form of specifications.

One of the challenges facing project managers and their teams is to complete projects not just on time and not just within budget, but also according to customer specifications. The specifications—along with the budget and all applicable deadlines—constitute the success criteria for a project. In other words, the success criteria for projects are as follows:

- Complete the project on time
- Complete the project within budget
- Complete the project according to specifications

Projects Are Process-Oriented

A Guide to the Project Management Body of Knowledge (PMBOK Guide) is recognized as the authoritative reference for practicing project managers. Published by the Project Management Institute, the *PMBOK Guide* makes the important point that the work of projects is completed through processes.[1] It describes the processes of project management in terms of three components:[2]

- Inputs (documents, plans, designs, specifications)
- Tools and techniques
- Outputs (documents, products, services)

The *PMBOK Guide* encompasses more than 40 different processes that fall into one of five broad process groups as follows:[3]

- Initiating
- Planning
- Executing
- Monitoring
- Closing

These five process groups encompass the major responsibilities of project managers and describe what they do. The work of project managers revolves around initiating, planning, executing, monitoring, and closing out projects. The knowledge needed to carry out the processes of project management falls into nine knowledge areas. In other words, the 40 plus processes recognized in the *PMBOK Guide* fall into five process groups and nine knowledge areas. These knowledge areas are as follows:[4]

- Integration management
- Scope management
- Time management
- Cost management
- Quality management
- Human resource management
- Communications management
- Risk management
- Procurement management

Figure 1.2 is a chart that shows how the five process groups and nine knowledge areas of project management are related to each other. It also shows some of the processes that must be completed during the course of a project and which process group/knowledge area each falls into. The five process groups of project management can be seen on the left side of the matrix while the knowledge areas read from left to right across the top. As an example of how to read the matrix, find the process group labeled *planning* and the knowledge area labeled *scope*. At the intersection of these two headings—a process group and a knowledge area—two processes are listed: (1) scope development and (2) work breakdown structure development. The various processes shown in Figure 1.2 represent the basics that students and professionals who want to be project managers must learn.

It is important for students of project management to understand processes as a concept, the three components of a process, and the specific processes that are used in completing projects. The most important and most widely used of these processes are explained in this book. The better project managers become at using these processes, the more effective they will be as project managers.

KNOWLEDGE AREAS

Process Groups	Integration	Scope	Time Management	Cost Management	Quality Management	HR Management	Communi-cation Management	Risk Management	Procurement Management
Initiating	• Project charter (Contract, drawings, and specifications)	–	–	–	–	–	• Identify stakeholders	–	–
Planning	• Project management plan	• Scope development • Work breakdown structure development	• Estimate time and duration of activities • Develop schedule	• Estimate costs • Establish budget	• Plan quality	• Develop HR plan	• Develop communi-cation plan	• Identify and analyze risks • Plan risk management	• Develop procurement plan
Executing	• Project execution	–	–	–	• Assure quality	• Establish build/lead project	• Communicate with all stakeholders regularly	–	• Procure needed resources
Monitoring/ Controlling	• Monitor, track progress, control • Adjust as changes occur	• Control scope	• Control schedule	• Control costs	• Control quality	• Monitor team performance	• Report on progress and performance	• Monitor and control risks	• Manage the procurement process
Closing	• Close	–	–	–	–	–	–	–	• Close procurements

FIGURE 1.2 Phases and elements of the project management process.

NEED FOR PROJECT MANAGERS

Engineering and technology firms receive contracts from other entities that need something designed, developed, tested, and/or produced. These contracts become projects that are undertaken by project teams. Engineering and technology firms also undertake internal projects on the basis of their own initiative to improve on some aspect of their performance or to add new capabilities. External and internal projects in engineering and technology firms are undertaken by project teams. A project team in an engineering and technology firm is like a symphony orchestra: It has a lot of different players—each with a specific instrument and role. Without a conductor to lead, coordinate, and facilitate, the symphony members are more likely to produce more noise than music. Like orchestras, project teams in engineering and technology firms are often cross-functional in nature. This means they are composed of individuals from various different functional units or departments.

A firm that provides both engineering and manufacturing services would be composed of several different departments or functional units including engineering/design, manufacturing, quality, purchasing, human resources, sales and marketing, and accounting (see Figure 1.3). Project teams in firms such as the one shown in this figure are often cross-functional in that their members come from all or at least several of the various departments. This is not always the case, but it often is.

Regardless of whether they are cross-functional or composed of individuals from just one department, project teams, like an orchestra, need a conductor who can meld the members into one coherent, mutually supportive, well-coordinated team and keep them on task and on time. That conductor is the project manager. Without project managers who can step across departmental boundaries and transform a disparate group of people into a well-coordinated, mutually supportive team, an engineering and technology firm will tend to operate as a collection of disjointed, disconnected, autonomous departments.

In a competitive business environment, engineering and technology firms excel by completing the projects they undertake on time, within budget, and according to specifications. Those that cannot meet these basic success criteria soon lose business to other firms

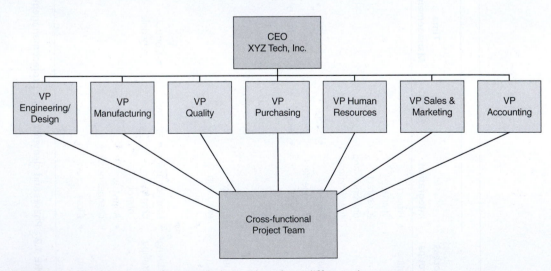

FIGURE 1.3 Project teams often get their members from different departments.

that can. Ensuring that a project is completed on time, within budget, and according to specifications is the job of the project manager. Without a conductor, orchestras are likely to make more noise than music. Without a project manager, project teams are likely to make more problems than progress.

Just as the best orchestra conductors know the music, the parts of all of the various players, and the capabilities of each individual player, the best project managers know their projects, the processes required to complete them, and the capabilities of the members of project teams.

Project Management Scenario 1.1

Why Do We Need a Project Manager?

Dale Cartwright is the Quality Director for ABC Technologies, Inc. He has been given the challenge to form a team for establishing performance benchmarks for all of ABC's manufacturing processes. Further, he has been asked to be the project manager. Cartwright agrees that the processes need to be benchmarked, but does not understand why the company needs to be so formal. "Why do we need a project manager? Why not just assign a process to each team member and let them get it benchmarked? These are all good people. They know what to do." Cartwright's supervisor tried to explain that there would be more to the project than just making work assignments, but Cartwright still did not seem to understand why there needed to be a project manager.

Discussion Question

In this scenario, Dale Cartwright has been assigned to serve as project manager for an internal project. But he thinks his company is going overboard by forming a project team and asking him to be the project manager. He does not understand the need for a project manager. If he asked you to explain why a project team needs a project manager, what would you tell him?

ESSENTIAL ELEMENTS OF A PROJECT

Regardless of size and complexity, all projects require the following: (1) charter/scope/plan, (2) schedule, (3) resources, and (4) leadership. These four essential components are all interrelated and interdependent. It is important for project managers to understand each of these components as well as how one can affect the other.

Project Charter/Scope/Plan

Every project undertaken by an engineering and technology firm begins with a need. For external projects—those originated by a customer—the customer needs a certain product or service. To secure the product or service it sends out a request for proposals (RFP) or a request for quotes (RFQ) to which engineering and technology firms respond. The response that best meets the needs of the customer is selected and a contract is awarded to an engineering and technology firm to provide the product or service. The contract gives the engineering and technology firm a *charter* to complete the project.

For internal projects—those originated by an engineering and technology firm to enhance its competitiveness in some way—the process is different. Higher management in the firm decides that an improvement of some type needs to be made. It develops a *charter* that describes the needed improvement and everything a project manager will need to lead a project ream in making the improvement. The project team undertakes the improvement project on the basis of the internally developed project charter.

The project—whether it is external or internal—has what is known as a *scope*. The scope of a project is the entirety of all work that must be accomplished to complete the project. In some instances—as with external projects—the scope is spelled out in the contract. With internal projects the scope is spelled out in the charter document(s) provided to the project manager. In any case, the scope of a project encompasses everything the members of the project team will need to do to complete the project on time, within budget, and according to specifications.

Regardless of whether the project in question is external or internal and regardless of how detailed the contract is, the scope of a project should be made clear for the project team through the development of a written *scope statement*. The scope statement for an engineering and technology project should answer the following questions:

- What result is the project team committing to (e.g., purpose of the project, expectations of the customer, objectives, deliverables, features and functions of the products and/or services)?
- What resources and other types of support does the project team need to fulfill its commitment (e.g., funds, people, time, facilities, technology)?
- What assumptions have been made by the project team's organization concerning the terms and conditions under which the work of the project will be completed by the project team (e.g., assumptions, restrictions, and constraints that apply)?

A practical and effective outline to use when writing a scope statement for an engineering and technology project is as follows (Figure 1.4):

- **Project overview.** In this element of the scope statement, the following types of information should be completed: (1) how the project came into being, (2) why the project is being undertaken, (3) scope of the work to be completed, (4) how the project might

---

Typical Outline for a Project Scope Statement

- Project overview
- Deliverables
- Features and functions descriptions
- Acceptance criteria
- Restrictions/constraints
- Uncertainties

FIGURE 1.4   This is a practical outline for a project scope statement.

affect other activities in the organization, and (5) how the project might be affected by other activities in the organization.

- ***Deliverables.*** In this element of the scope statement, the products and/or services the project team will produce are listed. Some project managers write this element as a list of objectives for the project (e.g., to produce the next generation of the XYZ radar system, etc.).
- ***Features and functions descriptions.*** In this element of the scope statement the required features and functions of the product or service to be delivered are described.
- ***Acceptance criteria.*** This element of the scope statement describes the process that will be used for customer acceptance of the product or service delivered and the criteria that will be used for determining what is or is not acceptable.
- ***Restrictions/constraints.*** This element of the scope statement describes all constraints that might restrict or inhibit what the project team is able to produce and when. For example, assume that the project team can produce the expected deliverables on time only if it is able to procure a certain material that is in worldwide shortage at the time. This constraint would be explained in the restrictions section of the scope statement. Common constraints on engineering and technology projects are time, cost/budget, personnel, material, equipment/technology, and quality expectations.
- ***Uncertainties.*** In any project there will be areas of uncertainty. For example, if the design of a certain product depends on the outcome of a specified test and the results of the test are uncertain, this issue would be explained. This element in the scope statement would also explain how the project team plans to deal with the uncertainty. This explanation might take the form of an "if-then" statement or several if-then statements.

The project charter is broader than just the project scope. The project charter is what empowers the firm to undertake the project in question. It provides all of the information needed by the project manager, including the scope of the project. With external projects, the project charter can take any one of several forms. For example, it is typically a package of project documents including the customer's contract, specifications, and the information contained in the firm's response to the customer's RFP or RFQ. The project scope statement can be developed on the basis of the customer's RFP/RFQ or using material contained in the contract, specifications, and RFP/RFQ response. It contains the more specific information shown above concerning the actual undertaking and proposed approach to completing the project.

The project plan or project management plan, as it is sometimes called, is a comprehensive package of smaller plans that documents how the project is going to be completed and how all aspects of the project are going to be managed over the course of the project. The project management plan typically consists of several smaller plans, including the scope statement, budget, schedule, quality plan, human resource plan, communications plan, risk management plan, and procurement plan.

Project Schedule

Every project has a beginning and an ending date. However, it is wise to also establish intermediate dates for the work of the project. For example, assume that a given task is to begin on January 5 and must be completed by June 1. The project manager should set intermediate target dates for when the task should be 25 percent, 50 percent, 75 percent, and 100 percent complete. A complete project schedule will show all of the activities that must be completed and the sequence in which they must be completed. It will also show the planned duration of each activity. Of course, few projects are linear in their makeup. Consequently, activities often overlap in terms of their starting and ending dates. There are a variety of scheduling tools available to help project managers with this element of the scope statement (Chapter Four).

Project Resources

In the context of project management, a resource is any asset necessary to complete a project on time, within budget, and according to specifications. The most common resources necessary for engineering and technology projects are people, money, time, technology, facilities, and materials. Time is such a critical resource for project managers that it is covered in detail in three chapters of this book. Chapter Four explains how to develop a schedule for an engineering and technology project. Chapter Ten explains how to monitor the schedule, and Chapter Twelve explains how project managers can manage their own time. All of the resources needed to complete a project on time, within budget, and according to specifications must be considered when developing a bid in response to an RFP or RFQ and in developing the project budget.

Leadership

One can manage budgets and control processes, but it is necessary to lead people. Project teams are made up of people, all of whom have their own personalities, agendas, motivations, and points of view. The essential ingredient in melding a disparate group of people into a well-coordinated, mutually supportive team whose members are committed to the mission is leadership. Leadership is the act of inspiring people to make a wholehearted commitment to the mission that brought them together. Leadership is so important a people skill for project managers that Chapter Twelve is devoted to this topic.

INTERNAL VERSUS EXTERNAL PROJECTS

Most projects in engineering and technology firms are initiated by a contract from a customer. After all, completing development, design, testing, and manufacturing projects is the purpose of engineering and technology firms. Customers of engineering and technology firms are other organizations that need a product designed, developed, and/or tested or a service provided. Engineering and technology firms bid to provide the needed product or service. The firm that wins the bid receives a contract and the contract is converted into a project. Projects that are initiated in this way are external projects because their source comes from outside of the firm. Projects can also be internally initiated.

Internal projects are initiatives undertaken by engineering and technology firms to enhance their competitiveness. In a competitive environment, engineering and technology firms sustain themselves in two ways: by providing superior value to their customers (superior value is a combination of superior quality, superior cost, and superior service); and by innovating (anticipating new and emerging market needs and being first or best in meeting them).

The most competitive engineering and technology firms self-initiate internal projects on a continual basis to enhance value. These projects involve continually improving the quality of products and services while simultaneously lowering their costs (think of the microcomputer chip). Competitive engineering and technology firms also self-initiate internal projects to develop innovative new products and services that will give them a competitive advantage in the marketplace, allow them to keep up with the competition, or help them catch up with market leaders that have pulled ahead of them. For example, a project to institute best practices such as Six Sigma, Lean Six Sigma, or a zero-tolerance safety program would be a self-initiated internal project.

Challenging Engineering and Technology Project

MANAGING THE NITROGEN CYCLE

While global warming is debated continually, another environmental phenomenon gets little attention but is an equally important issue. In fact, this particular issue—managing the nitrogen cycle—is one of the most important challenges facing engineering and technology professionals. The nitrogen cycle plays a critical role in the production of food. Interrupt or alter this cycle and it may be impossible to maintain a sustainable global food supply without sever damage to the environment.

The nitrogen cycle occurs naturally. Microbes that live in the soil use enzymes to convert nitrogen from the atmosphere into forms that plants can use. The process is known as *fixation*. Plants then turn the fixed nitrogen into organic nitrogen, a process that is essential to plants and the animals that eat the plants. The nitrogen cycle is completed when—through a process known as *denitrification*—nitrogen molecules are returned to the atmosphere. Historically, fixation was the only way for nitrogen to be pulled from the atmosphere for use by living organisms. However, human production of fertilizer and burning of fuels have begun to interrupt the natural nitrogen cycle. When nitrogen is removed from the air by human processes, environmental problems such as the greenhouse effect, protecting the ozone layer, smog, acid rain, and contaminated drinking water are worsened.

The challenge to engineering and technology professionals is to find ways to maintain a sustainable food supply worldwide without causing serious environmental problems by developing methods for remediating human disruption of the nitrogen cycle. These methods will have to include: 1) ways to increase denitrification, 2) ways to improve efficiency when applying fertilizer so that more of the nitrogen goes into the plants and less into the environment, and 3) better ways to recycle human waste.

Meeting these challenges will require creative thinking, innovation, and persistence on the part of engineering and technology professionals. It will also require effective project management. *Before proceeding with this chapter, stop here and consider how managing the nitrogen cycle will require the various process and people skills of project management.*

Source: Based on *National Academy of Engineering.* http://www.engineeringchallenges.org

Internal projects are often known as "loss-leader" projects. This means that the engineering and technology firm invests funds on the front end of internal projects in anticipation of earning them back over time. The "loss" in these cases should be temporary and "lead" to profits in the future. For example, an engineering and technology firm might invest in developing an innovative new product that leaders in the firm believe will open up a whole new market for them. The same firm might also undertake a variety of quality initiatives to make key processes more productive in the manufacture of existing products or delivery of existing services.

The same principles apply when managing internal projects as when managing external projects. Project managers need to be skilled at initiating, planning, executing, monitoring/controlling, and closing out projects as well as at building, leading, and motivating project teams. All of the process skills and people skills explained in this book apply to both external and internal projects. Project managers in engineering and technology firms should be prepared to lead both internal and external project teams.

SPEAKING THE LANGUAGE OF PROJECT MANAGEMENT

One of the most basic skills needed by project managers is speaking the language of project management. There is an extensive glossary of project management terms at the end of this book, and all of the terms defined in it are important. However, there are certain terms that are used so frequently in project management that they need to be understood before proceeding through the remaining chapters of this book. These terms include the following:

- *Project.* A fully coordinated group of interdependent tasks that are completed by people using resources and processes. Projects have definite starting and ending dates and success criteria.
- *Program.* A series of projects. For example, a college degree can be considered a program and the courses of which it is comprised can be considered projects. For another example, consider a software package called the XYZ Package. The overall package as it is updated over time might be referred to as the XYZ Program. Each time it needs to be upgraded to the next higher version, that version would be considered a project. All past, current, and future versions of the XYZ software make up the XYZ Program.
- *Goal.* The goal of a project is its overall purpose. The goal of an external project is defined by the contract the engineering and technology firm receives from the customer. The goal of an internal project is defined by the charter given to the project manager by the firm's higher management team.
- *Objective.* Collection of tasks that must be completed to produce a deliverable. For example, assume that one objective of an internal project is to develop a comprehensive manual for a training program on the concept of statistical process control (SPC). The deliverable is the comprehensive training manual. One task would be to develop a course description accompanied by a set of broad learning goals. Another task would be to develop specific lesson plans for satisfying the broad learning goals. Another task would be to develop a comprehensive list of all instructional and instructional-support materials needed. Another task would be to develop an evaluation process for determining the extent to which trainees have learned the material. Tasks are, in turn, broken down into activities (see Figure 1.5). Once all of these tasks and others that would be required have been completed, the objective has been accomplished and the deliverable is ready for use. Projects often have more than one objective.
- *Deliverable.* The actual product or service developed and provided by the project team. It is the deliverable(s) in a project that must be completed on time, within budget, and according to specifications. In the previous definition of the term *objective,* the deliverable was a comprehensive training program. Projects often have more than one deliverable.
- *Scope.* The scope of the project is a comprehensive definition of the project. It explains the following types of information: project overview, list of deliverables (often stated

FIGURE 1.5 Objectives are broken down into tasks which are broken down into activities.

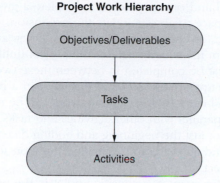

Project Work Hierarchy

as objectives), features and functions of the deliverables, the customer's acceptance criteria, restrictions and constraints that impinge on the successful completion of the project, and uncertainties that could affect completing the project on time, within budget, and according to specifications. With external projects, the project scope is the most important connection between the project team and the customer. With internal projects, the project scope is the most important link between the project team and the firm's higher management team.

- *Tasks and activities.* Objectives consist of tasks, and tasks consist of activities. Objectives are broken down into tasks as shown in the earlier explanation, and tasks are broken down into more finite activities. For example, in the earlier definition of the term *objective,* one of the stated tasks was to develop lesson plans for the training manual for an SPC course. That task would be broken down further into such activities as: (1) developing a description of the scope and purpose of each individual lesson, (2) developing specific learning outcomes for each individual lesson, (3) developing instructional strategies for each individual lesson, (4) developing a list of all instructional and support material and equipment needed to present each individual lesson, and (5) developing a test or some other way to assess participant learning for each individual lesson.

- *Duration.* The time from beginning to end that is required to complete an activity, task, or objective. There can be, and often is, a difference between the actual time an activity, task, or objective eventually takes to complete and the scheduled time. The duration is the scheduled time as established during the planning stage of the project.

- *Constraint.* Any factor affects the project manager's ability to complete a project on time, within budget, and according to specifications. Common constraints include time, cost/budget, quality expectations, and personnel. Other constraints that sometimes come into play are technology/equipment and facilities. Time, cost/budget, and quality expectations are almost always constraints. Customers typically want the needed product or service as soon as they can possibly get it at the lowest price and highest quality possible. Being able to satisfy these demands is what separates the engineering and technology firms that excel from those that never perform beyond the level of mediocrity. Personnel can become a constraining factor when the organization: (1) has insufficient personnel, (2) lacks the personnel with the knowledge and skills required by the project, or (3) has the needed personnel tied up on other pressing projects that require all of their time. Facilities can become a constraining factor when there is a need to add personnel and/or equipment to complete a given project and there is no space or insufficient space to accommodate the expansion.

- **Schedule.** A timetable for completing all activities, tasks, and objectives in a project. The schedule runs from the starting date of the project until the deadline for completion. All work that must be completed in between these two dates is put on a schedule with target completions dates. Some of the work in a project must be completed before other work can begin. However, some work can be undertaken simultaneously and be completed in parallel with other activities and tasks. A comprehensive project schedule will reflect not just the beginning and ending date for the overall project, but the beginning date, ending date, and estimated duration for all activities, tasks, and objectives for the project.
- **Milestone.** A significant event or a specified point in the project. Often the engineering and technology firm is able to bill the customer for a percentage of the overall price of completing the project when selected milestones have been achieved.
- **Resource.** A resource is any asset needed to complete a project on time, within budget, and according to specifications. Resources that typically concern project managers the most are time, personnel, money, material, technologies/equipment, and facilities.
- **Processes.** The work of engineering and technology firms is completed by people using processes. A process is a series of standardized steps used over and over to produce a given result. A process is not an end result. Rather it is a series of events that lead to a predictable end result. Project management processes in engineering and technology firms include the project initiation, planning, execution, monitoring/control, and closeout. Products are produced and services are delivered through standard processes. The better a process works, the higher the quality of the product produced or service delivered. This fact coupled with the demands of global competition form the basis for the need for continual improvement of processes in engineering and technology firms.

PROJECT SUCCESS CRITERIA

Every engineering and technology project comes with a built-in set of success criteria. Three broad success criteria apply to all projects. These criteria are to complete the project on time, within budget, and according to specifications (quality expectations). Another way to describe these criteria is schedule, budget, and quality. These basic success criteria are emphasized throughout this book. These broad success criteria are broken down into more specific terms by the contract and specifications that accompany them.

Quality expectations can be found in the RFP or RFQ issued by the customer to solicit bids from engineering and technology firms, and the specifications that accompany the contract when a bid is awarded. When an entity that has a need for a product or service sends out an RFP or RFQ, the documents specify not just what is needed but the customer's quality expectations for the product or service in question. These quality expectations often determine whether or not a firm can even submit a bid. Engineering and technology firms that cannot meet the customer's quality requirements need not even bid on the contract. This is why having capable processes that are continually improved as well as properly credentialed, talented personnel are so important for engineering and technology firms. Firms that cannot meet exacting quality requirements have fewer opportunities to bid on contracts. For those firms that can satisfy the requirements, the requirements become success criteria.

Time is always of the essence with engineering and technology projects. When a customer awards a contract to an engineering or technology firm, a hard and fast deadline for completion is specified in the contract. That deadline becomes the fabled *immovable object*. Project managers must develop a schedule that will ensure the project is completed and closed out by this deadline. The schedule for a project will show the beginning and ending dates for all required tasks—as well as the duration of each.

The deadlines for individual tasks in a project schedule may be revised as adjustments are made during the course of the project, but the overall deadline typically cannot. In fact, some contracts have built-in penalty clauses that apply for every day a project extends beyond the specified deadline. There are exceptional instances in which the specified deadline can be revised. However, this happens only if specific conditions apply that are worked out in advance with the customer.

Cost is always in issue with engineering and technology projects. One of the key factors in any contract awarded to an engineering or technology firm is the amount the customer has agreed to pay the firm for completing the resulting project. In accepting the contract, the firm has agreed to produce the product or deliver the service for the specified price. This means the organization must be able to cover its costs for the project and still make a profit without exceeding the contract price. A budget for the project will be established that factors in the firm's costs and profit margin. The project manager will participate in developing this budget and then become responsible for completing the project on time and according to specifications without exceeding the budget.

PEOPLE SKILLS IN PROJECT MANAGEMENT

Up to this point, all of the project management concepts and skills explained have been process oriented. The other side of project management is the people side, a side that requires specific people skills. To be successful as a project manager, engineering and technology professionals must be effective at: (1) building projects teams; (2) leading project teams; (3) motivating members of project teams; (4) communicating with team members, customers, and colleagues; (5) managing time—that of the project as well as their own; (6) managing change; (7) managing diversity; and (8) leading project teams through periods of adversity.

The need to be able to function at a high level of effectiveness in terms of both the process and people sides of the job is what makes excelling as a project manager a challenge. Some engineering and technology professionals have excellent process skills but struggle with the people side of the job. Other engineering and technology professionals have outstanding people skills but struggle when it comes to managing processes. Professionals at either end of the process/people continuum will find it difficult to excel as project managers. Only those who commit to developing both the process and the people skills required will excel.

The good news is that with effort and commitment both types of skills can be developed. Engineering and technology students and professionals can learn to initiate, plan, execute, monitor/control, and close out projects. At the same time they can learn to lead, manage, motivate, and communicate with people. Part Two of this book is devoted to helping engineering and technology students and professionals develop the process skills of project management. Part Three is devoted to helping them develop the people skills needed by project managers.

Project Management Scenario 1.2

How Do We Know If We are Successful?

On her first day as a project manager, Janet Carson was asked a question she was not sure how to answer. One of her team members—himself new to serving on a project team—asked Carson an interesting question: "How do we know when we are successful?" Carson was taken aback by the question. In fact, she had been so focused on selecting team members, establishing a budget, and developing a schedule that she had not even thought about how her team's performance would be evaluated. To herself Carson said, "That's a good question, and I need to know the answer." To her team member she said, "Good question. Let's talk about it later today." As soon as she could get away, Carson went to the office of a colleague who was an experienced project manager and placed the question before him.

Discussion Question

In this scenario, Janet Carson is a new project manager who is feeling a little overwhelmed by her new responsibilities. When asked a fairly basic question by a team member, she is unable to answer it. If Carson came to you with this question, how would you recommend she answer it?

SUMMARY

A project is a fully coordinated group of interdependent tasks that are completed by people using resources and processes. Projects have definite starting and ending dates as well as success criteria. The broad success criteria for all projects are to complete them on time, within budget, and according to specifications. Project managers are needed in engineering and technology firms for the same reason conductors are needed in orchestras. Projects are composed of a number of separate but interdependent tasks all of which must be planned, scheduled, budgeted, staffed, and fully coordinated if they are going to be completed on time, within budget, and according to specifications. Project managers serve the same purpose as orchestra conductors in making this happen.

There are four main components of any project: scope, schedule, resources, and leadership. These four components are all interrelated and interdependent. Every project must have a project scope that summarizes everything that the members of the project team need to know to understand the project. An outline for a project scope contains the following elements: project overview, deliverables, features and functions, acceptance criteria, restrictions/constraints, and uncertainties.

The schedule for a project contains beginning and ending times for and the duration of all tasks that must be done to complete the project on time. Project resources include any and all assets necessary to complete the project on time, within budget, and according to specifications. Typical project resources are people, time, technology, facilities, and material. Internal projects are initiated by engineering and technology firms higher for the purpose of enhancing the firm's competitiveness. External projects are initiated by customers for the purpose of satisfying the need for a product or service.

Important terms and concepts in the language of project management include the following: project, program, goal, objective, deliverable, scope, task, activity, duration, constraint, schedule, resource, and processes. Every project has five distinct phases: initiation, planning, execution, monitoring/control, and closeout. People skills needed by project managers include team-building, leadership, motivation, communication, time management, change management, dealing with diversity, and leading teams during times of adversity.

KEY TERMS AND CONCEPTS

Project
Fully coordinated interdependent tasks
People processes, and resources
Starting date
Ending date
Success criteria
Project scope
Project schedule
Project resources
Leadership
Project overview
Features and functions descriptions
Acceptance criteria
Uncertainties
Internal projects

External projects
Program
Goal
Objective
Deliverable
Tasks and activities
Duration
Constraint
Processes
Initiation
Planning
Implementation
Closeout
People skills

REVIEW QUESTIONS

1. Define the term *project*.
2. What is meant by *fully coordinated interdependent tasks*?
3. What are the three success criteria that apply to all projects?
4. Explain why project managers are needed by engineering and technology firms.
5. List and explain the main components of a project.
6. Summarize the elements that should be contained in a project scope.
7. Distinguish between an internal and an external project.
8. What is the difference between a program and a project?

9. What is the difference between a goal and an objective in a project?
10. Give an example of a project deliverable for an engineering or technology firm.
11. Give an example of a constraint that might affect a project in an engineering or technology firm.
12. Explain the term *process*.
13. List the phases of a project and the corresponding processes associated with each of these phases.
14. Think of an assignment you have had to complete in a college class. What were the success criteria your professor established for the project?
15. List the people skills needed by project managers.

APPLICATION ACTIVITIES

The following activities may be completed by individual students or by students working in a group:

1. Contact an engineering and technology firm and identify an individual who is willing to cooperate in

helping complete this project. Ask this individual to identify a project his or her firm has completed. Ask this individual to provide the following information about this project: (a) the main tasks that had to be completed in the project, (b) number of people assigned to the project and their roles, (c) starting and ending date for the project, (d) deliverables for the project, and (e) success criteria for the project.

2. Assume that you are required to develop a class project covering Part Two of this book

that culminates in a class presentation. You will have 10 people in your group including yourself, and you are to be the project manager. Explain how you would initiate, plan, execute, monitor/control, and close out the project.

ENDNOTES

1. Project Management Institute, *A Guide to the Project Management Body of Knowledge,* 4th ed. (Newtown Square, Pennsylvania: Project Management Institute, 2008), 37.

2. Ibid.
3. Ibid., 6.
4. Ibid., 43.

CHAPTER 2

Roles and Responsibilities of Project Managers

Chapter One explained what project management is. This chapter explains what a project manager does. To be effective project managers, professionals in the fields of engineering and technology must develop both process and people skills. Process skills allow project managers to provide leadership in the initiation, planning, execution, monitoring/control, and closeout of projects. People skills allow project managers to mold disparate groups of individuals into effective project teams and lead them to peak performance in carrying out projects. Both sets of skills are essential for engineering and technology students and professionals who want to be effective project managers.

THE PROJECT MANAGER'S FUNCTIONS

When an engineering and technology firm receives a contract to provide a product or service for a customer, a project is established and a project manager is appointed. When an engineering and technology firm decides to undertake an internal project to enhance its competitiveness, a project charter is developed and a project manager is appointed. Once an engineering or technology professional has been appointed to serve as the project manager for either an external or internal project, that individual will play a well-defined role and have specific responsibilities. The overall responsibility of a project manager is to ensure that projects are completed on time, within budget, and according to specifications. In carrying out this overall responsibility, project managers have a number of specific functions to perform that fall into two categories: process functions and people functions. These functions are explained in the next two sections.

PROCESS FUNCTIONS OF PROJECT MANAGERS

When a project has been established and a project manager assigned, a certain prescribed list of responsibilities come with the assignment. Regardless of whether the project is internal or external, the project manager is responsible for the following process functions:

PROCESS FUNCTIONS OF PROJECT MANAGERS

1. **Project initiation**
   - Develop project charter
   - Identify stakeholders

2. **Project planning**
   - Develop the project schedule
   - Develop the cost estimate/budget
   - Develop the quality, human resource, communication, and risk management plans

3. **Project execution**
   - Direct and manage project work
   - Assure quality
   - Conduct procurements

4. **Project monitoring/control**
   - Control changes
   - Control the scope, schedule, costs, qualify, performance, and risk

5. **Project closeout**
   - Close procurements
   - Close all other project activities

FIGURE 2.1 Project managers provide the leadership in carrying out these process functions.

project initiation, planning, execution, monitoring/control, and closeout (see Figure 2.1). The process function is actually a group of processes as can be seen in Figure 2.1.

Project Initiation Group

The project initiation group consists of the processes necessary to define the project, identify project stakeholders, and authorize the project. Ideally, project managers are named before project initiation begins so that they can lead or at least participate in the initiation processes. This allows the project manager to interact with the customer in ways that: (1) establish a clear understanding of customer expectations, (2) establish a positive working relationship from the outset, and (3) become familiar with all of the project's stakeholders. The Project Management Institute identified the following two processes that must be completed during project initiation: (1) development of the project charter and (2) identification of project stakeholders.[1]

PROJECT CHARTER. The project charter is a document that authorizes the project. It also summarizes the general requirements of the project. A project charter should contain at least the following information:[2]

- Purpose of the project
- Project objectives and success criteria
- General requirements
- General project description and product/service characteristics
- Summary schedule showing project milestones

- Project approval requirements (how success is defined and who decides)
- Project manager (including responsibilities and authority)
- Name and responsibility of the individual(s) authorizing the project

PROJECT STAKEHOLDERS.   Stakeholders are all individuals and/or organizations that have a stake (interest) in the project. The process of identifying project stakeholders is important because project managers need to know what individuals and organizations have expectations and influence concerning the project. Part of project initiation involves beginning the process of gaining the support and assistance of key stakeholders. Consequently, it is important for project managers to not just identify stakeholders for their projects, but to create a stakeholder *register*.

The registry should contain the names and positions of all stakeholders as well as their interest in and expectations of the project. Notes should be made in the registry concerning any influence—positive or negative—each stakeholder has relating to the project. Project managers use their stakeholder registers for more than just knowing who the players are relating to their projects. They also use them to develop strategies for gaining buy-in and assistance from key stakeholders. For example, assume that a project will require a substantial amount of precision machining work. The machine shop supervisor would be an important stakeholder who the project manager would want to enlist as an ally.

Project Planning Group

The project planning group is the most extensive group of processes project managers are responsible for and the ones most widely associated with project management. Processes in this group include the following:

- Defining the scope of the project
- Developing the work breakdown structure for the project
- Developing a schedule for the project showing all activities including their sequence and durations
- Developing the cost estimate and budget
- Developing other individual plans that are part of the larger project management plans including the quality, human resource, communications, risk management, and procurement plans

SCOPE OF THE PROJECT.   If the project scope statement was not developed as part of the initiation process for the project, it must be developed at this point. The scope statement should contain at least the following information: (1) project overview, (2) deliverables, (3) features and functions of the product or service to be provided, (4) acceptance criteria including who has the authority to approve work as being acceptable, (5) restrictions, and (6) uncertainties.

WORK BREAKDOWN STRUCTURE.   In developing the work breakdown structure or WBS for a project, the project manager begins with the deliverables that are to be produced or provided and works backwards. This process—known as *decomposition* breaks the deliverables down into smaller components. It results in a group of *work packages* that represent the activities and tasks that must be accomplished to complete the project. The work package

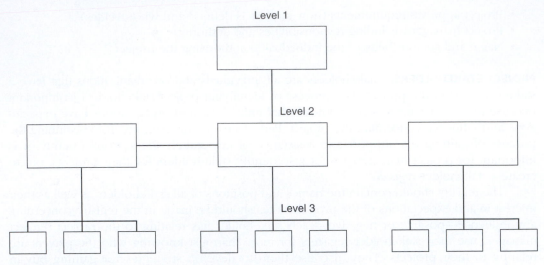

FIGURE 2.2   Typical format for a simple work breakdown structure.

is the lowest level of activity in a work breakdown structure. Work packages become the basis for estimating the duration and cost of the project. Figure 2.2 is an example of a work breakdown structure.

PROJECT SCHEDULE.   Scheduling is part of project planning and was introduced in the previous section. However, it is such a specialized planning function that it is treated separately here. Project managers are responsible for either developing comprehensive schedules for all of the work that must be completed in their projects or for working with a professional scheduler to do so. There is a great deal of software available to assist project managers with this critical and often complex responsibility. However, even if using scheduling software, project managers must be able to develop and interpret Gantt charts and PERT charts, identify critical paths and dependencies in WBSs, develop network logic diagrams, and understand the types and amounts of resources that must be allocated to project activities.

COST ESTIMATE AND BUDGET.   Few things are more important to a successful project than beginning with a well-thought out, realistic and accurate budget. The first step toward arriving at a realistic budget is to accurately estimate the duration and cost of all of the activities in the work breakdown structure. In addition, cost estimators should consider other cost inducing factors such as the schedule, human resource plan, risks, and any miscellaneous factors that might affect the cost of the project. These miscellaneous factors are sometimes referred to as *enterprise environmental factors.*[3] Enterprise environmental factors can be anything that might affect the cost of the project including the organization's corporate culture and capabilities of its processes, government regulations, availability and capabilities of human resources, market conditions, and the level of risk stakeholders are willing to endure. Once all applicable factors have been considered, the cost estimate becomes the basis for the project's budget.

QUALITY PLAN.   The quality plan sets forth how the team will meet or exceed specifications and customer expectations. The underlying philosophy on which the quality plan is built

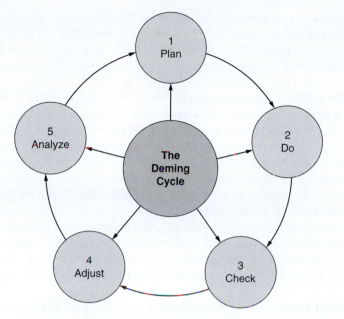

FIGURE 2.3 The Deming Cycle developed by quality pioneer
W. Edwards Deming.

should be continual improvement—improvement of the performance of processes and people.
A well-developed quality plan will incorporate the Deming Cycle developed by quality pioneer
W. Edwards Deming[4] shown in Figure 2.3 and the Ten Steps to Quality Improvement advocated
by quality pioneer Joseph Juran:[5]

- Build awareness of the need for improvement and create opportunities for improvement.
- Set goals for improvement.
- Organize to meet the improvement goals.
- Provide training.
- Implement problem-solving projects.
- Report on progress on a regular basis.
- Give recognition where it is deserved.
- Communicate results.
- Keep score (establish benchmarks for process and people performance and measure
  improvements against the benchmarks).
- Maintain momentum by building improvement into the firm's regular systems, proc-
  esses, and procedures (improved practices become the firm's new best practices).

HUMAN RESOURCE PLAN. One of the keys to success with any project is the human element. In the final analysis, even the most advanced technology firms must depend on the
performance of people as a major determinant of productivity and quality. The human
resource plan is the project manager's plan for properly staffing his or her project team. In
developing a human resource plan, project managers identify all of the positions that will
be needed to complete the project on time, within budget, and according to specifications.

The plan documents: (1) all positions that must be filled for the project as well as the roles and responsibilities of the personnel who will fill these positions, (2) qualifications required for each position in the project (i.e., knowledge, skills, experience, and credentials), and (3) reporting relationships of all personnel assigned to the project.

COMMUNICATION PLAN.   Communication is critical to project managers. All stakeholders in a project must be kept fully informed and up to date throughout the duration of the project. The communication plan documents how this is going to be achieved. *A Guide to the Project Management Body of Knowledge* describes communication management for projects as follows: ". . . the processes required to ensure timely and appropriate generation, collection, distribution, storage, retrieval, and ultimate disposition of project information."[6]

If a project can be viewed as a machine, communication is the oil that keeps its gears lubricated and working well. Project managers communicate within their teams, between their teams and the larger organization, and with external stakeholders. Because of this they have to become adept at communicating with people of different backgrounds, levels of education, cultures, interests, personalities, and needs. The communication plan documents how the project manager will ensure effective communication throughout the duration of the project.

RISK MANAGEMENT PLAN.   Every project comes with a certain amount of risk. Project risk consists of factors that might inhibit the team's ability to complete the project on time, within budget, and according to specifications. The risk management plan documents the processes that will be used for identifying, analyzing, minimizing, monitoring, and controlling the risks associated with the project. The risk management plan documents how the team will minimize negative risk factors and exploit positive risk factors. Figure 2.4 lists some of the more common risk factors associated with engineering and technology projects. The risk factors are categorized according to source. Figure 2.4 is neither an exhaustive list nor it is intended to be. Rather, it lists risk factors that are present with almost every project.

PROJECT RISK FACTORS

TECHNICAL FACTORS
- Specifications that are difficult to comply with
- Technology requirements that are difficult to meet
- Quality expectations that might exceed current capabilities

EXTERNAL FACTORS
- Supplies that might be difficult to obtain
- Government regulations that are difficult to comply with
- Market forces that change quickly

INTERNAL FACTORS
- Priority the firm gives the project
- Availability of the necessary resources: human, financial, and technical
- Quality of project management (i.e., planning, executing, monitoring/controlling, and closing)

FIGURE 2.4   Every project comes with risk.

PROCUREMENT PLAN.   Procurement is the process of acquiring the materials, services, and other outside resources needed to complete the project on time, within budget, and according to specifications. The procurement plan documents the approved procedures for carrying out the procurement process. It also documents the procedures for developing contracts with material suppliers and service providers. The procurement plan documents how: (1) purchasing decisions will be made, (2) the procurement process will be carried out, (3) procurement relationships will be managed, (4) procurement changes and errors will be managed, and (5) procurements will be closed out.

Project Execution Group

The project execution group consists of the processes for establishing and managing the project and executing the quality, human resource, communication, and risk management plans. Project managers rarely have the luxury of just selecting anyone they want to serve on their project teams. Some personnel may already be assigned to other projects. Supervisors might not want to release certain personnel to serve on the project team. Consequently, it is often necessary to negotiate with the supervisors of personnel project managers want for their teams (see Chapter Eleven).

When a team is first established, it must is likely to be nothing more than a collection of disparate individuals, all with their own personalities, motivations, egos, agendas, and ambitions. Consequently, teambuilding and development should begin as soon as the team is established so that the group of disparate individuals can be transformed into a well-coordinated group of people who are committed to a common mission and will mutually support each other in trying to achieve it. That mission, of course, is to complete the project on time, within budget, and according to specifications. As the team comes together, the project manager's job is to lead it in achieving peak performance and continual improvement.

In the planning phase of the project, a communications plan was developed that outlined the reporting procedures to be followed. In this step, those procedures are put into action as the communication plan is implemented. A procurement plan was also developed during the planning phase of the project. In this phase, that plan is executed so that all of the resources needed to complete the project on time, within budget, and according to specifications are obtained in a timely manner.

Project Monitoring and Controlling Group

The project monitoring and controlling group consists of the processes for monitoring and controlling the project scope, schedule, costs, quality, risk, and procurements. Monitoring and controlling the project scope amounts to managing the change order process that the engineering and technology firm does not find itself doing more than it is being paid to do. In managing the change order process, project managers will find themselves negotiating with the customer over the causes of changes and who is responsible for paying for the extra work they generate. Effective change order management will prevent *scope creep,* the process of adding to the project's scope without properly compensating the engineering and technology firm.

Monitoring and controlling the project schedule is a matter of regularly checking the actual status of the work on the project against schedule projections and taking the appropriate corrective action if the work falls behind schedule. This is a critical process for ensuring that the project is completed on time. Monitoring and controlling costs is a similar process of regularly checking actual expenditures against budgeted expenditures. It involves monitoring the performance of the project team, identifying the causes of cost variances and making the appropriate corrections, and managing the change order process effectively.

Monitoring and controlling quality amounts to ensuring that the work is being completed according to specifications. Quality specialists use regular inspections, sampling, and statistical process control to ensure that poor quality work is detected and corrected before it goes to the customer. Monitoring and controlling risks involves carrying out the risk management plan developed earlier, identifying new risks that are introduced after the work of the project is being done, and taking the necessary steps to eliminate or minimize risk. Monitoring and controlling procurements amounts to administering contracts that have been signed, managing relationships with suppliers, subcontractors, and service providers, and managing the procurement change/correction process.

Project Closing Group

The project closing group consists of the processes for closing out projects. These processes include closing out the project itself and closing out procurements for the project. Closing a project amounts to completing all processes associated with the project in all of the process groups (i.e., initiating, planning, executing, monitoring and controlling, and closing). Closing procurements amounts to finalizing all contracts and ensuring that all work contracted for has been completed according to specifications.

It is important to ensure that projects are properly closed out so that there are no lingering responsibilities or liabilities that apply to any project stakeholders. Further, the engineering and technology firm depends on a proper close out of the projects as a prerequisite to receiving final payment for its work. In addition, how a project is closed out can have a bearing on the customer's decision to award future contracts to the firm.

PEOPLE FUNCTIONS OF PROJECT MANAGERS

In addition to the process functions of project management, there are also people functions. The people functions have to do with applying people skills in ways that encourage peak performance from the members of project teams while also motivating them to improve continually. The people functions of project managers are important because peak performance and continual improvement from team members are essential to successfully completing projects on time, within budget, and according to specifications. No matter how well a project is initiated, planned, executed, monitored and controlled, or closed out the performance of team members is still the major delimiting or enabling factor in successfully completing projects. The people functions of project managers are as follows (Figure 2.5):

- *Leadership function.* Project managers must be able to inspire—by their example—team members to make a total commitment to achieving the team's mission. Hence, project managers must be good leaders. Leadership skills are covered in Chapter Twelve.
- *Teambuilding function.* Effective project teams do not just happen. They must be built. Consequently, project managers must be good team builders. Teambuilding skills are covered in Chapter Eight.

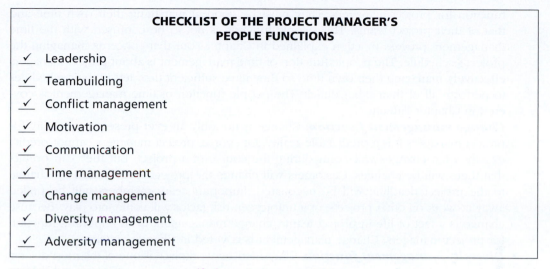

**CHECKLIST OF THE PROJECT MANAGER'S
PEOPLE FUNCTIONS**

- ✓ Leadership
- ✓ Teambuilding
- ✓ Conflict management
- ✓ Motivation
- ✓ Communication
- ✓ Time management
- ✓ Change management
- ✓ Diversity management
- ✓ Adversity management

FIGURE 2.5   People functions are half of a project manager's job.

- **Conflict management function.** Any time people work together in groups there is the potential for conflict. Spirited debates over the best way to do something should be encouraged in project teams, but spirited debates can quickly become counterproductive conflicts if not handled properly. When conflict occurs it can polarize a project team and render it ineffective as time and energy that should be devoted to getting the job done is diverted to petty squabbles. Consequently, project managers must be good conflict managers. Conflict management is covered as part of Chapter Eight.

- **Motivation function.** People come to project teams with different agendas, outlooks, and attitudes. Plus, people have their good days and bad. Regardless of all of this, the project manager's responsibility to ensure peak performance and continual improvement from team members does not change. Consequently, project managers must know how to motivate people to give their best efforts. Motivation is covered in Chapter Thirteen.

- **Communication function.** Effective communication is critical in project teams. First, communication is the oil that lubricates the gears of human interaction. People who work in teams need to know what is expected of them, when it is expected, why it is expected, and how they will be evaluated against expectations. They also need to be kept up to date on the daily information flow, on changes that affect them, and on anything else that will help them perform at peak levels and improve continually. Further, they need someone in a position of authority—in this case the project manager—who will listen when they have concerns, complaints, ideas, or recommendations. All of this means that project managers must be effective communicators, which includes being good listeners. Communication is covered in Chapter Fourteen.

- **Time management function.** Time is always an important consideration for project managers. When a contract is received for an external project or a charter is provided for an internal project, time will be one of the most prominent of the various success criteria. Remember, that completing projects on time is always a success criterion on engineering and technology projects. Consequently, time management is an important function for project managers. They must be good at managing their own time and that of their project teams. This people function is not to be confused with the time management process function explained in Chapter Four that concerns managing the project's schedule. The people function of time management is about project managers effectively managing their own time so they have sufficient time left in their schedules to perform all of their other duties. The people function of time management is covered in Chapter Fifteen.

- ***Change management function.*** Change is not only an ever-present probability for project managers it is a predictable reality. Of course, project managers cannot predict exactly what changes will occur during the course of a project, but they can predict that there will be changes. Customers will change the project specifications, a change to the project deadline will be negotiated, important team members will be pulled away to work on other projects, or a unforeseen risk factor will come into play. Because change is a fact of life in project teams, change management is an important function for project managers. Change management is covered in Chapter Sixteen.
- ***Diversity management function.*** The American workplace is one of the most diverse in the world, and project teams in engineering and technology firms are likely to reflect this diversity. It is predictable that the members of project teams will have different backgrounds, genders, races, cultures, worldviews, perspectives, levels of education and experience, agendas, and attitudes. Molding a diverse group of people into a well-coordinated, mutually supportive team is an important function of the project manager. Depending on how the project manager handles it, diversity can strengthen the team or to cause it to devolve into counterproductive factions separated by their diversity. Ensuring that diversity is an asset in project teams is the responsibility of the project manager. Diversity management is covered in Chapter Seventeen.
- ***Adversity management function.*** Project teams will go through times of adversity. Like change, adversity is a predictable fact of life for project managers and the members of project teams. Individual team members may suffer some form of loss or crisis that affects their work. The engineering or technology firm may fall on hard times. Changes in project specifications or deadlines may make it difficult to complete the project on time, within budget, and according to specifications. Consequently, leading their teams through adversity is an important function of project managers. The best project managers are those who can give their team members hope and inspiration on the days when completing the project successfully seems like a remote possibility at best. Adversity management is covered in Chapter Eighteen.

Project Management Scenario 2.1

I Don't Like Dealing with People

Danny Cutter is a project manager for a mid-sized engineering and technology firm. He does an excellent job of planning, scheduling, cost estimating, procuring resources, tracking progress, reporting, and handling risk. In other words, as a project manager, Cutter has strong process skills. In spite of this, his project teams never seem to get beyond the level of mediocrity. His teams complete their assigned projects, but never without a steady stream of problems. There always seems to be a lot of conflict and confusion among his team members. The firm's personnel are vocal about not wanting to be assigned to teams led by Cutter. In addition, customers complain frequently.

After the firm finally got through Cutter's latest project, an especially difficult project plagued by problems from the outset, the company's vice president for engineering decided she had to get to the bottom of the problem. In a meeting with Cutter, she came straight to the point. "Danny you are the best planner, scheduler, estimator, and cost controller we have, but your projects never seem to go well. Where is the disconnect here?" Cutter was equally frank in his reply. "I can plan projects with my eyes closed and manage the process aspects of projects with my hands tied behind my back. But when it comes to dealing with customers, my team members, and higher management it's another story. I don't like dealing with people."

Discussion Questions

In this scenario, Danny Cutter has excellent process skills but lacks people skills. This short-coming causes problems when he is assigned to lead project teams. Have you ever worked in any setting with someone who was technically competent, but lacked people skills? If so, did this shortcoming cause problems? Elaborate on the situation. Unless something changes, will Danny Cutter ever be an effective project manager?

CHARACTERISTICS OF AN EFFECTIVE PROJECT MANAGER

Serving as a project management can be both challenging and rewarding. Those project managers who best meet the challenge and, as a result, enjoy the most rewards exhibit the following characteristics on a consistently basis:[7]

- **Strong process skills.** This characteristic may appear obvious and perhaps it is, but it is important enough to deserve special emphasis. The most effective project managers are those who develop advanced process skills—they make sure that they become thoroughly accomplished at all the processes required to effectively initiate, plan, execute, monitor and control, and close out projects. Process competence is a strength shared by the most effective project managers.
- **Strong people skills.** This characteristic, like process skills, may appear obvious, but it should never be taken for granted. Molding a diverse group of people into a high-performing team and leading them in ways that ensure that projects are completed on time, within budget, and according to specifications is never easy. Even the strongest process skills will not be sufficient for success as a project manager if they are not coupled with strong people skills.
- **Intellectual curiosity.** The most effective project managers are not blindly conforming robots. Rather, they are intellectually curious. They want to know the *why* behind all factors that affect their projects. They also want to understand why things are done a certain way and if there might be a better way. By understanding why certain factors apply, they are better prepared to make the right decision when questions arise or when problems occur. By refusing to accept doing things the way they have always been done just because that is how they have always been done, they are able to achieve continual improvement.

- **Commitment.** The most effective project managers are committed to completing their projects on time, within budget, and according to specifications and they are willing to sacrifice to make this happen. They maintain a can-do attitude that can be summarized as follows: *I will do everything necessary—within the limits of legality and ethics—to successfully complete my projects.* Project managers who are committed go beyond just trying or even trying their best. They are determined to get the job done right. Consequently, they approach problems, changes, and other inhibitors not as insurmountable barriers, but as roadblocks to go around, over, or through.
- **Vision and insight.** The most effective project managers have both vision and insight. In other words, they understand the big picture—their firm's and the customer's—and where their project fits into it (vision). They also understand the details of their project and what must be done to ensure that they come together in a way that satisfies the vision of their firm and the customer (insight).
- **People orientation.** The most effective project managers understand that even the best planned, best scheduled, best budgeted, and most carefully monitored project is ultimately dependent on people for its success. Team members, the firm's higher management, the customer's representatives, suppliers, and subcontractors are all people with their own needs and interests in the project. Yet they must work cooperatively and in a mutually supportive manner if the project is to be completed on time, within budget, and according to specifications. Consequently, effective project managers learn to view people as assets, treat them with respect, and give credit where credit is due. They understand that the way they treat stakeholders will shape how the stakeholders view the projects they manage.
- **Character.** The most effective project managers are trusted by all stakeholders of their projects (i.e., team members, higher management, customer representatives, suppliers, subcontractors). They understand that people who do not trust them will not follow their lead in doing what is necessary to ensure that projects are completed successfully. Consequently, effective project managers make a point of being honest, dependable, and ethical in all of their dealings with stakeholders. Further, they insist that stakeholders do the same.

In addition to these characteristics, Duncan Brodie of ProjectSmart suggests that effective project managers also have the following characteristics:[8]

- **Focus on solutions.** When problems arise, as they surely will, effective project managers focus on solutions. They do not let themselves become frozen by fear of the consequences of failure. Rather, they approach problems systematically, analytically, and with a mind to solving them.
- **Participative and decisive.** Project managers lead project teams. Consequently, when decisions must be made, they ask for input from the team members who will have to carry them out. This does not mean they ask for a vote. Rather, it means they get input from those who will be affected by the decision before making it. Once they have collected the input of their team members—the participative aspect of decision making—effective project managers make what they believe to be the best possible decision under the circumstances they face. This is the decisive aspect of their decision making.
- **Focus on the customer.** The best project managers have a customer focus. This is important because, ultimately, it is the customer they must please. This fact applies

regardless of whether the customer in question is an external or an internal customer. How would the customer want us to do this? This question is always at the forefront of a good project manager's thoughts. In fact, there will be times when project managers will be forced to play the role of advocate on behalf of customers. A project that satisfies the customer in all of its aspects is a firm's best marketing tool for generating future work.

- **Focus on win-win outcomes.** The customers who generate projects have specific needs relating to the project. Engineering and technology firms that take on the projects have needs of their own. The best project managers are those who can keep the needs of both entities at the forefront of their thinking and concentrate on finding approaches and solutions that allow both parties to benefit. Project managers who think they benefit when their firm wins and the customer loses are shortsighted. Win-lose solutions dampen relationships and make the possibility of future contracts with the same customer unlikely. Further, disgruntled customers make their dissatisfaction known to potential customers. Win-win solutions, on the other hand, build relationships help generate future contracts.

- **Lead by example.** The best project managers expect their team members to "do as I do" rather than "do as I say." Good project managers are good leaders. Good leaders inspire team members to give their best by consistently exemplifying everything they expect of team members. Members of project teams are just like anyone else. They are more likely to follow the leader's example than his words. Consequently, the best project managers make sure that their actions reinforce their words.

- **Get the best from all stakeholders.** Project management is a team sport. In order to succeed, project managers must be adept at getting the best from all of a project's stakeholders. This means they take full advantage of the strengths of their team members while minimizing their weaknesses. It also means they do what is necessary to pull the individual team members together into a unified whole that performs better than the sum of its individual members.

FUNCTIONAL, MATRIX, AND PROJECT-BASED ORGANIZATIONS

The organizational structure of an engineering or technology firm can have an effect on project managers and the composition of their teams. Consequently, students and professionals in the fields of engineering and technology need to understand the various ways organizations might structure themselves. The three most common organizational structures for engineering and technology firms are functional, matrix, and composite.

Functional Structure

The traditional and still most common organizational structure for engineering and technology firms is the functional structure. The functionally structured organization is sometimes referred to as a *line organization*. This is because functional organizations have clear lines of authority in terms of who reports to whom. Figure 2.6 is an example of a firm organized according to the functional structure. The company is subdivided into four broad functional areas: engineering, manufacturing, sales/marketing, and accounting/finance. Each of these functional areas is led by a vice president who reports to the firm's chief executive officer. Each vice president of a functional area has a staff of personnel who do the work associated with that area.

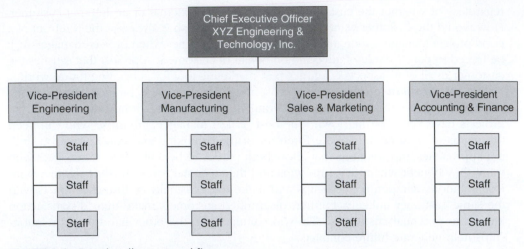

FIGURE 2.6 Functionally structured firm.

The boxes in the organizational chart in Figure 2.6 under "Vice President Engineering" contain the various positions and personnel who do the firm's engineering work. This would include engineers, CAD technicians, administrative support personnel, and project managers. The box labeled "Manufacturing Vice President" contains all of the positions and personnel who do the firm's manufacturing work. This would include quality professionals, various technology professionals, manufacturing technicians, assemblers, machinists, administrative support personnel, and project managers.

Engineering and technology professionals in functionally structured organizations may be called on to lead project teams staffed by personnel solely from their function area or cross-functional teams that draw their members from multiple functional areas. For example, assume that the Engineering and Technology firm in Figure 2.6 wins a contract to provide design services, but not to manufacture the product in question. A project manager from the engineering department might lead a team consisting of engineers, CAD technicians, and administrative support personnel—all of whom are also from the engineering department. In this instance, the project manager might have line authority over the members of the project team. The word "might" is used here because even when all team members come from the same functional department, the project manager does not necessarily have line authority over all team members.

Now assume that the same firm receives a contract to design, prototype, test, and manufacture a given product. The project manager selected might come from either the engineering or manufacturing departments. This individual would lead a team of personnel drawn from both engineering and manufacturing at a minimum, but possibly from sales/marketing and accounting/finance too. In a case such as this, the project manager would not have line authority over all of the team members.

There simply is no guarantee that project managers will have line authority over all individuals selected to serve on their project teams. Consequently, even in organizations with the functional structure, project managers may be called upon to lead teams of personnel who report to other managers or supervisors. This is why leadership skills and the other characteristics explained in this chapter are so important for those who want to be project managers. Project managers do not always have the luxury of leading project teams in which all of the members report directly to them, even in a functional structure.

Matrix Structure

The matrix structure is one in which people from similar backgrounds—engineering, manufacturing, sales/marketing, accounting/finance—are grouped into their own departments and report to a manager in that department. This aspect of the matrix structure is similar to the functional structure shown in Figure 2.6. Each of these departments in a matrix structure is viewed as a pool of assets to be drawn from when forming project teams. The work of a matrix organization is done by project teams while the administrative support tasks are done within the functional departments. Administrative support tasks include such things as completing performance appraisals, approving leave requests, making recommendations for salary increases and promotions, selecting personnel for special recognition, and administering discipline.

Figure 2.7 is an example of a matrix structure. Notice that it is similar to the functional structure in Figure 2.6 except that it has project team boxes added on the left side of the organizational chart. Each of these project teams draws it members from the functional departments as needed. In this way, project teams in matrix organizations are always cross-functional in nature. Because of this, project managers in matrix organizations do not have line authority over their team members. This means they must work cooperatively and collaboratively with the supervisors of their team members in motivating and leading project team members. Project managers in matrix organizations must be strong leaders, but they must also be good diplomats, effective negotiators, and dedicated relationship builders.

Project Structure

Occasionally organizations will be structured around project teams. In the project structure—as shown in Figure 2.8—project managers have line authority over their

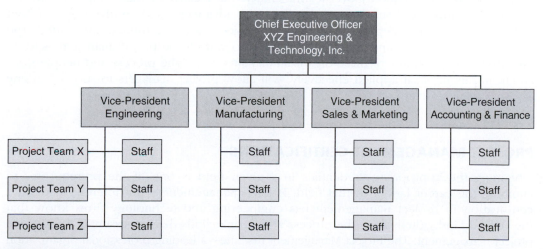

FIGURE 2.7 Matrix structured firm.

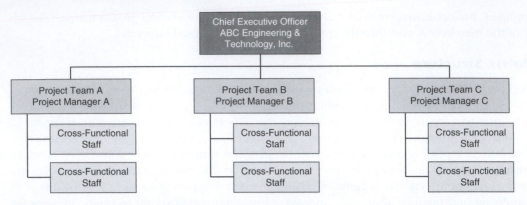

FIGURE 2.8 Project structured firm.

team members. This means that in addition to supervising their work, project managers complete performance appraisals, approve leave, make recommendations for salary increases and promotions, single personnel out for special recognition, and administer discipline. In other words, in the project structure project managers perform both traditional management duties as well as those associated with project management.

The major challenge faced by project managers in the project structure is competition between and among project managers for the resources needed to complete their projects. In Figure 2.8, Project Manager A is going to want the best personnel, most up-to-date equipment, and most advantageous facilities for his team. Project Manager B will want the same, as will Project Manager C. All three project managers will want their projects to have budget priority. This constant competition for resources and priority can turn project-structured firms into warring factions if not handled well.

Each organizational structure has its advantages and disadvantages for project managers. The matrix structure promotes cross-functional communication and cooperation but gives the project manager limited authority over team members. The project structure gives the project manager full authority over team members but creates internal competition for resources. The functional structure can replicate the advantages and disadvantages of both of the other structures depending on how the project teams are formed and staffed. The message in this for students and professionals who want to be project managers is to be prepared to work in any of these organizational structures. The process and people skills taught in this book, if applied effectively, will allow project managers to succeed in any organizational structure.

PROJECT MANAGEMENT CERTIFICATIONS

Obtaining the appropriate credentials in a given field is one of the prerequisites to climbing the career ladder in that field. Project management is no different. Earning a certification in project management lets engineering and technology firms know that the certified individuals have the process and people skills that are essential to effective project management. The Project Management Institute—a leading professional organization for project managers—offers a variety of professional certifications. These include the following:[9]

- **Project Management Professional (PMP)**®. This is the highest level of project management certification. This certification is for individuals who meet certain specified knowledge and experience requirements. Consequently, this certification is for experienced project managers.
- **Certified Associate in Project Management (CAPM)**®. This is an entry-level certification with no experience requirement. Professionals in the fields of engineering and technology who have just been named project managers may wish to test for this certification.
- **Program Management Professional (PgMP)**®. Programs are made up of projects. Therefore, program management is a broader and higher level job than project management. This certification is for experienced program managers who oversee multiple projects that, taken together, are considered a program.
- **PMI Scheduling Professional (PMI-SP)**®. This is focused certification for project management professionals who specialize in the scheduling component of project management. In large firms that handle multiple large projects simultaneously, scheduling is a complex task requiring focused attention and advanced skills. Professionals who work or plan to work primarily as schedulers may wish to test for this certification.
- **PMI Risk Management Professional (PMI-RMP)**®. All project managers are responsible for assessing risk in the projects they manage. However in organizations that handle a number of large and complex projects simultaneously, risk management is a specialized skill. Professionals who work or plan to work primarily as risk managers may wish to test for this certification.

Project Management Professional Examination

To be certified as a Project Management Professional (PMP), engineering and technology professionals must meet specific education and experience requirements and pass an examination. The education and corresponding experience requirements to sit for the PMP examination are as follows:[10]

- High school diploma or an associate degree and 60 months of experience in project management. The 60 months must encompass 7,500 hours of experience.
- Baccalaureate degree with 36 months of experience in project management. The 36 months must encompass 4,500 hours of project management experience.
- Formal education in project management consisting of a minimum of 35 contact hours. This requirement applies to all candidates regardless of their level of education and length of experience.
- Information about professional certification for project managers is available from the Project Management Institute at www.pmi.org. The Project Management Institute is the professional organization of project managers. In addition to professional certification, the Project Management Institute provides opportunities for project managers to keep their knowledge and skills up to date and interact with other professionals in their field.

Project Management Scenario 2.2

You Just Need to Learn Diplomacy

Sherry Johns is one frustrated project manager. Before accepting her current position at ABC Technologies, Inc., she had been a project manager at a competing firm, Manu-Tech, Inc. The senior managers at ABC Technologies had given Johns a substantial salary increase and better benefits to make the jump to their firm. Her former colleagues at Manu-Tech think Sherry Johns made a wise career move, but lately Johns doesn't feel that way. In fact, she is beginning to regret the move and think it was a mistake. Her problem is that ABC Technologies is structured as a matrix organization. Manu-Tech was organized as a traditional functional organization.

At her former firm, Johns always had direct line authority over most of the members of her project teams. She chose them, completed their performance appraisals, and recommended them for promotions and salary increases. In short, Sherry Johns had complete control over her project teams when she worked at Manu-Tech, Inc. In her mind, "borrowing" team members from other managers and supervisors as she is required to do at ABC Technologies is absurd. She has no control and ABC's managers and supervisors are not very cooperative in providing the team members she needs for her projects.

On the verge of resigning because she cannot seem to get the level of commitment and cooperation she wants out of her team members, Johns decided to talk the situation over with the executive who hired her. After listening patiently, the executive said, "Sherry, it's not as bad as you think. The matrix structure takes a little getting used to, but eventually you will come to appreciate its benefits. In the meantime, you just need to learn diplomacy."

Discussion Questions

In this scenario, Sherry Johns is accustomed to having line authority—hence greater control—over the members of her project teams. Now that she works in a matrix organization, she feels she has no control. Have you ever been in a situation in which you had to depend on others over whom you had no control to get a job done? If so, expound on the situation. Did you find the situation difficult? How did you handle the situation? What did the executive mean when he told Johns she needed to learn diplomacy?

SUMMARY

Project managers perform both process and people functions. Process functions fall into the following groups: initiating, planning, executing, monitoring and controlling, and closing out projects. People functions include leadership, teambuilding, motivation, communication, time management, change management, diversity management, and adversity management. To be an effective project manager an individual must have knowledge and skills in both process and people functions.

Project management can be a step up the career ladder toward higher management positions. A common career path for engineering and technology professionals is project manager, program manager, senior corporate executive, and chief executive officer.

Effective project managers have the following characteristics: advanced process skills, advanced people skills, intellectual curiosity, commitment, vision, insight, people orientation, and character. In addition to these characteristics, Duncan Brodie of ProjectSmart adds the following: (1) focus on solutions, (2) participative and decisive, (3) focus on the customer, (4) focus on win-win outcomes, (5) lead by example, and (6) elicit the best from all stakeholders.[11]

Project managers may work in firms that have a functional, matrix, or project-oriented organizational structure. The functional structure is hierarchical. Functionally structured organizations are sometimes referred to as line organizations. In functional organizations, project managers may be called upon to lead teams that are staffed by personnel over whom they have direct line authority or teams that are cross-functional and draw their members from a number of different disciplines and/or departments. Project managers typically do not have line authority over cross-functional teams.

In a matrix organization, people of similar backgrounds are grouped together in their own departments—engineering, manufacturing, quality, sales/marketing—and supervised by the managers of their respective departments. Each department is considered a pool from which members are drawn as needed to staff project teams. Project managers in matrix organizations do not have line authority over their team members. Hence, they must be accomplished diplomats.

In organizations that adopt the project structure, everything revolves around project teams. Project managers in these types of organizational structures have complete line authority over their team members. The major challenge for project managers in project-oriented firms is competition between and among project teams for the resources needed to complete projects on time, within budget, and according to specifications.

Project management certifications are available from the Project Management Institute for individuals who meet the education and experience requirements. The various levels of certifications, candidate requirements, and examination information are available from the Project Management Institute at www.pmi.org.

KEY TERMS AND CONCEPTS

Planning projects
Scheduling projects
Estimating the cost and duration of projects
Procuring project resources
Tracking progress and reporting
Handling risk
Leadership function
Teambuilding function
Conflict management function
Motivation function
Communication function
Time management function
Change management function
Diversity management function

Adversity management function
Advanced process skills
Advanced people skills
Intellectual curiosity
Commitment
Vision and insight
People orientation
Character
Functional structure
Matrix structure
Project structure
Project management certifications
Project management professional examination

REVIEW QUESTIONS

1. List and briefly summarize the process functions of project managers.

2. List and briefly summarize the people functions of project managers.

3. List eight skills that make up the skill set for project managers.

4. What are the characteristics of an effective project manager?

5. Explain the concept of the functional organization structure and how it can affect the project manager's job.

6. Explain the concept of the matrix organizational structure and how it can affect the project manager's job.

7. Explain the concept of the project-oriented organizational structure and how it can affect the project manager's job.

8. Explain the various project management certifications that are available to engineering and technology professionals.

9. What are the requirements to sit for the Project Management Professional certification examination for an individual who holds a Baccalaureate degree?

APPLICATION ACTIVITIES

The following activities may be completed by individual students or by students working in groups:

1. Identify a project manager in an engineering or technology firm who is willing to cooperate and ask this individual the following questions: (a) How did he or she become a project manager? (b) What advice would he or she give a student who wants to become a project manager? and (c) Does he or she recommend becoming certified in project management?

2. Review the process and people functions of project management. Which ones will represent the biggest learning challenge for you? Which ones will come more easily to you? Review the characteristics of an effective project manager. Which of these characteristics do you already have? Which ones you will have to develop will come most easily to you? Which ones you will have to develop will be the most difficult for you?

ENDNOTES

1. Project Management Institute, *A Guide to the Project Management Body of Knowledge,* 4th ed. (Newtown Square, Pennsylvania: Project Management Institute, 2008), 45–46.

2. Ibid., 351.

3. Ibid., 14.

4. W. Edwards Deming, "The Deming Cycle." Retrieved from www.balancedscorecard.org/?TabId=112.

5. Joseph, Juran, "Total Quality Management." Retrieved from http://totalqualitymanagement. wordpress.com/2009/06/07dr-joseph-juran/ on February 17, 2012.

6. Project Management Institute, *A Guide to the Project Management Body of Knowledge,* 243.

7. David Litten, "How to Become a Project Manager." Retrieved from http://projectsmart.co.uk/ how-to-become-a-project-manager.html on January 14, 2012.

8. Duncan Brodie, "7 Habits of Brilliant Project Managers." Retrieved from http://projectsmart. co.uk/7-habits-of-brilliant-project-managers.html on January 14, 2012.

9. Project Management Institute, "Which Certification is Right for You?" Retrieved from http:// www.pmi.org on January 15, 2012.

10. Project Management Institute, "Project Management Professional (PMP)." Retrieved from http:// www.pmi.org/Certification/Project-Management-Professional-PMP.aspx on January 20, 2012.

11. Duncan Brodie, January 14, 2012.

CHAPTER 3

Project Initiation

Projects in engineering and technology firms may be initiated from external or internal sources. The external sources of projects for engineering and technology firms are typically customers, but they might also include agencies and organizations that provide grant funding for specific high-priority projects or any other source of external funding. Internal sources of projects are typically members of the engineering and technology firm's management team who are interested in improving the firm's competitiveness in some way (e.g., quality improvement project, addition of a new product or service, expansion of a facility, equipment upgrade, software implementation).

Project initiation is the first phase of a project, the other phases being planning, execution, monitoring and controlling, and closing (Figure 3.1). It is during the project initiation phase that an engineering and technology project is conceptualized—where the idea for the project is given form and substance. It is also where the decision is made to commit the engineering and technology firm's resources to a project. No project should go forward without a definite commitment from the firm's top executives. The project initiation processes are about identifying, compiling, and summarizing the information needed by the firm's executives to decide whether or not to proceed with a proposed project.

The outcome of the project initiation phase is a comprehensive description of the purpose of the project; an explanation of the projects' feasibility; a summary of project outcomes and deliverables; a description of the scope of the project; by whose authority the project is undertaken; and the project's critical success factors. If the project initiation phase is completed properly, all stakeholders will fully understand and agree on the purpose, goals, and objectives of the project. Projects that begin without this level of understanding and agreement are guaranteed to encounter problems—problems that can undermine the satisfaction of the customer and the reputation of the engineering and technology firm.

FIGURE 3.1   Project phases.

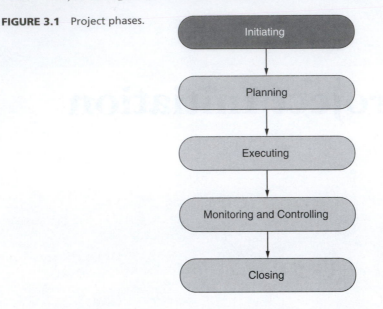

OUTCOMES OF THE PROJECT INITIATION PHASE

Like all phases of project management, project initiation is process-oriented. This means there are inputs, methods/procedures, and outcomes. Inputs can include such things as an RFP/RFQ, a statement of work, a business case, or a signed contract. Information from these various sources, as well as others, is used to produce the various products or outcomes of project initiation. These products/outcomes include the following (Figure 3.2):

- Project description
- Feasibility analysis report
- Concept document
- Project charter
- Stakeholder register
- Project kickoff meeting

Checklist
OUTCOMES OF PROJECT INITIATION

✓   Project description (product, process, and/or service)

✓   Feasibility analysis report

✓   Concept document

✓   Project charter

✓   Stakeholder register

✓   Project kickoff meeting

FIGURE 3.2   Products/outcomes of the project initiation phase.

Project Description
DEVELOP NEW EXERCISE MACHINE

Elderly Fitness International (EFI) requires a new exercise machine designed specifically for elderly individuals who are trying to prevent recurrences of lower back pain. The machine must be easily accessible to elderly individuals of limited mobility and provide sufficient support and built-into safeguards to ensure against injury from improper use. This machine has the potential to be purchased in multiple copies for all of EFI's outlets worldwide.

FIGURE 3.3 Sample project description.

PROJECT DESCRIPTION

The project description is a brief but comprehensive overview of the project. The project description is a high-level description of the product, process, or service to be developed and provided by the engineering and technology firm. The project description summarizes the following essential information: (1) what the project involves (e.g., developing or providing a product, process, or service), (2) who the project is for (e.g., external or internal customer), and (3) why the project is important. Figures 3.3 and 3.4 are examples of project descriptions that contain all three elements of essential information.

These examples contain the three elements that should be summarized in a project description. In Figure 3.3 an external customer named Elderly Fitness International wants to contract with the engineering and technology firm to develop—design, prototype, test, and manufacture—a new exercise machine that can be used to help prevent lower back pain. The project is important for two reasons: (1) the help it will provide elderly people and (2) the size of the potential market is sufficient to cover development costs and produce an acceptable profit. In Figure 3.4 an external customer—the Florida Department of Education—wants to contract with an engineering and technology firm to develop a Web-enabled software application for doing online audits. The software application has the potential to save state government millions of dollars annually.

PROJECT FEASIBILITY ANALYSIS

Before undertaking a project, it is important for engineering and technology firms to determine the feasibility of doing so. There are two sides to project feasibility. The first and most obvious side deals with the question of cost versus benefit. This side answers

---

Project Description
WEB-ENABLED SOFTWARE APPLICATION

The State of Florida's Department of Education requires a Web-enabled software application that will allow state auditors to perform financial and performance audits for school systems, colleges, and universities online rather than in person. The software application must be secure, convenient, and easy to use for auditors who are not IT professionals. This software application could save the State of Florida millions of dollars annual in travel costs for auditors.

FIGURE 3.4 Sample project description.

the question: Do the benefits of undertaking this project outweigh the costs? In other words, should the engineering and technology firm even pursue the project? The second side of project feasibility deals with the question of approach. This side answers the question: If the project is feasible from a cost-benefit perspective, what is the best approach to use in completing it?

In deciding if it should undertake a given project, an engineering and technology firm must answer several pertinent questions: (1) Is the firm is already operating at capacity? (2) Does the project fall within the firm's set of core competencies? (3) Is the potential return on investment sufficient (cost-benefit analysis)? and (4) Is the customer financially able to meet its contractual obligations. When analyzing a project for cost-benefit and other go/no-go feasibility factors, engineering and technology firms should attempt to identify all potential constraints and problems associated with a project. In other words, it is important to approach potential projects objectively and realistically. If the project appears to be sufficiently feasible to proceed, the next phase of project analysis must be completed.

Once it appears to be advisable to proceed with a project, the engineering and technology firm must decide how best to approach the project. This aspect of the project feasibility analysis proceeds in five steps as follows:

- Analyze the problem the project is supposed to solve (business problem)
- Decide what approach is best for solving the business problem
- Develop potential solutions to the business problem
- Identify and compare the advantages and disadvantages of the potential solutions
- Make recommendations concerning how best to proceed with the project

To understand how this step in the feasibility analysis process works, consider the example of the project description in Figure 3.3. In this example of a project description, a customer proposes to have an engineering and technology firm design, develop, and manufacture an exercise machine for elderly people. The purpose of the machine is to help elderly individuals strengthen their lower back muscles as a way to prevent lower back pain. The business problem is to produce a machine that will do what is expected but that will have sufficient safeguards to prevent injuries from improper use.

Project Management Scenario 3.1

Who Cares If the Project is Feasible—We Need the Work

ABC Engineering and Technology, Inc., is like most other firms. In recent years, it has found that global competition and a soft economy have thinned out the market. With more and more jobs going overseas, there seem to be fewer jobs available to bid on and more companies bidding on every job that is available. Consequently, the executive management team at ABC Engineering and Technology has begun taking on jobs that fall outside of the firm's core competencies. In fact, they are considering one such job right now. However, unlike other jobs the firm has taken on, this potential project falls well outside of ABC's core competencies. So far, in fact, that the firm's CEO is balking on submitted a bid claiming, "This project is just not feasible." His vice president for marketing disagrees. "Who cares if the project is feasible—we need the work."

Discussion Question

In this scenario, ABC Engineering and Technology, Inc., is struggling to stay afloat in an ocean of global competition. To do so, the firm is taking on projects that stretch its capabilities somewhat. Now the vice president for marketing wants the firm to take on a project that will do more than stretch the firm's capabilities. It will take the firm into areas where it has no capabilities. If you were the project manager asked to conduct the feasibility analysis for this project, what would you recommend the firm do?

Engineering and technology firms interested in this project would have to ask themselves several important questions about this business problem including:

- Does our firm has the expertise to design, develop, and manufacture the machine so that it satisfies all expectations?
- Is our firm willing to take the risk of producing a machine for elderly individuals who could be injured while using it in spite of built-in features designed to prevent improper use?

If the firm determines it can produce the desired machine and that it is willing to do so, the next concern is how to go about it. There are usually a number of different ways to solve a business problem. This step in the feasibility analysis would involve developing several different designs as well as examining the potential for adapting an existing design. The advantages and disadvantages of the various designs as well as those of the adaptation option would be identified and recorded. The firm would then compare the advantages and disadvantages of the different potential approaches and select the one that appears to be the best solution. At this point, recommendations for how best to proceed with the best approach would be developed. These recommendations will provide guidance for planners in the next phase of the project.

PROJECT CONCEPT DOCUMENT

The project concept document is a comprehensive description of the project in question. If properly prepared, this document will provide executive-level decision makers in the engineering and technology firm with sufficient information to make an informed decision concerning whether or not to proceed with the proposed project. Before developing the project concept document, the engineering and technology firm must do the following:

- *Select the project manager.* If the project manager has not been selected by now, he or she should be selected at this point. The project manager is responsible for developing the project concept document and for carrying out the remaining initiation processes (e.g., developing the project charter, helping establish the project team, working with concept input partners, developing the stakeholder register, and conducting the kickoff meeting).
- *Select the members of the project team.* The ideal time to select the members of the project team is prior to development of the project concept document so that its members can engage in the development process. The earlier in the initiation phase the project manager and project team members are named, the better their understanding of the project will be and the more ownership they will feel in the project.

- ***Identify concept input partners.*** In developing the project concept document, the project manager and project team members must work with several different entities that should have input into the document. These entities are individuals and departments within the engineering and technology firm that can help define the project and establish its direction.
- ***Identify key stakeholders.*** It is important to know who all the players associated with a project are. These players or stakeholders are identified as part of the development of the project concept document. Later, following development of the project charter, the project manager and project team annotate the list of stakeholders developed here to create a more comprehensive *stakeholder register*. The stakeholder register is explained in more detail later in this chapter.

Developing the Project Concept Document

With the project manager and project team selected and with concept input partners and key stakeholders identified, the project concept document can be developed. The document should contain at least the following information:

- Overview of the project (this can be the project description statement developed earlier)
- Purpose statement
- Goals and objectives of the project
- Selected approach and strategies for implementing the project
- Success factors
- Financial information and resource requirements
- Schedule information
- Risk information

The information needed to develop these various components of the project concept document is obtained by examining the project description and the feasibility report; interviewing concept input partners; and conducting brainstorming sessions and other meetings. It might also be necessary to meet with the customer, important suppliers, and other external stakeholders to gain their input and to verify that the project concept as explained in the draft project concept document is accurate. Once the project concept document has been completed and the final determination to proceed with the project has been made by the engineering and technology firm's executive management team, the next step in the initiation phase of the project is to develop the project charter.

Challenging Engineering and Technology Project

PREVENTING NUCLEAR TERROR

Concerns over nuclear terror began well before the tragedy of September 11, 2001. A terrorist attack using a nuclear weapon result in the loss of hundreds of thousands of lives and the loss of thousands of square miles of land—land that would be rendered useless for any kind of productive use. The downside of nuclear power is that the materials needed for making a nuclear weapon are a by-product of the process used to create the power. In addition to nuclear power plants, there are nuclear reactors in research laboratories throughout the world that also produce weapons grade nuclear material. Further, the directions for making nuclear bombs have been widely published.

Because the directions for making a nuclear weapon are so readily available, the primary obstacle facing a terrorist group wishing to make a nuclear bomb is obtaining a sufficient supply of fissile material. Less than ten kilograms is needed to make a nuclear weapon and there are more than two million kilograms extant in the world today. Consequently, nuclear security is a critical challenge facing all developed countries in the world. Engineers and technologists have a key role to play in meeting the challenge of preventing nuclear terror.

There are five major challenges facing the world in attempting to prevent nuclear terrorism: 1) developing foolproof methods for securing fissile materials, 2) developing better methods for detecting nuclear material at the greatest possible distance, 3) developing better methods for disarming a nuclear device, 4) developing better method of emergency response, communication, and clean up in the aftermath of a nuclear explosion, and 5) developing better methods for identifying the perpetrators of nuclear incidents. Not all of the solutions are engineering and technology solutions, but some are. These engineering and technology solutions include the following:

- Development of a passive device located near a nuclear reactor that could detect immediately any removal of fissile material and transmit real-time data concerning the removal to appropriate authorities.
- Development of a passive device that can determine if a nuclear reactor is being operated in a way that will maximize the production of plutonium rather than electrical power.
- Development of better methods for detecting nuclear material that is hidden in containers being transported by ground, rail, or seaborne transports. One solution to this challenge that is currently being worked on is a sophisticated scanning system that conducts what is called a *nuclear carwash*.

Meeting the challenges associated with preventing nuclear terrorism will require creative thinking, innovation, and persistence on the part of engineering and technology professionals. It will also require effective project management. *Before proceeding with this chapter, stop here and consider how preventing nuclear terrorism will require the various process and people skills of project management.*

Source: Based on *National Academy of Engineering.* http://www.engineeringchallenges.org

PROJECT CHARTER

At this point, as a result of the information contained in the project concept document, the engineering and technology firm's executive management team has made a decision to move forward with the project. As a result, it is necessary to develop a project charter. The project charter is the first official document that says *there is a project and our firm is going to undertake it*. The project charter authorizes work on the project to begin and is used by the project manager and project team as the basis for the planning phase of the project.

Different firms use different formats for their project charters. However, regardless of format project charters should contain at least the following information (Figure 3.5):

- General information
- Project overview (business problem and project objectives)

Project Charter Format

- General information

- Project overview (business problem and project objectives)

- Project scope, milestones, and deliverables

- Authority and responsibility

- Project organization

- Disaster recovery methodology

- Resources and funding

- Signatures

FIGURE 3.5   This information must be covered in the project charter regardless of the format chosen by a given firm.

- Project scope, milestones, and deliverables
- Authority and responsibility
- Project organization
- Disaster recovery methodology
- Resources and funding
- Signatures

General Information

This section of the project charter contains general information including the project title and contact information. The contact information element should include the executive in the engineering and technology firm who is higher management's representative for the project and the individual with the authority to commit resources to the project. This is the individual the program manager goes to for support, decisions beyond the authority of the project manager, and for advice concerning problems that occur during the course of the project. Also included in the contact element are the project manager, the customer's representative(s), and any other pertinent stakeholder the project manager may have to work with on a regular basis throughout the project.

Project Overview

The project overview provides a snapshot of the project including the project description developed earlier for the project concept document, the business problem the project will solve, a summary of feasibility information, and the objectives of the project. The overview might also contain a brief summary of methodologies to be used to deploy the necessary processes and technologies.

Project Scope Statement

The scope of any project is the totality of the work to be done on the project. Consequently, the scope statement contained in the project charter (see Figure 3.6) must contain a list of

Project Charter Format

General Information

Project Title _____

Contact Information

Position	Title/Name/Organization	Phone	E-mail
Responsible Executive			
Project Manager			
Customer Representative(s)			
Other			

Project Overview

Business Problem

Project Objectives

Project Scope (product description, acceptance criteria, deliverables, exclusions, constraints, and assumptions).

FIGURE 3.6 Standard format for a project charter.

Major Milestones/Critical Success Factors

Milestone	Deadline

Authority and Responsibility

Authorization (executive authorizing the project)

Project Manager (authority and responsibility)

Roles and Responsibilities of other Stakeholders

Project Organization

Organization Chart for the Project

FIGURE 3.6 (Continued)

Disaster Recovery Methodology

Resources and Funding

Resources	Source	Funding
Executive Support		
Project Team and Staff		
Facilities		
Equipment		
Software		
Materials		

Signatures

Position/Title	Signature	Date
Responsible Executive		
Project Manager		
Other Stakeholders as needed		

FIGURE 3.6 (Continued)

the project deliverables—the products or services to be produced or provided—and the tasks that must be performed to produce or provide those deliverables. The project scope statement should ensure that all stakeholders understand exactly what the project is as well as what it is not. It should rule out any future disputes among stakeholders about what was supposed to be done. Where this becomes important is when changes are requested during the execution phase of the project.

Change orders are a normal part of most projects. However, the critical question that always arises with change orders is who is responsible for their cost. A well-written scope statement can prevent disputes between the engineering and technology firm and the customer concerning who is responsible for bearing the cost of change orders. If the work that is requested in a change order is part of the project scope, the engineering and technology firm is responsible for performing the work and bearing its cost. However, if a change order asks for work that is not part of the project scope, the customer is responsible for the cost.

Because it can be used to prevent or settle disputes over money and deadlines, the project scope should be comprehensive and detailed. A well-written project scope statement will contain the following information:[1]

- **Product/service description.** This element of the scope statement lists the characteristics of the product(s) or service(s) to be produced or provided. For example, if the product to be produced is 100 copies of a specified conference table, the product description would stop with just *conference table*. It would list all of the characteristics of the conference table such as material, height, width, length, color, type of paint or stain to be used, weight, type of legs or supports to be used, and so on.

- **Acceptance criteria.** This element of the scope statement lists the criteria for determining if the product produced or the service provided is acceptable to the customer. It also explains the process that will be used for making acceptance determinations. Returning to the example of the conference table as the product to be produced, the acceptance criteria would match the desired characteristics set forth in the customer's requirements/specifications and reiterated in the *product scope description* (e.g., Proper material used? Proper length, width, height, and weight?). The process for making acceptance determinations in this case would probably be an inspection of the first article. The engineering and technology firm would be required to produce one perfect prototype that meets all of the acceptance criteria. Once the first article is approved by the customer, the firm would produce the other 99 copies just like it.

- **Deliverables.** This element of the scope statement lists all products and/or services to be delivered as well as supportive documents and any other materials that will be provided as part of the contract. In the example of the conference table, the deliverables would be 100 tables and an owner's manual that explains regular maintenance and upkeep procedures.

- **Exclusions.** This element of the scope statement lists all activities that are excluded from the project—work the engineering and technology firm is not responsible for performing. The scope is intended to ensure that all stakeholders understand what is included in the project. An effective way to ensure that this understanding exists is to list what is not included. By stating exclusions in writing, the scope of the project is clarified even beyond what is explained in the *product/service description* and the *deliverables* elements. Listing exclusions is a risk management strategy that helps project managers prevent the concept of *scope creep*—the gradual expansion of the scope over the course of the project. Any work beyond the scope of the project should be the result of a change order that has been approved by both the customer and the engineering and technology firm. In the example of the conference tables, delivery beyond the loading dock might be excluded from the scope of the work to be performed.

- **Constraints.** This element of the scope statement lists all factors that must be complied with by the engineering and technology firm. Common constraints include budget, deadlines, and specifications. With external projects, constraints are explained in the contract. Hence, they may be simply referred to in the scope statement or attached as an appendix. However, with internal projects the constraints should be included in the scope statement. In the example of the conference tables, all applicable constraints would appear in the contract and would include cost, deadlines, and quality specifications.

- *Assumptions.* This element of the scope statement lists all assumptions the engineering and technology firm makes in agreeing to complete the project on time, within budget, and according to specifications. For example, an assumption might be that a certain material that is difficult to obtain but is essential for completion of the project can be obtained in sufficient quantities and on time. This element also explains the ramifications of each assumption being wrong (e.g., increase in the cost of the project, inability to complete the project on time). Listing project assumptions is a risk management strategy as well as a way to ensure communication among stakeholders. In the example of the conference tables, an assumption might be that a difficult to obtain nontoxic, low-odor sealing compound specified by the customer can be delivered on time.

Milestones/Critical Success Factors

With every project undertaken by engineering and technology firms, the most fundamental success factors are time, cost, and quality. In other words, success ultimately means completing the project on time, within budget, and according to specifications. However, to ensure that these expectations are met, project milestones are established and deadlines for achieving the milestones are established. For example, it is common to set deadlines for 25, 50, 75, and 100 percent completion of the project. Each milestone is a success factor. Of course, with each time related milestone the product being developed or service being provided must meet all quality specifications.

Authority and Responsibility

Projects undertaken by engineering and technology firms require many decisions. To keep the project on schedule typically requires solving a multitude of problems. Because of this need to make decisions and to solve problems, it is important for all stakeholders in the project to understand their authority, roles, and responsibilities. The following issues of authority and responsibility must be made clear in the project charter:

- Who in the organization has the authority to commit the necessary resources to the project and the responsibility for making sure the project team has the resources it needs to complete the project on time, within budget, and according to specifications?
- What are the bounds of the project manager's authority for planning, executing, monitoring/controlling, and closing out the project? For example, when a change order is requested by the customer, does the project manager have the authority to commit the firm's resources or to establish a new completion date or does the project manager negotiate these issues and take them to higher management for approval. Does the project manager have the authority to pull other employees in the firm from their regular jobs to assist with some aspect of the project or does this type of action require the approval of higher management?
- Who does the project manager report to within the firm and seek assistance from for gaining the approval of higher management for decisions that must be made and problems that must be solved concerning the project?
- What are the roles and responsibilities of other key stakeholders in the project?

Project Organization

This section of the project charter provides an organizational chart for the project showing where it fits into the engineering firm and how it is organized. The top of the chart should be the customer. Under the customer is the responsible executive the project manager reports to for the project. Next in the hierarchy is the project manager and then the project team. Other key stakeholders are shown where they appear in the hierarchy of the project with either staff or reporting lines indicated.

Disaster Recovery Methodology

Because the work of engineering and technology firms has become so dependent on information technology (IT) systems, it is common to include a disaster recovery element in project charters. This element in the project charter explains how the firm is prepared to recover from a major IT problem so that it can complete the project on time, within budget, and according to specifications in spite of the problem. The disaster recovery element of the project charter should describe how the firm is prepared to affect an IT disaster recovery. The following domains of concern should be included in this element:[2]

- Servers
- Storage
- Software and automation
- Networking and physical infrastructure
- Skills needed to operate the other domains of concern

This element should describe the disaster recovery methodology the firm has selected for each of these five domains.

Resources and Funding

This section of the project charter lists a summary of the various resources that will be committed to the project, the amount of funding required for each, and the source of the funding. Typical entries on the list of resources include the time of the responsible executive, the project team (including the project manager and support staff), facilities, equipment, software, and materials.

Signatures

The final section of the project charter contains the signatures of the key internal stakeholders for the project. At a minimum, there should be a signature for the responsible executive in the firm and the project manager. Other stakeholders may need to sing the charter depending on the extent of their commitment to the project.

STAKEHOLDER REGISTER

The stakeholder register is a directory of all individuals who have a stake in the project. People who have a stake in a project are those who are directly involved, indirectly involved, or able to have an impact on the successful completion of the project. *Impact* in the current context can be thought of as the result of power, influence, and interest.[3] Power is the ability to impose one's will in terms of the project. The customer and the firm's responsible executive are examples of stakeholders with power. Influence is the ability to shape decisions without exercising power. Important suppliers and critical internal support personnel are examples of stakeholders who have influence. Other stakeholders who can have an impact on the project include subcontractors who control any aspect of the project and representatives of applicable regulatory bodies.

There are sometimes stakeholders who have either power or influence over the outcome of the project that the project manager or the project team members may not be aware of. These *hidden stakeholders* can be identified by asking known stakeholders to identify others who might be able to affect the project in a positive or negative way. In every organization, there are individuals with hidden agendas. Occasionally, these hidden agendas will provide the motivation for these individuals to quietly work for or against the successful completion of the project. For example, the author has dealt with situations in which two project managers were competing for a promotion to program manager. In order to gain an edge in the competition, one of the project managers tried to quietly undermine the performance of the other's project teams. It is important for project managers to take development of the stakeholder directory seriously. Stakeholders can either help or undermine a project manager's efforts. Better to know from the outset who is on board and who is not.

The stakeholder register should contain all stakeholders who can be identified who have an interest in the project or power or influence over the successful completion of the project. The register contains the stakeholders' names, contact information, relationship to the project, and annotations concerning how they might affect the project team's ability to complete the project on time, within budget, and according to specifications. The overall purpose of the stakeholder register is to help the project manager understand which relationships are most important to maintain for the benefit of the project and where he or she might need to exert more effort in terms of building and maintaining relationships. Figure 3.7 is a sample page from a stakeholder directory.

Stakeholder Register
ABC ENGINEERING AND TECHNOLOGY, INC.

Stakeholder Name _____

Position _____

Contact Information
_____
_____
_____
_____

Potential Impact on Project
_____
_____
_____
_____

Miscellaneous Comments
_____
_____
_____
_____

FIGURE 3.7 Entries in the stakeholder register should be thorough.

Agenda
KICKOFF MEETING: XYZ PROJECT

Date: September 27, 9:00 AM, 3rd Floor Conference Room

Project Manager: Mark Wheland

AGENDA ITEMS

1. Welcome by Mark Wheland

2. Introductions of team members

3. Distribution and discussion of the project charter:
   a. Project overview
   b. Assumptions
   c. Project scope with milestones and deliverables
   d. Authority and responsibility
   e. Project organization
   f. Roles and responsibilities
   g. Disaster recovery
   h. Resources and funding

4. Stakeholder register
   a. Known stakeholders
   b. Hidden stakeholders

5. Next steps

6. Around the room
   a. Questions
   b. Concerns

FIGURE 3.8 Project managers should develop a comprehensive agenda for the kickoff meeting.

PROJECT KICKOFF MEETING

The final step in the initiation of a project before the planning phase begins is the project kickoff meeting. Up to this point, although project team members have been involved in various aspects of the initiation phase the entire team may not have met together. Even if they have, the meetings have been to move some aspect of the initiation phase forward. The purpose of the project kickoff meeting is to formally begin the project. The project manager uses the kickoff meeting to get all members of the team oriented to the project so that they understand the project's big picture as well as their individual and respective roles in it.

Before convening the kickoff meeting, the project manager should develop an agenda and collect any back-up materials that will be distributed during the meeting. The agenda for the kickoff meeting should contain at least the following items (Figure 3.8):

- Welcome
- Introductions
- Distribution and discussion of the project charter
- Discussion of stakeholders

- Discussion of next steps
- Questions and concerns from the team

Welcome

The project manager begins the kickoff meeting by welcoming the team, introducing him or herself, and explaining what is to be accomplished during the kickoff meeting. If the team members and project manager do not already know each other, the project manager should spend some time explaining his or her background to begin establishing credibility. It is important that project managers use the welcome to show their enthusiasm for the project and to set a positive tone from the outset. If the project manager is not positive and enthusiastic about the project, why should the team members be?

Introductions

After the welcome, the project manager should ask each team member to introduce him or herself. Introductions should include name, area of expertise, and specific assignment on the project in question. The project manager should be prepared to draw members out who are reluctant to talk about themselves and to rein in those who are not. The introductions give the project manager an excellent opportunity to set the right tone for team meetings—to encourage team members to participate but not dominate. Teambuilding begins with the kickoff meeting and continues for the duration of the project. To promote teamwork from the outset, the project manager should wait until all team members have introduced themselves and explained their individual responsibilities on the project. Then he or she should explain that a project team is like a baseball team in that all members have their own positions to play but that they are also responsible for mutually supporting each other.

Distribution and Discussion of the Project Charter

It is important that all team members understand the project's big picture, their individual and respective roles in successfully completing the project, and the roles of their fellow team members. The project charter is the best tool for establishing this understanding. Consequently, the project manager should lead the team through an explanation of each individual element of the project charter giving team members opportunities to ask questions for clarification. Every team member should leave the kickoff meeting understanding the purpose of the project, project objectives, project scope, deliverables, constraints, assumptions, milestones, and any special concerns the project manager may have about completing the project on time, within budget, and according to specifications.

Discussion of Stakeholders

Project team members need to know who the key stakeholders in the project are as well as any special concerns about stakeholders. On the other hand, they can be an excellent source of information for helping the project manager identify *hidden stakeholders*. In fact, project team members can be invaluable in helping the project manager identify people within the organization who can either help or hinder the project's successful completion.

Typically, the stakeholder register will have to be revised following the kickoff meeting. Of course, this is true of all project documentation. No project document is really complete until the project itself is complete.

Discussion of Next Steps

Because the entire project team is present, the kickoff meeting gives project managers an opportunity to discuss next steps. Of course, the immediate next step is the planning phase of the project. Project managers should discuss the planning phase, including which members of the project team will be involved in the planning and what their assignments will be. They should also discuss in broad terms the executive, monitoring/controlling, and closeout phases of the project. The *next steps* item on the agenda gives project managers the opportunity to make sure all team members understand the project's big picture as well as where they fit into it.

Questions and Concerns from the Team

Perhaps the most important item on the agenda of the kickoff meeting is *questions and concerns*. This is because asking their specific questions and stating their specific concerns allow team members to gain clarity on their individual issues and concerns relating to the project. Although it is important for all members of the project team to understand the big picture, it is only human for them to be more concerned about their individual place in it. Consequently, project managers should give as much time to this agenda item as necessary to ensure that project team members are clear about both the big picture for the project and their individual roles in helping complete it on time, within budget, and according to specifications.

Project Management Scenario 3.2

We Don't Need a Kickoff Meeting—Let's Just Get to Work

Melissa Fountain had wanted to be a project manager since she was in college and took her first project management course. She liked the idea of being responsible for taking a project from birth to maturity. Consequently, now that she had been named a project manager, Fountain intended to jump in with both feet, and no wasting time on preliminaries. Consequently, when one of her team members asked when she planned to have the project kickoff meeting, Fountain responded, "We don't need a kickoff meeting—let's just get to work." Seeing that her team member was taken aback, Fountain asked him to explain why he thought they needed to have a kickoff meeting.

Discussion Question

In this scenario, Melissa Fountain is so anxious to get started on her project that she plans to skip preliminary activities such as the kickoff meeting. One of her team members who is surprised at her decision has been asked to justify the need for a kickoff meeting. Put yourself in this team member's place. How would you explain the need for a kickoff meeting?

SUMMARY

The products/outcomes of the initiation phase of a project are a project description, feasibility analysis report, concept document, project charter with scope, stakeholder register, and the project kickoff meeting. The project description summarizes what the project involves, who the project is for, and why the project is important. The feasibility analysis should answer several questions: (1) Is the firm already operating at capacity? (2) Does the project fall within the firm's core competencies? (3) Is the potential return on investment sufficient? (4) Is the customer financially able to meet its contractual obligations?

The project concept document is a comprehensive summary of the project in question. Before developing the project concept document, engineering and technology firms should: (1) select the project manager, (2) select the members of the project team, (3) identify concept input partners, and (4) identify key stakeholders. The project concept document should contain the following information: overview of the project, purpose statement, goals and objectives, selected approach and strategies for implementing it, success factors, financial information and resource requirements, schedule information, and risk information.

The project charter is a more detailed document than the project concept document. It encompasses the information in the concept document and includes additional information. Regardless of the specific format used by a given firm, the following information should be included in a project charter: general information, project overview, assumptions, project scope, milestones, deliverables, authority and responsibility, project organization, roles and responsibilities of all stakeholders, disaster recovery methodology, resources and funding, signatures.

The project stakeholder register is a directory of all individuals who have a stake in the project. The registry contains information on all stakeholders who can be identified in the following manner: names, contact information, relationship to the project, and annotations concerning how each stakeholder might affect the team's ability to complete the project on time, within budget, and according to specifications. The project kickoff meeting should cover the following agenda items at a minimum: welcome, introductions, distribution and discussion of the project charter, discussion of the stakeholder, discussion of next steps, and questions/concerns from the team.

KEY TERMS AND CONCEPTS

Project description
Feasibility analysis report
Concept document
Project charter
Stakeholder register
Project kickoff meeting
Assumptions

Project scope
Milestones
Deliverables
Authority and responsibility
Project organization
Roles and responsibilities of stakeholders
Disaster recovery methodology

REVIEW QUESTIONS

1. List and explain the outcomes of the initiation phase of a project.

2. What four things must be accomplished in preparation for developing the project concept document?

3. List and explain the information items that should be contained in the project concept document.
4. What is a project charter?
5. What information should be contained in a project charter? Explain each item of information.
6. List and explain the information that will be contained in a well-written project scope statement.
7. Explain the term "milestone."
8. Explain the concept of disaster recovery as it relates to project management. What are the domains of concern when developing a plan for disaster recovery?
9. What are *hidden stakeholders*? How can project managers identify hidden stakeholders?
10. What is the purpose of the project kickoff meeting? List the items that should be on the agenda of the project kickoff meeting.

APPLICATION ACTIVITY

The following activity may be completed by individual students or by students working in groups. However, this activity is best done in groups. Assume that your professor has assigned a class project in which you are to prepare a PowerPoint presentation covering this chapter on project initiation. In addition to the PowerPoint presentation, you are to submit a written paper that expounds further on your presentation material. THIS IS NOT YOUR PROJECT. Your project is to develop the project initiation outcomes for the assignment: (1) project description, (2) feasibility analysis report, (3) concept document, (4) project charter, (5) stakeholder register, and (6) agenda for the project kickoff meeting. If this activity is undertaken by a group of students, the group should select one student to be the project manager. The other group members will constitute the project team.

ENDNOTES

1. Project Management Institute, *A Guide to the Project Management Body of Knowledge,* 4th ed. (Newtown Square, Pennsylvania: Project Management Institute 2008), 115.
2. Cathy Warrick and John Sing, *A Disaster Recovery Solution Selection Methodology* (New York: IBM Corporation, 2004), 2.
3. Project Management Institute, *A Guide to the Project Management Body of Knowledge,* 249.

CHAPTER 4

Project Planning

The Schedule

Once a project has been initiated and the kickoff meeting completed, it is time to move into the next major phase: *planning and scheduling* (Figure 4.1). During the initiation phase milestones were established. These milestones represent specific success criteria for the project. In addition to the milestones, there are the general success criteria that always apply: time, cost, and quality. Time means completing the project on time. Cost refers to completing the project within budget, and quality means completing the project according to specifications. It is important for project managers to understand that none of these criteria—specific or general—can be met on a project, large or small, without *thorough planning* and *careful scheduling*. The planning and scheduling phase of the project involves developing a Work Breakdown Structure (WBS), using the WBS to develop a schedule, and developing a project management plan.

The three principal outputs of the planning and scheduling process explained in this chapter are the WBS, project schedule, and the project management plan. The WBS is a visual representation of the project's deliverables broken down into more finite tasks and activities. The schedule shows the starting and ending dates and duration for the project as well as for all of the tasks and activities in the WBS. It also shows project milestones and the critical path for the tasks and activities. The project management plan is the compilation of the various subsidiary plans that are developed by the project manager and project team (e.g., schedule, cost management/budget, quality management, human resource, communications, procurement, and risk management plans). Development of the project management plan begins in this chapter and continues through Chapter 10.

BENEFITS OF PLANNING AND SCHEDULING

Even a small project is a complex undertaking with a lot of parts that have to fit together in the right way and at the right time. Because of this, the project manager's job is akin to that of the orchestra conductor. If properly conducted, an orchestra will produce beautiful

FIGURE 4.1 Project phases.

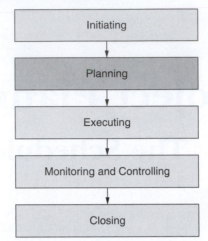

music. But if poorly conducted the orchestra will produce only disjointed noise. The same is true of project managers. If they do a good job of leading their project teams and conducting their work, the work will be completed properly, on time, and according to specifications. This is the project manager's equivalent of making beautiful music.

The more complex the project, the more difficult the *orchestration* becomes. The most important tool in making all of the moving parts fit together properly and at the right time is a comprehensive, well-planned schedule. Ensuring that a project has a comprehensive, well-planned schedule is the responsibility of the project manager. The quality of the project's schedule will have a major effect on the success of the project. Of course, just developing a well-planned schedule is just one side of the coin. The other side is in keeping the project's work on schedule (monitoring and controlling—Chapter 10).

Time and Cost-Related Benefits of Scheduling

The old saying that time is money certainly applies to engineering and technology projects. Anything that adversely affects one adversely affects the other. For example, poorly planned projects have an uneven workflow. With unsuccessful projects, it is often the case that members of the project team do not get serious about their work until they realize that a deadline for a project milestone is drawing near. Then, all of sudden, team members go into the hurry-up mode. Invariably, when project teams go into hurry-up mode quality and safety begin to suffer. The schedule suffers because hurry-up mode produces bottlenecks that throw the workflow into disarray which, in turn, causes work to fall behind schedule. When work falls behind schedule, the budget suffers because unbudgeted overtime must be approved in order to meet pressing deadlines. When a project team goes into hurry-up mode quality also suffers. The scrap rate increases as more products are rejected because of quality problems and rework time increases as more work must be redone to meet specifications. A well-planned schedule, on the other hand, enables an even workflow that helps prevent emergencies, bottlenecks, unplanned overtime, quality shortcomings, and safety violations.

Another problem with poor scheduling is that the uneven workflow it creates results in project team members working on top of each other and getting in each other's way as they try to do their respective aspects of the work for the project. Worse yet, it can result in one team member standing around doing nothing while waiting for another to complete his or her work. Often, the work of one team member is dependent on that of another in engineering and technology projects. This type of situation undermines productivity, and productivity is a key factor in completing a project on time and within budget. An even workflow has the opposite effect. It gives team members the room they need to do their jobs without undue interference and without standing around doing nothing while waiting for another person to finish his or her work.

A well-planned schedule has the benefit of providing a yardstick for measuring progress on a project. There is a reason that yard lines are painted on a football field: They allow all stakeholders to know if progress is being made and, if so, how much. The best way to ensure that progress is being made is to measure it.

Another time and cost-related benefit of developing a schedule is that it encourages critical thinking. Project managers need to be able to mentally develop the projects they are responsible for. They need to be able to think critically about all aspects of their projects. Part of the critical thinking process involves envisioning the scope of the project and all activities within it. Developing a schedule requires the project manager and members of the project team to mentally develop a project from start to finish. The critical thinking skills developed and honed when planning the schedule for a project will be invaluable when the time comes to execute, monitor/control, and closeout a project.

A final time and cost-related benefit of a well-planned schedule is that it promotes more effective communication among the many stakeholders on a project. Consider the following scenario. Even a small project has a long list of stakeholders: customer, higher managers within the engineering and technology firm, project manager, individual members of the project team, suppliers, subcontractors, and various colleagues within the firm. Ineffective communication among these stakeholders can create disorganization, disruption, and chaos. On the other hand, effective communication will allow all of the various players involved in a project to be fully informed so they can ensure that the work of the project is completed on time, within budget, and according to specifications. A well-developed schedule is as much a tool for the project manager as the baton is for an orchestra conductor.

Quality-Related Benefits of Scheduling

The customer's quality expectations are translated into practical terms by the project's specifications. Consequently, completing all work in strict accordance with the specifications—doing quality work—is of paramount importance. A well-developed schedule promotes quality work. To do quality work, team members need time, resources, and support from the project manager and higher management. They also need the work that must be completed before their work can start to be completed on time and according to specifications. Ensuring that this happens requires a well-planned schedule. Without such a schedule the work environment becomes disorganized and rushed. People cannot do their best work in an environment of hectic disorganization and rushed deadlines.

Safety and Environment-Related Benefits of Scheduling

Just as quality goes down when work is disorganized, hectic, and rushed, safety and environmental concerns also become an issue. When team members are rushed unrealistically they respond by taking shortcuts, cutting corners, and neglecting proper work practices. When this happens, safety and environmental concerns are put aside for the sake of making up time and getting the work completed by the deadline. For example, when working with powered equipment it can take more time to observe the various safety precautions. Team members might simply ignore safe operating procedures and take dangerous shortcuts.

Another example is the proper disposal of toxic materials. It can take time to go through all the steps necessary to properly dispose of toxic waste materials. Team members who are rushed might simply pour the material into the firm's sewer system or throw it in a dumpster to save time. These types of unsafe and environmentally unsound practices can be avoided by having a well-developed schedule that keeps work flowing properly and gives team members the time they need to do their work properly. One of the leading causes of accidents in the workplace is the neglect of approved safety and health procedures by individual employees who feel compelled to ignore them in order to complete the assigned work on time. Team members cannot help the team from a hospital bed. Consequently, project managers must be concerned about the safety ramifications of scheduling.

Project Management Scenario 4.1

I Don't Need a Schedule—It's in My Head

Mack Gainer has worked in engineering and technology positions for more than 25 years and is excellent at what he does. Consequently, Gainer is usually selected to serve on the project teams for his company's most important projects. The firm's latest project is the most important in its history. If the firm performs well on this project, it will receive a follow-on contract 10 times the size of the current contract. Naturally, Mack Gainer will play an important role in the project team. The firm is depending on Gainer and his fellow team members—all of whom have been handpicked—to complete the project on time, within budget, and according to specifications so as to make a favorable impression on the customer.

Consequently, the project manager—Gerald Caldwell—was shocked when he asked Gainer to work with him to develop a tight, well-planned schedule for the job. Gainer responded, "I don't need a schedule—it's in my head." Caldwell has been with the firm only six months, but he was brought in to help professionalize the company's project management. He is concerned that Mack Gainer—in spite of his obvious skills—might not make the desired impression on the customer in this case. He needs to convince Gainer of the value of having a well-planned schedule on paper.

Discussion Question

In this scenario, a talented employee operates by the seat-of-the-pants method. The plans for his work on projects are in his head but not laid out on paper. Gerald Caldwell, the project manager in this case, has been brought in to help correct unprofessional practices such as this no matter how talented the people who use them may be. If you were in Caldwell's place, how would you explain the importance of having a well-planned, comprehensive schedule available for all stakeholders to use?

THE PLANNING AND SCHEDULING PROCESS

Like so much of project management, developing a schedule is a process. Processes have three components: (1) inputs, (2) tools/techniques/methods, and (3) outputs. The primary inputs for the scheduling process are the project scope and WBS. Commonly used tools, techniques, and methods include bar charts, network logic/critical path diagrams. The output of the planning/scheduling process is a well-planned schedule that encompasses all of the activities that have to be completed to finish the project on time, within budget, and according to specifications. The schedule then becomes a major component in the project management plan.

Project planning and scheduling can be a complex undertaking, but there are numerous scheduling aids available. Scheduling software is readily available and much of it is excellent. However, regardless of whether a project is planned and scheduled using appropriate software—which is typically the case—or by hand, the process involves putting the WBS activities in sequence, computing/estimating and charting the duration of WBS activities, identifying milestones, developing the network diagram and determining the critical path, and transforming all of this information into a comprehensive schedule.

DEVELOPING THE WORK BREAKDOWN STRUCTURE

Project managers use the concept of deconstruction to break the entire project down into its components parts and then identify all deliverables that must be produced or provided for each component. Deliverables for a project may be broken down into three levels as shown in Figure 4.2. Level 1 is called a *parent node*. It is the parent to the boxes in Level 2 (i.e., 2a, 2b, and 2c). The boxes in Level 2 and the box in Level 1 when taken together represent 100 percent of the scope of the project in question. The boxes in Level 2 are parent nodes to the three boxes in Level 3 under them. The three boxes under 2a, for example, represent 100 percent of the scope of the parent node (box 2a). This type of WBS uses what is called the *deliverables format*.[1]

There are several other formats that can be used for developing a WBS for a project. The more commonly used of these other formats include the following:[2]

- *Verb-oriented WBS.* This type of WBS format is action-oriented rather than deliverables-oriented. The WBS consists of the actions that must be completed to produce or provide each deliverable for the project. Consequently, the first word in each element of the WBS is an action verb (e.g., plan, design, test, manufacture).

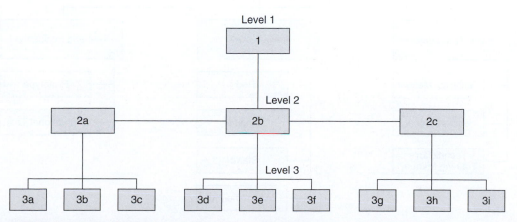

FIGURE 4.2 Typical format for a deliverables-oriented WBS.

- ***Noun-oriented WBS.*** This type of WBS is deliverables-oriented in which project work is defined in terms of the components that make up a deliverable. The first word in each element of the WBS is a noun that names that component of a larger deliverable (e.g., Subassembly A, Component B, Part C).
- ***Time-phased WBS.*** This type of WBS breaks a project down into phases. It is used for projects that are uncommonly long in duration. The first phase of the project is planned in detail to begin the project. Each subsequent phase is planned in detail only after sufficient progress has been made in the preceding phase to allow for informed planning. The time-phased approach makes use of a concept known as *rolling wave planning*. When a deliverable will be produced so far in the future that it has not yet been defined in sufficient detail to allow for accurate decomposition, it is not broken down until later when more accurate information is available.
- ***Miscellaneous WBS formats.*** There are a variety of other formats that can be used for developing a WBS. These include organizational, geographical, cost, and profit-center formats. Engineering and technology firms do not necessarily use any one type of WBS format exclusively. It is not uncommon for a firm to develop two or more WBS formats to satisfy specific planning needs within the organization.

Regardless of which format is chosen for the WBS for a project, it is important that the WBS fully encompass the entire scope of the project. The WBS should encompass all deliverables and all work that must be accomplished during the project. This is sometimes called the *100 percent rule*, a concept advocated by the Project Management Institute.[3]

A WBS can be developed on the basis of deliverables, on the basis of phases of work, or a combination of the two approaches. Figure 4.3 is an example of a WBS developed on the basis of project deliverables. Figure 4.4 is an example of a WBS developed on the basis of phases of work. The author recommends the deliverables format for developing a WBS unless the duration of the project is so extended that it must be divided into phases and the detailed planning for later phases put off until more and better information is available to planners.

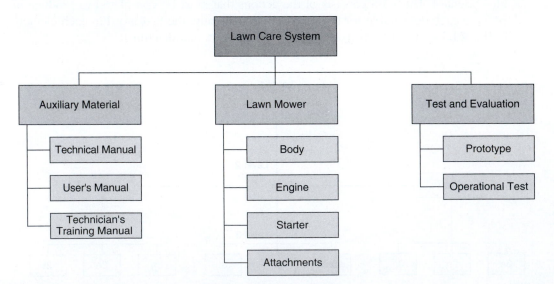

FIGURE 4.3 WBS based on major deliverables for a project.

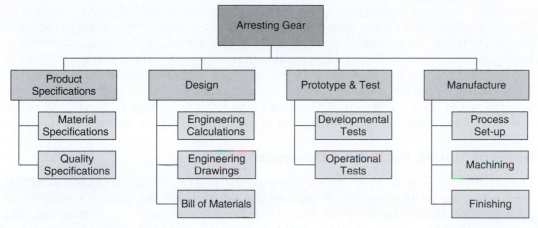

FIGURE 4.4   WBS based on phases.

Regardless of what approach is selected for developing a WBS, building the WBS around outcomes is recommended. Michael D. Taylor makes the point that "A well-designed WBS describes planned outcomes instead of planned actions. Outcomes are the desired ends of the project, such as a project, result, or service and can be predicted accurately. Actions, on the other hand, may be difficult to predict accurately. A well-designed WBS makes it easy to assign any project activity to one and only one terminal element of the WBS."[4]

Levels in the WBS

When developing a WBS, there are can be several different levels. For example, the WBS in Figure 4.5 is decomposed into three levels. Level 1 is the project—in this case a motorcycle

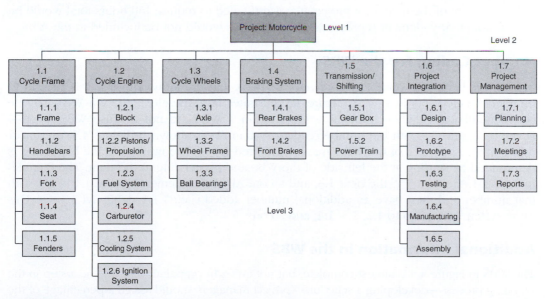

FIGURE 4.5   WBS for producing a motorcycle with numbering.

that will be designed, prototyped, tested, manufactured, and assembled. Level 2 consists of the major deliverables for the project—including project management and project integration: cycle frame, cycle engine, cycle wheels, braking system, and transmission/shifting. Level 3 consists of the subdeliverables for each major deliverable. These subdeliverables are called *work packages*. For example, the cycle frame—a major deliverable—is decomposed further into several work packages (i.e., the frame, handlebars, fork, seat, and fenders). The items shown in Level 3 represent 100 percent of the work to be done in Level 2. All of the deliverables in Level 2 represent 100 percent of the work to be done for the project.

Rules of Thumb for Project Decomposition

The question of how far to go in the decomposing process when developing a WBS is always an issue for planners. There are several rules of thumb that project managers can apply to answer this question. These rules of thumb are the outcome rule, 40-hour rule, and 4-percent rule.[5] These are explained as follows:

- *Outcome rule.* The question of how far to decompose a project is easily answered if planners are developing the WBS on the basis of outcomes rather than actions. A good stopping point when decomposing a project is when it is no longer possible to define an element in the WBS as an outcome. At a certain point, the next step in the decomposition process would result in actions rather than outcomes. Once this point is reached, the decomposition process should be terminated.
- *40-hour rule.* The 40-hour rule applies to the amount of direct labor a given element in a WBS requires. According to this rule of thumb, elements that require less than 40 hours of direct labor to complete are too finite to include in the WBS.
- *4-percent rule.* The 4-percent rule—attributed to Gary Heerkens[6]—states that the lowest level of decomposition has been reached when an element in the WBS represents 4 percent or less of the total project in terms of either time or cost. For example, 4 percent of the time for a project that is estimated to require 150 hours total would be 6 hours. Any element representing less than this would not be included in the WBS.

Number Coding for the WBS

Just as it is helpful to number the pages in a book and the items in a list, it is helpful to number the elements in a WBS. Figure 4.5 illustrates how the elements of a WBS may be coded numerically. When numbering elements in a WBS, do not confuse WBS levels—Levels 1, 2, and 3—with the number coding system. As is shown in Figure 4.5, the major deliverables in Level 2 of the WBS are all numbered 1.___. The number in the blank line is determined by starting at the left side of the WBS and proceeding to the right. The first element is 1.1, the next 1.2, the next 1.3, and so on. All of the elements under 1.1 begin with that number and then have an additional number added (i.e., 1.1.1, 1.1.2, 1.1.3 . . .). This same system is applied to 1.2, 1.3, 1.4, and so on.

Additional Information in the WBS

The WBS in Figure 4.5 is almost complete, but not quite. In preparation for the next step in the planning process—developing a schedule—project managers should assign a percentage of the total time for the project to each major deliverable as shown in the WBSs in Figures 4.6 and 4.7.

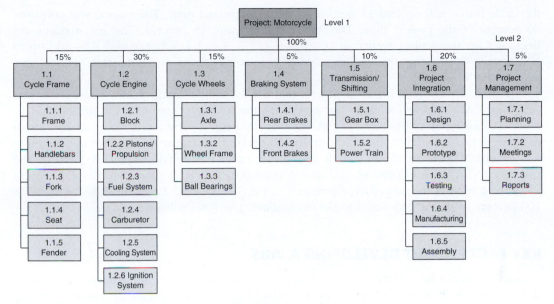

FIGURE 4.6   WBS for producing a motorcycle with time allocations and numbering.

Later, when planning the budget, the WBS can also be used for assigning a percentage of the overall budget to each major deliverable.

In Figure 4.6, the project box at the top of the WBS represents the project—a motorcycle. Therefore, the percentage of time assigned to this element is 100 percent. The major deliverables shown in Level 2, taken together, must add up to 100 percent. In Figure 4.6,

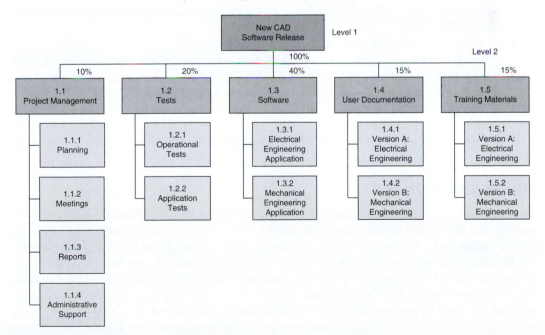

FIGURE 4.7   WBS for developing a new CAD software release with time allocations and numbering.

the cycle frame will require 15 percent of the total project time. The engine will consume 30 percent of the project time. The wheels will require 15 percent. The seven major deliverables for the project shown in Level 2—1.1, 1.2, 1.3, 1.4, 1.5, 1.6, and 1.7—add up to 100 percent. Notice that the time devoted to project management—element 1.7—must also be included in the WBS.

Figure 4.7 is the WBS for a project that involves developing a new computer-aided design or CAD software release. The new software package represents the project and is, therefore, assigned 100 percent of the time for the project. The major deliverables shown in Level 2, taken together, must add up to 100 percent. In Figure 4.7, project management will require 10 percent of the overall time of the project, tests will require 20 percent, the software will require 40 percent, and user documentation and training materials will require 15 percent each. Taken together, elements 1.1, 1.2, 1.3, 1.4, and 1.5 of the WBS account for 100 percent of the time devoted to developing the new CAD software release.

KEY FACTS ABOUT DEVELOPING A WBS

Michael D. Taylor has identified a list of key facts that summarize critical information about developing a WBS that all project managers should know. The facts are as follows:

- The top element of the WBS (Level 1) should be the overall deliverable for the project.
- Level 2 of the WBS should be a set of outcomes that collectively add up to 100 percent of the project's scope.
- Element in a WBS are best defined as outcomes—desired ends—rather than actions. The outcomes for a project will not change, but the actions taken to secure those outcomes might.
- Each element in a WBS should have a code number which identifies where it belongs in the WBS.
- All project deliverables must be included in the WBS.
- There should be no overlap in scope definition between different elements in the WBS.
- A WBS is not a plan, schedule, or chronological listing of the work to be done in a project. It is a tool used to develop the schedule for the project.
- A WBS has been sufficiently decomposed when further decomposition would identify actions rather than outcomes.
- A WBS is a picture of the project's scope not a comprehensive list of all work to be done.

THE WBS DICTIONARY

The WBS dictionary is a document that contains more detailed information about each element in the WBS. This information might include a code of account identifier, the organization or individual responsible for the element, a description of the work for the element, quality requirements, required resources, and acceptance criteria. The WBS dictionary can be an invaluable tool for project team members provided sufficient time is invented in making it comprehensive. This is a tool and an outcome of the planning process that is sometimes overlooked by project teams, but it should not be overlooked. The WBS dictionary can be as valuable to project managers and members of project teams as a regular dictionary is to a writer.

ENTERPRISE ENVIRONMENTAL FACTORS

The planning process for a project can be affected by enterprise environmental factors. These are factors—internal and external—that can influence the success of the project in ways both good and bad. Consequently, it is important to identify enterprise environmental factors so that they can be either accommodated or exploited during the planning process. Common enterprise environmental factors include:[7]

- Market conditions
- Organizational culture
- Organizational structure
- Organizational processes
- Government regulations
- Outside standards for quality, ethics, and professionalism
- State of existing facilities and equipment
- Human resource support
- Risk tolerance
- Organizational communication channels
- Quality of the workforce
- Organizational databases and information technology systems

These are just a few of the types of enterprise environmental factors that can affect the ultimate outcome of a project. The key to ensuring that these and other factors do not inhibit the success of a project is for project managers and their teams to identify as many enterprise environmental factors relating to the project in question and build their mitigation or elimination into the planning process.

ESTIMATING ACTIVITY DURATION AND SEQUENCING ACTIVITIES

Once the WBS had been completed and the project has been decomposed to the level of work packages, the duration of the activities in the project must be estimated. Estimating activity duration involves estimating the amount of work to be done and how long it will take for each work package in the WBS. For example, refer back to Figures 4.6 and 4.7. In Figure 4.6 planners have estimated that the cycle frame will require 15 percent of the overall time for the project in question. This overall figure was determined by estimating the duration for each work package under the "cycle frame" element of the WBS (i.e., 1.1.1, 1.1.2, 1.1.3, 1.1.4, and 1.1.5). In Figure 4.7, for example, 40 percent of the time for the project is estimated to be spent developing the software (element 1.3). This figure was arrived at by estimating the duration of the work packages coded 1.3.1 and 1.3.2.

Estimating the Duration of a Work Package

The duration of a work package must be estimated as accurately as possible. Estimating the duration of a work package involves answering the following question: *How many people working how many hours will be required to complete this work package?* Underestimating the duration of work packages will ensure that the project is not completed on time. Overestimating the duration of work packages will increase the project's budget and could price an engineering and technology firm's work too high which, in turn, will make it less competitive.

Estimations should be informed by past experience, records of the productivity of the firm's processes and people, the expert opinions of experienced personnel in the firm, and any other source of reliable information pertaining to the project in question.

Some planners arrive at their duration estimates by averaging a best case and worse case estimate. The actual estimate for the work package in question then becomes this average. Others establish the most accurate estimate they can arrive at and then adjust it based on the enterprise environmental factors identified earlier in the planning process. For example, the quality of the workforce that will perform the work required in a given work package might lengthen or shorten the duration of that work package.

Sequencing Project Activities: Activity List, Activity Attributes, and Project Milestones

To sequence project activities, it is necessary to create an activity list, activity attributes, and project milestones. An activity list is simply a list of all the activities within the scope of the project. This is the list of activities that will be sequenced. Activity attributes are details about each activity on the activity list, such as responsible people, assumptions, constraints, and the location where the activity occurs. Project milestones are important events that occur during the course of the project.

In any project, there will be activities that can be completed simultaneously and others that must be completed sequentially. Sequencing project activities involves carefully determining what is referred to as *precedence relationships* among activities. In project planning, a precedence relationship is one that depends on a given activity coming first and another following it. An activity that must precede another activity has a precedence relationship with that follow-on activity. An activity that must follow another activity has a precedence relationship with its predecessor activity. In sequencing activities, planners determine which activity must come first, second, third, and so on, as well as which can be accomplished simultaneously with others.

As stated above, milestones are critical events in a project. Milestones are typically viewed as critical progress points. It is not uncommon for the contract issued by the customer to specify that certain milestones must be achieved by certain specified dates. For example, milestones in Figure 4.6 might include completion of 1.1 the cycle frame, 1.2 the cycle engine, 1.3 the cycle wheels, and so on. These events would be viewed as critical progress points in the overall project: developing a motorcycle. Some projects state milestones as percentage of completion of the project (e.g., Milestone 1 = 25% completion, 2 = 50% completion, 3 = 75% completion). However, most use specific events or deliverables as project milestones.

Determining the Critical Path of Activities

When all activities have been sequenced and assigned durations, they form *paths*. The critical path is the longest path. By totaling the durations of all activities along the critical path, the overall duration of the project can be determined. Shorter paths outside of the critical path have extra time known as *float*. Critical path and float are important scheduling concepts that are explained in greater detail later in this chapter.

DISPLAYING THE PROJECT SCHEDULE

Once the WBS has been completed and the activity durations have been estimated, planners can chart the project schedule. The most commonly used methods for charting project

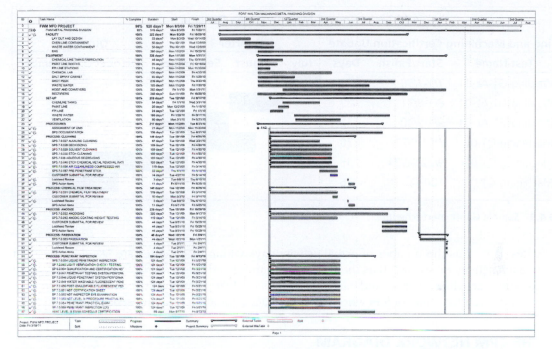

FIGURE 4.8   Combined WBS & Gantt chart schedule format. Courtesy of Fort Walton Machining, Inc.

schedules are the Gantt chart (also known as a *bar chart*), the critical path method network diagram (CPM network diagram), and a combination WBS/Gantt chart. Because it lends itself so well to scheduling software, the combination WBS/Gantt chart method is emerging as the most widely used method for displaying project schedules. Leading project management software packages such as Microsoft Project typically use this method (see Figure 4.8).

Displaying Project Schedules on Gantt Charts

Gantt charts, also known as bar charts, provide a simple and easy-to-understand tool for graphically displaying WBS activities and their respective durations. With the bar chart, all activities can be listed in the proper sequence along with a starting and ending date. The bar between these two extremes represents the duration of the activity in question. In addition, the bar chart can show which activities are interdependent and must, therefore, be performed sequentially as well as which can be performed concurrently.

Figure 4.9 is a bar chart showing the schedule for an internal project in which ABC Technologies will update its certifications in certain welding and machining processes. The welding recertification tests begin in July and run through the end of September. Notice that the number of days allocated in each month does not include Saturdays and Sundays. In the middle of August the machining recertification tests will begin and overlap with the welding tests through the middle of September. The overlap continues through the end of September. Then, there is overlap in October between the QT-03 and QT-04 machining tests.

Figure 4.9 demonstrates the advantages of a Gantt or bar chart. The activities scheduled are listed down the left side of the chart in approximate sequential order. Starting and finishing

ABC Technologies: Manufacturing Division								
Welding and Machining Certification Update				**Quarter 3**			**Quarter 4**	
ID	Test Name	Start/Finish	Days	July	August	September	October	November
1	Welding-GT-AL-01	7/1-7/30	21					
2	Welding-GT-AL-01	8/1-8/31	23					
3	Welding-GT-AL-01	9/1-9/30	22					
4	Machining-QT-01	8/11-9/21	30					
5	Machining-QT-01	9/22-10/31	35					
6	Machining-QT-01	10/6-11/10	26					

FIGURE 4.9 Sample Gantt chart schedule.

dates as well as the number of actual activity days within those dates are listed to the right of each activity on the list. Bars then show the starting and finishing dates graphically as well as where activities overlap and can be accomplished simultaneously. Gantt charts are easy to read, even by stakeholders who may not be accustomed to reading schedules.

THE CPM NETWORK DIAGRAM

Although Gantt charts have their advantages, they also have some shortcomings. For example, they do not display the concept of *slack*. Most projects will have activities in them that contain slack. This problem can be overcome by using the CPM network diagram for displaying project schedules (see Figure 4.10). A CPM network diagram displays each activity in the project as a coded box. The formats for the activity box can vary, but the one shown in Figure 4.11 contains the minimum information that an activity box must display: activity name, early start, early finish, late start, late finish, duration

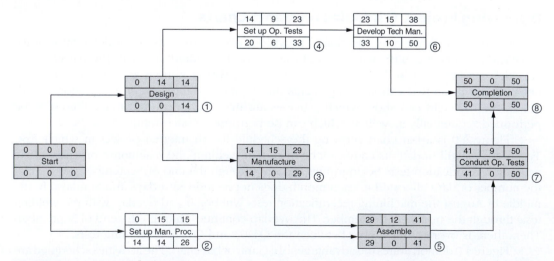

FIGURE 4.10 CPM network diagram.

FIGURE 4.11 Commonly used format
for the activity box.

Early Start	Duration	Early Finish	
	Activity Name		#
Late Start	Slack	Late Finish	
	Responsible person (optional)		

of the activity, and slack. Some firms choose to add the name of the person responsible for the activity in the box. When this is done, the name can be written over, under, or beside the activity box.

To understand slack, consider that every activity in an engineering project has an early start and early finish date as well as a late start and late finish date. When the early start and late start and early finish and late finish dates are the same, there is no slack (see Figure 4.10, Activity Box 1: Design). But when these dates differ, the activity has slack build into it. Calculating the difference between the early start and late start dates will show how much slack—room for error or room to maneuver—there is in the activity (see Figure 4.10, Activity Box 2: Set up manufacturing processes). This activity has 21.12 days of slack built into it. Consequently, if the activity is started late it can still be completed by the late finish date without throwing other dependent activities in the diagram off schedule.

The critical path in Figure 4.10 is shaded in grey. These activities represent sequentially dependent activities. For example, activity number 3 (manufacture) cannot be undertaken before activity number 1 (design) is completed. But Activity 2 (set up manufacturing processes) can be undertaken simultaneously with activity number 1. Figure 4.10 shows the benefits of using a CPM network diagram to display a project schedule. There are also some disadvantages.

Advantages and Disadvantages of the CPM Network Diagram

CPM network diagrams offer several benefits to project managers including the following: (1) give a picture of how the overall project fits together, (2) identify the most critical activities in the project—those around which all other activities revolve, (3) give the project manager a basis for setting priorities, (4) make determining the consequences of change orders easier by showing how the change can have a ripple effect throughout the project, and (5) allow the project manager to experiment on paper with different sequences to determine the optimum sequence.

The principal disadvantage of the CPM network diagram is that becoming expert in using this method can take time. One other disadvantage is that this type of schedule can require a lot of space. This final disadvantage can be turned into an advantage. By reproducing the CPM network diagram in a large format, it can be taped to the wall of the project manager's office. In this enlarged format, the project manager and other stakeholders can use it as a manual tracking tool that offers a quick visual display of progress. Some project managers actually color in the cells as progress in the project moves across the diagram from left to right.

The critical path for the activities is easy to observe, dependencies between and among activities are shown graphically, and the activity boxes contain the most critical information needed by the project team and other stakeholders.

Using CPM network diagrams for scheduling was a less popular method prior to the advent of powerful personal computers and scheduling software. Many considered it too labor intensive and difficult to learn. In those manual scheduling days, the activity boxes were typically 3 × 5 cards that were tacked or taped to a wall. However, with personal computers and scheduling software such as Microsoft Visio now readily available, CPM network diagrams have become more popular. However, project managers—present and future—should understand that even with personal computers and scheduling software mastering CPM network diagrams for scheduling still requires a great deal of work, persistence, and experience. It also requires focused study in the area of project scheduling. Students should not walk across the stage at graduation expecting to be experts at using CPM network diagrams for scheduling. It will take some time on the job before that level of competence will emerge. However, this section provides sufficient explanation of the concept to give students a good start on their journey.

Developing the CPM Network Diagram

Students might find it helpful to make a copy of Figure 4.10 to refer to while reading this section. To understand how to develop such a diagram, it is helpful to begin with a finished product and work backwards. The CPM network diagram provides a lot of information in a relatively small amount of space. The best place to start with is the activity box.

Examine the activity box key in Figure 4.11. This is one of the more commonly used formats for activity boxes in CPM network diagrams. There are other formats, but they all contain the following information or similar information:

- Activity number
- Activity duration in days
- Activity name
- Early start day or date
- Early finish day or date
- Late start day or date
- Late finish day or date
- Slack in days
- Name of the person responsible for the activity (optional)

The activity boxes in Figure 4.10 that are shaded (activities boxes 1, 3, 5, 7, and 8) represent the critical path for the project. These boxes represent activities that are sequential in nature and must be completed within the specified duration for each activity. The activity boxes that branch off from the critical path boxes represent activities that can be undertaken concurrently with another activity or activities. Arrows point to successive activities. For example, Activities 3 and 4 follow Activity 1, but Activity 4 may be undertaken at the same time as Activity 3. Further, Activity 6 may be undertaken concurrently with Activity 5 but must precede Activity 8.

BEGIN BY REVIEWING THE WORK BREAKDOWN STRUCTURE.   Before attempting to lay out a CPM network diagram such as the one on Figure 4.10, the project manager should review the WBS in which all of the activities in the project have already been sequenced and for which the durations have been estimated. These steps were explained earlier in this chapter. The project manager may have also developed a simple bar chart to display the WBS and durations graphically—a step that simplifies the development of the CPM network diagram. However, a WBS in any format—graphic or outline form—that shows the activities sequenced and with durations can serve as the starting point.

IDENTIFY SEQUENTIAL, CRITICAL, AND CONCURRENT ACTIVITIES. By examining the work breakdown structure for the project, the project manager is able to determine which activities must be completed sequentially and which can be completed concurrently with other activities as well as which activities must precede which other activities. In the example in Figure 4.10, the sequence of activities turned out to be as follows:

1. Design (critical)
2. Set up manufacturing processes (concurrent)
3. Manufacture (critical)
4. Set up operational tests (concurrent)
5. Assemble (critical)
6. Develop technical manuals (concurrent)
7. Conduct operational tests (critical)
8. Completion (critical)

LAYOUT, NUMBER, AND LABEL THE ACTIVITY BOXES. In this step, the activity boxes are laid out beginning with the critical path boxes which are placed in order of sequence from left to right as shown in Figure 4.10. As can be seen from the previous step, all activities that will be included in the CPM network diagram have numbers and the numbers are in sequence. Notice that the critical path activity boxes skip numbers in the sequence as concurrent activities are worked into the sequence at appropriate places.

Once the critical activity boxes are laid out, they are numbered and labeled to indicate the activity they represent. Then the concurrent activity boxes are numbered, labeled, and laid out. This step requires some thought. The concurrent activities boxes must be placed on the diagram in locations that will allow the scheduler to indicate that they precede certain activities and follow others. In addition, there must be room to show their relationships—by arrowed lines—not just to the critical path activities but to each other. This step can require some *juggling* of concurrent activity boxes to find the simplest, least cumbersome, and most descriptive arrangement. Once the activity boxes are laid out, the relational arrows are added.

ADD START AND FINISH DAYS, DURATION, AND FLOAT. The duration of each activity was estimated in an earlier step in the scheduling process. Hence the first task in this step is to indicate the duration above each activity box. For example, in Figure 4.10 duration for various activities can be determined by studying the information in the respective boxes. The next task in this step of developing the CPM network diagram is to calculate early start/early finish days—also referred to as the *forward pass*—and late start/late finish days—also referred to as the *backward pass*. The forward pass is calculated first and the backward pass second. The forward and backward passes are calculated as follows:

• *Calculating the Forward Pass (Early Start/Early Finish Days)* The early start/early finish days for each activity are determined using what is called *the forward pass*. The forward pass is accomplished by beginning at the start of the project and reading the network diagram from left to right. The durations for each activity box were determined earlier when developing the WBS. These durations will be used to calculate the early start/early finish days for each activity. Beginning with the first activity in Figure 4.10—Design—place a zero in the early start corner of the box. The zero represents the beginning of the day on the first day of work. Add the duration—14 days in Activity 1—to the zero to determine the early finish day for Activity 1. The early finish day for Activity 1 becomes the early start day for successive activities. When a single activity has two or more predecessor activities, use the larger of the early finish dates of the predecessor as the early start date for the successor activity. This process is repeated from left to right until the early start/early finish days have been determined for all activity boxes in the CPM network diagram.

- ***Calculating the Backward Pass (Late Start/Late Finish Days and Slack)*** The late start/late finish days and slack are determined using what is called the *backward pass*. The process for calculating the late start/late finish days and slack is called a backward pass because the calculations begin at the end of the project and proceed backwards through the network from right to left (just the opposite of the forward pass). This is the major difference. The calculations are performed in the same way. Beginning with the last activity in the CPM network diagram—Activity 8 in Figure 4.10—copy the early finish day into the late finish corner of the box. Subtract the duration for that activity from the late finish day to determine the late start. Proceed in this way through the entire schedule using late finish days and the duration to determine late start days. As the scheduler moves backwards through the schedule, the late start day of a successor activity becomes the late finish day of its predecessor. When two or more successor activities back up into a single activity, the earliest late start day becomes the late finish day for the single activity.

- ***Calculating Slack*** Slack is the difference between the early start and the late start days for a given activity in a CPM network diagram. If the difference is zero, there is no slack. When there is no slack, the activity is on the critical path. For example, refer to Activity 3—manufacture—in Figure 4.10. The late start day is day 14 and the early start day is day 14. This means there is no slack in activity 3 and it is on the critical path of the project. Now examine Activity 6: develop technical manuals. The late start day is 33 and the early start day is 23. Therefore, this activity has 10 days of slack in it. Slack calculations should be performed for all activities in the CPM network diagram.

SCHEDULING SOFTWARE AND COMBINED SCHEDULE FORMATS

Computer software has simplified the process of scheduling projects markedly over the old days when schedules had to be developed by hand and tacked to the wall. One of the more advantageous aspects of the better scheduling software is that it will produce schedules in bar chart form that contain more information than the typical bar chart (e.g., milestones, work package summaries, internal tasks, and external tasks). See Figure 4.12 for an example of such schedules. There are a number of scheduling software packages that are widely used. Of these, the leading package is Microsoft Project.

Figure 4.12 shows the type of schedule that can be produced using Microsoft Project and other leading scheduling software packages. Tasks are numbered and named on the left side of the schedule. In the middle of the schedule there are different columns: (1) a percent complete column that is updated throughout the project, (2) duration column, (3) start date column, and (4) finish date column. The lower side of the schedule shows task durations,

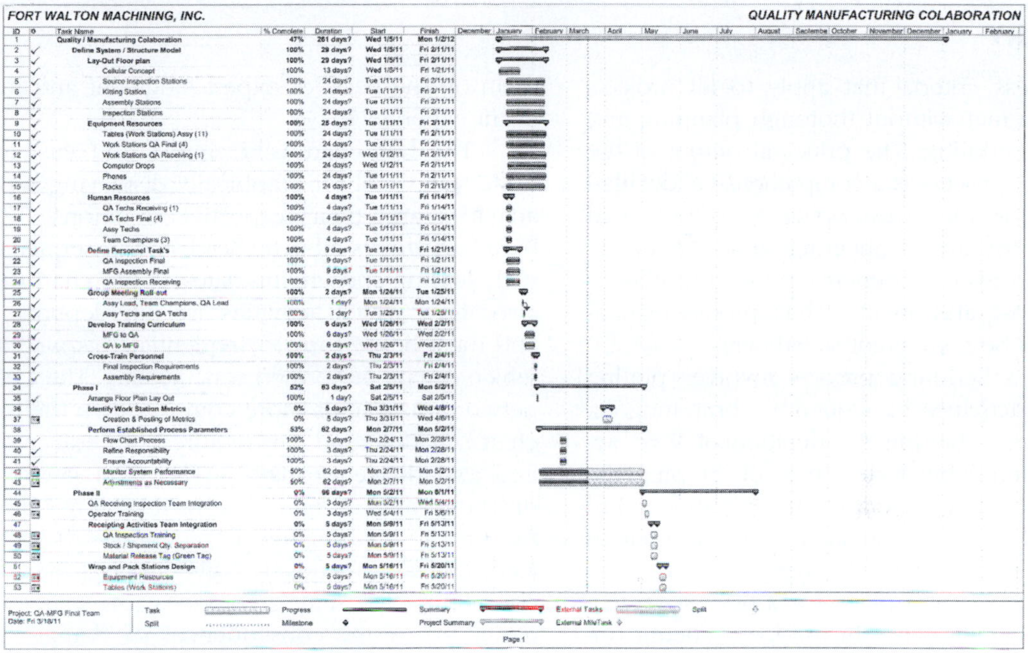

FORT WALTON MACHINING, INC.

QUALITY MANUFACTURING COLABORATION

FIGURE 4.12 Sample schedule.

milestones, intermediate summaries, a project summary, and the months over which the project will be completed. Project management students can expect to use various scheduling software packages as a normal part of their jobs.

Project Management Scenario 4.2

Let's Just Use the Bar Chart Schedule—CPM is Too Complicated

Danny Forester has just been named project manager for his firm's new aircraft refurbishment project. Magna-Tech is a technology company that specializes in updating and refurbishing military aircraft and the technologies in them. The company is thriving and has just received a contract to refurbish 50 C-130 aircraft for the Air Force. Danny Forester is the project manager for the C-130 project. Naturally, his company wants to do a good job and make a favorable impression on the Air Force. Consequently, when Forester's boss—Amanda Parker—asked to see his tentative schedule for the ten aircraft, she became concerned when the project manager produced only a rudimentary bar chart. When Parker asked if he had started laying out the CPM network diagram, Forester replied: "Let's just use this bar chart—CPM is too complicated."

Discussion Question

In this scenario, Danny Forester does not want to take the time to develop a comprehensive CPM network diagram. He would rather just use a simple bar chart. If you were Amanda Parker, would you approve this recommendation? Explain why or why not.

SUMMARY

The success criteria that apply to all projects cannot be met without thorough planning and careful scheduling. The principal output of the scheduling process is a comprehensive, detailed project schedule. Developing a well-planned schedule for an engineering and technology project can result in benefits in the critical areas of time, cost, and quality. Thus, project managers should be competent schedulers.

The scheduling process involves putting the WBS activities in sequence, computing/estimating and charting the duration of WBS activities, identifying milestones, developing the network diagram and determining the critical path, and transforming all of this information into a comprehensive schedule.

The goal for a project is a brief statement that summarizes why the project is being undertaken. Project manager uses the concept of deconstruction to break a project down into its various component parts and then identify all activities that must be performed for each component. The process is used to develop a WBS for the project. Different firms do this differently, but all should develop a WBS to use when developing the actual schedule for the project.

Sequencing involves putting the activities identified in the WBS in the order they will be completed on the job site. Once project activities have been identified and sequenced, their durations must be computed or estimated. In addition to computing activity durations on the basis of hard data, project managers can estimate them on the basis of expert judgment and relevant experience.

The bar chart is a simple and easy-to-understand tool for graphically displaying WBS activities and their respective durations. The bar chart shows all activities listed in sequence with starting and ending dates. A bar chart can also show which activities are interdependent and must be performed sequentially as well as which can be performed concurrently. The CPM network diagram is more complex than the bar chart schedule and can require both education and experience to master. The CPM network diagram offers several benefits: (1) shows how the project fits together, (2) identifies the project's critical activities, (3) gives project managers a basis for setting priorities, (4) makes it easier to see the consequences of change orders, and (5) allows the project manager to experiment with different work sequences to determine the optimum sequence. The principal disadvantage of the CPM network diagram is that becoming expert in using this method can take time.

Scheduling software has simplified somewhat the task of developing project schedules. Scheduling software such as Microsoft Project can produce a schedule that lists and numbers all tasks, displays continually updated progress percentages, the duration of each task, start and completion dates, milestones, summaries, external activities, internal activities, and other critical information about a project.

KEY TERMS AND CONCEPTS

Time and cost-related benefits of scheduling
Quality-related benefits of scheduling
Safety and environmental benefits of
    scheduling
Work Breakdown Structure (WBS)
Sequencing
Duration
Critical path

Bar chart
CPM network diagram
Concurrent activities
Activity box
Float
Forward pass
Backward pass
Project milestone

REVIEW QUESTIONS

1. What are the time and cost-related benefits of scheduling?
2. What are the quality-related benefits of scheduling?
3. What are the safety and environmental benefits of scheduling?
4. List and briefly explain the steps in the planning and scheduling process for a project.
5. What is a Work Breakdown Structure?
6. Explain the concept of sequencing.
7. Explain how a project manager can determine the duration of a project activity.
8. Describe the types of information project managers can include on a bar chart schedule.
9. What are the advantages or benefits of the CPM network diagram?
10. What is the major disadvantage of the CPM network diagram?
11. What information is typically included on a CPM network diagram?

APPLICATION ACTIVITIES

The following activities may be completed by individual students or by students working in groups:

1. Contact an engineering and technology firm in your region that will work with you in completing this activity. Ask to see examples of schedules of projects the firm has completed or that are in progress. Does the firm use bar charts, CPM network diagrams, or both? Does the firm use scheduling software? What software package does the firm use?
2. Do the research necessary to identify the leading scheduling software packages available to engineering and technology firms. Develop a chart that compares all pertinent features, advantages, and disadvantages of the various packages. On the basis of your research, which package would you recommend that an engineering and technology firm uses?
3. Assume that you have been selected to serve as the project manager for a class project that must be presented on the last day of the current term. The presentation is to be given in a PowerPoint format with a written paper provided to all class members. You will have five class members as part of your project team to work with you in developing and making the presentation. Your task in this activity is NOT to actually make the presentation and write the paper. Rather it is to develop a CPM network diagram for the project. As an alternative, you may develop a computer-generated schedule for the project if you have access to scheduling software. You may choose the topic of the hypothetical presentation.

ENDNOTES

1. Michael D. Taylor, "How to Develop Work Breakdown Structures." (Systems Management Services: 2009), 4. Retrieved from http://www.projectmgt.com on January 14, 2012.
2. Ibid.
3. Project Management Institute, *A Guide to the Project Management Body of Knowledge,* 4th ed. (Newtown Square, Pennsylvania: Project Management Institute, 2008), 121.
4. Michael D. Taylor, "How to Develop Work Breakdown Structures," 3.
5. Michael D. Taylor, "How to Develop Work Breakdown Structures," 6–7.
6. Gary R. Heerkens, *Project Management* (New York: McGraw-Hill, 2002), 103.
7. Michale D. Taylor, "How to Develop Work Breakdown Structures," 9–10.

CHAPTER 5

Project Planning
The Cost Estimate and Budget

Accurately estimating the cost of a project and then translating that estimate into a well-considered, realistic budget are two of the most important planning activities project managers will be involved in. The extent of the project manager's involvement depends on the size and composition of the engineering and technology firm as well as timing. Some larger firms have individuals or even departments dedicated to preparing cost estimates and budgets. Further, the cost estimate is often developed in response to a request for proposal (RFP) or request for quote (RFQ) issued by a customer. When this is the case, the cost estimate is developed even before the project is initiated.

In this situation, project initiation is typically dependent on whether or not the firm's response to the RFP or RFQ is accepted by the customer who, in turn, issues a contract. However, it is not uncommon for a firm to begin the initiation phase of a project even before the contract is awarded, especially if the contract has a tight schedule. The project manager is involved in developing the budget more often than in developing the cost estimate. However, project managers can be called upon to participate in both processes. Consequently, future project managers should know the basics of developing cost estimates and budgets for their projects.

Preparing an accurate estimate is a matter of answering the following question: How much will it cost our firm to complete this project? This is an easier question to ask than to answer. However, once the answer to this question has been determined, the firm can add in its profit and any amount of *padding* or contingency that has been determined advisable. With these things accomplished, the firm has an estimate. The question then becomes one of accuracy: How accurate is the estimate? An estimate that is too high will lessen the firm's chances of winning the contract. However, it is not uncommon for an engineering and technology firm to build contingency funds into an estimate to compensate for information that is either not available at the time the estimate is developed or is not as complete as estimators would like it to be. On the other hand, an estimate that is too low will result in the firm losing money on the contract. Consequently, developing accurate estimates for projects that contain only minimal and essential contingencies is one of the most important activities project managers may be called upon to participate in.

Translating a cost estimate into an actual budget for a project is also an important activity. Completing projects within budget is one of the three basic success criteria that apply to all engineering and technology projects. Consequently, it is important for the budget to be both accurate and realistic. Although the cost estimate is the key input for developing a project's budget, the budget should be more accurate and more focused than the estimate, especially on larger projects.

In the time between completing the estimate and preparing the budget, new and more accurate information often becomes available on larger projects. The lack of information and incomplete information are why firms build contingencies into their cost estimates, as was just explained. Contingency dollars are built into an estimate when necessary to compensate for information that is either not available or is not sufficiently complete at the time of developing the cost estimate. Often, these informational shortcomings are cleared up by the time the budget is developed. Hence, the budget for larger projects is often based on better, more complete, more accurate information than the cost estimate. This benefit should be reflected in its accuracy. On smaller projects, the cost estimate and the budget are often the result of the same process and, in fact, are often the same product. Regardless of how they are developed, the cost estimate and budget become another component of the overall project management plan.

THE COST ESTIMATE

Businesses are in business to make a profit. Businesses that fail to make a profit eventually fail. Consequently, when an engineering and technology firm has an opportunity to produce a product or provide a service for a customer, the cost of doing so becomes a must-know factor. This fact makes accurate cost estimates essential to the success of engineering and technology firms. To be competitive, firms must be able to develop accurate cost estimates for the jobs they propose to do, add in a realistic profit margin, complete the work specified within the estimated cost.

A cost estimate is an informed prediction made at a given point in time based on the information available at that time of what it will cost to complete a given job. Every job undertaken by an engineering and technology firm will incur personnel, material, and overhead costs. Accurately predicting what these costs will be is essential to success. If an engineering and technology firm's estimate turns out to be less than it will actually cost to complete a project, the firm will lose money on it. Then, one might reasonably ask, "Why not just purposefully estimate high when estimating the cost of the personnel, materials, and overhead for a project?" In this way, the firm is sure to complete the project within budget. As has already been explained, this is known as *padding* an estimate or building in a contingency.

While it is wise to build a certain amount of contingency into an estimate for an engineering and technology project, padding an estimate too liberally will just ensure that some other firm wins the contract. When engineering firms are asked to submit competitive proposals or bids for a given job, the lowest bidder with the most realistic proposal is the one most likely to win the contract. An overly padded estimate is not likely to produce the lowest bid. Consequently, building in an unrealistic contingency to ensure against estimating too low is not an effective approach when developing cost estimates for projects. In point of fact, the key to success with estimating is to learn to estimate accurately. The ideal estimate is one that predicts exactly what it will cost the engineering and technology firm to complete the project in question. Of course, the perfectly accurate estimate has never been developed, but the more accurate an estimate is the better.

It is not uncommon for an engineering firm to have to submit a competitive proposal for a job before complete and accurate information about all aspects of the project are available. As long as all firms that want to submit proposals have access to the same information—no more and no less than each other—there is no problem with developing an estimate for the project because the playing field, though incomplete, is at least level. In fact, this type of situation occurs frequently in a competitive marketplace. Consequently, an engineering and technology firm's cost estimate for a project should not be viewed as a static product. Rather, it should be viewed as a living tool that is to be updated continually throughout the course of a project as better and more accurate information becomes available. In fact, comparing actual costs at the conclusion of a project to the original estimates made at the beginning of the project is an effective method for turning lessons learned into more accurate estimates in the future.

COSTS TO BE INCLUDED IN ESTIMATES

Different firms go about making estimates in different ways. There are direct and indirect costs associated with engineering and technology projects. Direct costs include those costs that are tied directly to the project in question such as the cost of personnel; materials that will be required; equipment that will have to be purchased, updated, or leased; facilities that will have to be added, leased, or renovated; services that will have to be contracted for; allowances for inflation; the cost of borrowing money (interest); and any contingency funds that will be built into the estimate to cover unknown factors that result from incomplete information. Indirect costs fall into a broad category known as *overhead*. Overhead costs are those associated with the everyday operation of a firm such as utilities and other bills that apply company-wide rather than to a given project.

Cost categories that are commonly included in cost estimates for engineering and technology projects include the following (Figure 5.1):

- Labor (personnel)
- Materials

Cost Categories for Estimates

- Labor (personnel)

- Materials

- Equipment/technology

- Services

- Facilities

- Cost of money

- Inflation allowance

- Contingency

- Risk

- Overhead (indirect)

FIGURE 5.1 Widely used cost categories for cost estimates.

- Equipment/technology
- Services
- Facilities
- Cost of money
- Inflation allowance
- Contingency
- Risk
- Overhead (indirect)

The list in Figure 5.1 contains both direct and indirect costs. Administrative support/overhead and contingency costs are considered indirect while the other costs in the list are direct. Indirect costs are not specific to a given project. Rather, they include generic costs associated with running the business (electricity, gas, water/sewer, insurance, and generic administrative costs). It is common practice to accommodate indirect costs by computing them as a predetermined percentage of the direct costs of the project. Different firms add different percentages to cover indirect costs. For example, a firm might estimate the direct costs of a project and then increase its estimate by 15 percent or some other predetermined percentage to accommodate indirect costs.

There are limited instances in which an engineering and technology firm might choose not to add indirect costs to its estimates for projects. For example, if the economy is in recession and the firm is trying to maintain enough work to just stay in business it might choose to make its proposals and bids more competitive by leaving off indirect costs. This is a common practice when an engineering firm is in the difficult position of having to lower its bid price or forfeit the opportunity to win a contract. However, this is a survival-mode approach to be used only when absolutely essential because indirect costs are real and must be paid from some source whether that source is projects or not.

The Project Management Institute lists several primary inputs for the cost estimating process. These inputs are the following (Figure 5.2):[1]

- Scope statement for the project in question including the Work Breakdown Structure (WBS) and WBS dictionary
- Schedule for the project
- Human resource plan for the project
- Risk register for the project
- Enterprise environmental factors for the project

Checklist of Resources for Developing Cost Estimates

- Scope statement with WBS and WBS dictionary for the project

- Schedule for the project

- Human resource plan for the project

- Risk register for the project

- Enterprise environmental factors for the project

FIGURE 5.2   Essential resources for the cost estimate.

Scope Statement, Work Breakdown Structure, and WBS Dictionary

The most fundamental questions that have to be answered when developing the cost estimate for a project are the following: (1) How much work has to be done? and (2) How many people with what types of expertise will it take to do the work? The best place to start in answering these questions is with the scope statement, WBS, and WBS Dictionary. These tools provide a wealth of information for the cost estimator, including a description of the product to be produced or service to be provided; deliverables; assumptions; constraints; and information about such pertinent concerns as health, safety, security, performance, insurance, intellectual property rights, licenses, and permits.[2]

Schedule for the Project

The schedule for the project will show all of the work that must be completed and the estimated duration of the work. This is invaluable information to cost estimators. For example, one of the costs that should be included in an estimate is the cost of money (interest to be paid on money borrowed to finance the project). The schedule can help cost estimators determine how long the interest will have to be paid. Another way the schedule is helpful to cost estimators is in determining the cost of time-sensitive aspects of the project. For example, the price quoted by a material supplier is good for a prescribed time period but increases after that date, cost estimators will have to factor the increase into their estimate if there is a good chance the materials will not be ordered by the specified date.

Human Resource Plan for the Project

If the human resource plan has been completed by the time the estimate is developed, it provides invaluable information about the labor aspects of the cost estimate. However, practically speaking the human resource plan is not always available when the cost estimate for a project is developed. It is not uncommon for the cost estimate to be developed in response to an RFP or RFQ even before the project initiation stage has begun. When this is the case, the human resource plan will not be available as an input for the cost estimate, although it will be ready as an input for developing the budget and is one of the best tools available to project managers who are developing budgets for their projects.

Risk Register for the Project

The risk register can be a valuable tool for cost estimators because risk minimization strategies come with a cost. The cost of the various strategies planned for minimizing project risk—purchasing insurance, building a contingency into the estimate, and other risk mitigation methods all have an associated cost that must be included in the direct costs of the project.

Enterprise Environmental Factors for the Project

Enterprise environmental factors include such things as government regulations, industry standards, process capabilities, process capacity, condition of the firm's project-support infrastructure, quality of human resources, marketplace conditions, and the political climate as well as other environmental factors that can affect the cost of a project. Consequently, it is important for cost estimators to identify all applicable enterprise environmental factors and assess their potential costs as part of developing the cost estimate.

Project Management Scenario 5.1

Why Don't We Just Pad Every Component of the Estimate?

John Markham is a new project manager. This is the first time he has ever had to prepare an estimate for a project, and he is nervous. Markham's problem is that he is afraid his estimate will be too low and the firm will lose money on his first project as a project manager. The Can-Com Project is not a big one—in fact it is quite small. Markham's company always starts its new project managers out on small projects. Nevertheless, Markham is determined to make the project a success. In discussing the preparation of the estimate with his project team, Markham has hit on the idea of padding the estimate to make sure that there will be no cost overruns. He began today's team meetings by asking, "Why don't we just pad every component of the estimate?" Some of his team members agree with him while others are opposed to the idea.

Discussion Question

In this scenario, John Markham is determined to ensure that there will be no cost overruns on his first project as a project manager. His idea is to simple pad every component of the estimate by a certain percentage. Several of his team members are opposed to the idea. Which side of the debate do you agree with? Explain and defend your reasoning.

ESTIMATING METHODS

Estimating project costs is both an art and a science. There are a number of different methods that can be used for developing project estimates, and different firms use different methods. Some use a combination of methods. The Project Management Institute lists the following cost estimation methods (Figure 5.3):[3]

- **Expert judgment.** Few factors contribute more to the accuracy of cost estimates than experience. Experienced professionals who are experts relating to the type of project in question are able to call on their insight and judgment to establish accurate projections for labor, material, and equipment quantities and costs. They will also have insight into such cost-related factors as inflation, risk, and indirect costs. The expert-judgment method of preparing cost estimates involves using historical data from similar past projects and then applying the insight, intuition, and wisdom gained through experience. The Achilles heel of the expert-judgment approach to cost estimating is the concept of *currency*. In preparing cost estimates, experts must have and use the most current data available for all cost factors (i.e., material, equipment, labor, inflation, risk). current data available for all cost factors (i.e., material, equipment, labor, inflation, risk).

- **Analogous estimating.** This cost estimating method compares the various components of the current project with analogous components of previous similar projects. For example, if the project involves developing the software for a certain application and the engineering and technology firm has developed similar software packages in the past, each component of the previous projects—material, equipment, labor, risk, inflation—can be used for arriving at an estimate for the analogous component in the current project. The challenge in using this method is accurately updating past data to current circumstances. The older the past analogous projects used as the basis for estimating, the more difficult and more critical the updating process is. As a student, you may have

---

COST ESTIMATION METHODS

- Expert judgment
- Analogous estimating
- Parametric estimating
- Bottom-up estimating
- Three-point estimating
- Reserve analysis
- Vendor bid analysis
- Estimating software

Source: Project Management Institute, *A Guide to the Project Management Body of Knowledge,* 4th ed. (Newtown Square, Pennsylvania: Project Management Institute. 2008).

FIGURE 5.3   Frequently used cost estimation methods.

used this method yourself in trying to estimate approximately what book, tuition, meals, rent, and other college-related costs will amount to for a future semester or quarter.

- ***Parametric estimating.*** This cost estimating method involves identifying all of the cost components of a given project and then comparing them to historical data for the same or similar components using statistics. A statistical relationship is established between the historical data and the current data. The more accurate the statistical model, the more accurate the estimate.

- ***Bottom-up estimating.*** This cost estimating methods involves breaking the entire project down into work packages, estimating the cost of each package, and then summarizing the cost of each individual work package to determine the cost of the project.

- ***Three-point estimating.*** This cost estimating method involves establishing the following three different estimates: (1) most likely, (2) optimistic, and (3) pessimistic. Estimators may use any of the other methods to arrive at these three points. The *most-likely* estimate is the one that cost estimators is most realistic. The *optimistic* estimate is based on the best-case scenario. The *pessimistic* estimate is based on the worst-case scenario. Some firms make a choice concerning which of the three estimates to use while others simply average the three and use the average as the estimate.

- ***Reserve analysis.*** This cost estimating method involves adding a contingency to the overall estimate or separate contingencies to each component of the estimate. Contingencies are sometimes referred to as *padding*. Padding involves adding a specified amount or a percentage to either the overall estimate or to selected components of the estimate to accommodate the potential for low estimates when insufficient information is available to cost estimators or when the information that is available is not 100 percent trustworthy. Adding a contingency is also a way to accommodate risk in a project. You have probably used this method as a college student. Assume that your parents have agreed to pay the cost of your books for the next semester and have asked you to give them an estimate. If you are not sure exactly what the books will cost, you can make your best estimate and then add a contingency. You might choose to add the contingency to each individual book or to the overall estimated costs of the books. The problem with adding contingency funds to estimates is that they increase the overall cost of the project, something that can cause the engineering and technology firm to be underbid by a competitor.

- **Vendor bid analysis.** This cost estimating method is used when the engineering and technology firm is serving as the prime contractor on a project and requesting bids from outside providers. Vendor bid analysis involves comparing the various bids received for specific components of the project. In using this method, it is important to ensure that all bids analyzed and compared come from providers that are *responsive* and *qualified*. A responsive bid is one that properly speaks to all criteria and specifications set forth in the RFP or RFQ. A qualified bidder is one that has demonstrated that it can actually perform the work in question.
- **Estimating software.** Cost estimating software does not replace any of the methods presented in this section, it just simplifies the process. Cost estimating software is widely available from numerous sources. Typical cost estimating software packages allow cost estimators to produce spreadsheets, use simulations, and apply a variety of helpful applications.

ESTIMATING PRODUCTS

Cost estimates for engineering and technology projects should result in two products: (1) a cost estimation summary and (2) cost estimation notes. The cost estimation summary is a summary of the estimate for each component of the project: labor, materials, equipment, facilities, services, inflation, interest rate, contingencies, and administrative support. Indirect costs can be added as a percentage for the overall project or as a percentage applied to each individual cost component. Figure 5.4 is an example of a summary form for a cost estimate.

Cost estimation notes provide detailed information documenting how the estimate was arrived at. This information should include the following as shown in Figure 5.5:[4]

- Explanation of how the estimate was developed
- Explanation of all assumptions upon which the estimate or component elements were based
- Explanation of all constraints and how they affected the estimate
- Explanation of any room for error applied to the estimate (plus or minus a percentage)
- Explanation of the cost estimator's level of confidence in the finished estimate

The cost estimation notes should explain what method was used for developing the estimate (expert judgment, analogous estimating, three-point estimating, combination of methods, etc.). Assumptions must be explained because they are a determining factor in arriving at an estimate. Returning to the earlier example of your parents asking for an estimate of what your textbooks will cost for the next semester, you would have to base your estimate on one or more assumptions. Since historical data show that your books have averaged a cost of $500 per semester for two years, you might reasonably assume that they will cost approximately $500 for next semester. Your estimate is based on this assumption. Therefore, you add a contingency of $50 to cover any factors you might have overlooked or are unaware of and give your parents an estimate of $550.

PROJECT ESTIMATION FORM
XYZ Engineering, Inc.

Prepared By: Date:

Approved By:

Estimate No.:

Deliverable or Component	Quantity	Material	Equipment	Labor Hours	Cost
		Subtotals			

Comments:

_____

_____

_____

Plus Indirect

Total

FIGURE 5.4 Typical cost estimation summary form.

COST ESTIMATION NOTES

• Explanation of how the estimate was developed

• Explanation of all assumptions upon which the estimate or component elements were based

• Explanation of all constraints and how they affected the estimate

• Explanation of any room for error applied to the estimate (plus or minus a percentage)

• Explanation of the cost estimator's level of confidence in the finished estimate

FIGURE 5.5 Explanatory notes for the cost estimate.

With every engineering and technology project, there are numerous potential constraints. Three that always exist are time, money, and quality. There are typically others. These constraints can have a direct effect on the cost estimate. For example, if a project has got to be completed by a specified date and it is going to take a certain number of labor hours, it might be necessary for the firm to subcontract some of the work out, hire additional employees, or build a certain amount of overtime into the estimate. The constraint of time and how it affected the estimate would be explained in the cost estimation notes, as would all other constraints.

It is not uncommon for an estimator to provide a final figure for a project and then apply a tolerance level or room for error percentage. For example, if the estimator thinks the project can be completed for $150,000 he or she might give himself or herself room for error of plus or minus 10 percent. When this approach is taken, it is not uncommon for the firm to submit the high end of the estimate as its bid but base its budget on the low end. Another way for estimators to hedge their bets is to attach a level of confidence to their estimates. For example, the estimate might be $150,000 with a level of confidence of 7 on a scale of 1 to 10. The level of confidence is used when an estimate must be based on incomplete information or information that is not completely clear.

Challenging Engineering and Technology Project

REVERSE-ENGINEER THE HUMAN BRAIN

Even the most advanced supercomputer cannot compare to the human brain. Consequently, for years some of the best and brightest in the field of engineering have been trying to build computers that could emulate human intelligence. Artificial intelligence has not come even close to emulating the capabilities of the human brain. Some scientists and engineers are beginning to think the reason for this is that designers have failed to pay close enough attention to human intelligence and how it works. Finding out how human intelligence actually works is the purpose behind the interest in reverse-engineering the human brain.

The more that scientists and engineers can learn about how the human brain actually functions, the more benefits to society they can produce. In fact, better, more capable computers is just one small benefit that could result. Other more important benefits include better biotechnology solutions to disorders caused by damaged nerve cells, disorders such as dementia and blindness. Technological innovations might be used to replace and do the work of the damaged nerve cells allowing blind people to once again see or crippled people to once again walk.

If scientists can determine how to make computers immolate the brain's ability to process multiple streams of information simultaneously rather than in the step-by-step manner of today's computers, the usefulness of computers will be enhanced by orders or magnitude. The key to making the next major breakthroughs in artificial intelligence is to decipher the brain's code for communicating. Nerve cells in the brain communicate by firing electrical impulses that release neurotransmitters. What makes deciphering the communication process a challenge is that every nerve cell receives messages from tens of thousands of other nerve cells, and circuits of nerve cells join together to form networks. Further, the firing of neurotransmitters through circuits and networks is not random. Rather, it appears that the cells fire in synchrony. *Before proceeding with this chapter, stop here and consider how reverse-engineering the human brain will require the various process and people skills of project management.*

Source: Based on *National Academy of Engineering.* www.engineeringchallenges.org/cms/8996/9109. aspx?printThis=1

Project Name: _____

Project Manager: _____

Project Task		Labor Hours	Labor Cost ($)	Material Cost ($)	Travel Cost ($)	Other Cost ($)	Total per Task
1	**Design**						
1.1	Broad Specifications						
1.2	Preliminary Specifications						
1.3	Detailed Specifications						
1.4	Acceptance Test						
	Subtotal						
2	**Development**						
2.1	Develop Components						
2.2	Develop Software						
2.3	Procure Hardware						
2.4	Integrate the Components						
2.5	Perform Integration Test						
	Subtotal						
3	**Delivery**						
3.1	Install System						
3.2	Train Customers						
3.3	Perform Acceptance Test						
	Subtotal						
9	**Project Management**						
9.1	Progress Meetings/Reports						
9.2	Interface with Vendors						
9.3	Interface with Internal Departments						
9.4	Quality Assurance						
	Subtotal						
10 - Other	Other Cost						
11 - Other	Other Cost						
	Sub-totals:						
	(Contingency):						
	TOTAL (*scheduled*):						

FIGURE 5.6 Sample budget summary forms.

THE BUDGET

The cost estimate is developed on the basis of the best information available at the time and used as the basis for responding to an RFP or RFQ. With small projects, the cost estimate often becomes the budget. But with larger projects, the budget is a more refined product than the cost estimate. It represents the actual amount of funds authorized for completing the project. The budget establishes the benchmark against which cost performance will be measured throughout the project. One of the differences between the cost estimate and the budget is that the estimate often contains contingencies. Some firms choose not to budget the contingencies but instead hold these funds in reserve to be used only if necessary.

Outputs of the budget development process are the budget summary and the cost performance baseline. Figure 5.6 is a budget summary form for a project that will involve design, development, and delivery of a product. Notice that the budget also includes line items for project management, a cost that should not be overlooked in a project budget. The cost performance baseline is used to monitor actual cost performance against a cost baseline throughout the project. Figure 5.7 is an example of a cost performance baseline. Notice in Figure 5.7 that the funding requirements are not uniform throughout the project. They start low—see the intersection of the X and Y axes on the graph—and build over time.

The cost baseline curve touches each of the X and Y intersection points of the funding requirements. To complete the project within budget the actual expenditures line should remain to the right of the cost baseline throughout the project. Of course, the two lines can temporarily intersect without the project running over budget in the final analysis, provided that actual expenditures stay to the right of the cost baseline for the majority of the project. However, the ideal condition is for the actual expenditures to stay to the right of the cost baseline throughout the project.

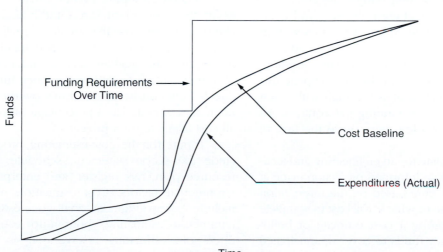

FIGURE 5.7   Cost baseline vs. actual expenditure graph.

Project Management Scenario 5.2

We Have an Estimate, Why Do We Need a Budget?

Amanda Perry is a new project manager for Ala-Tech Corporation. She began her career as an engineer, moved into quality management, and has recently been appointed a project manager after completing Ala-Tech's in-house training program for project managers with high marks. The Duncan Project is her first as a project manager. This morning her supervisor asked when he could expect to see the budget for the Duncan Project. Perry had answered saying, "No later than tomorrow morning." In truth, though, the question had confused her. Perry had planned to use the cost estimate for the Duncan Project for the budget. She decided to talk to a colleague before doing anything else. As they talked, Perry posed the question: "We have an estimate, why do we need a budget?"

Discussion Question

In this scenario, Amanda Perry wants to use the cost estimate for her project as the project's budget. However, her supervisor wants to see a budget. Is it realistic for Perry to assume she can use the cost estimate for the project as the budget? Explain your answer.

SUMMARY

Accurately estimating the cost of a project and then translating that estimate into a realistic budget are important planning activities for project managers. The extent of the project manager's involvement in developing an estimate and corresponding budget depends on a number of factors including the size and composition of the engineering and technology firm as well as timing. Preparing an accurate estimate is a matter of answering the following question: How much will it cost our firm to complete this project? An estimate that is too high will lessen the firm's chances of winning the contract. An estimate that is too low may cause the firm to lose money.

A cost estimate for an engineering and technology project is an informed prediction made at a given point in time based on the information known at the time of what it will cost to complete the project. *Padding* a cost estimate or building in contingency funds to cover unanticipated costs or estimating errors is common. However, too much padding can increase the size of the estimate to the point that the cost of the project is not feasible for the customer. An overly padded cost estimate is not likely to produce the lowest bid in a competitive bidding situation.

When developing a cost estimate, it is necessary to consider both direct and indirect costs. Direct costs are those that are tied directly to the project in question, including personnel, material, equipment, facilities, services, inflation, cost of money (interest), and contingency funds. Indirect costs consist of the firm's overhead and often are computed as a percentage of the overall cost estimate for a project.

Inputs for the cost-estimating process include the scope statement, schedule, human resource plan, risk register, and enterprise environmental factors. Cost estimation methods include expert judgment, analogous estimating, parametric estimating, bottom-up estimating, three-point estimating, reserve analysis, vendor bid analysis, and the use of estimating software.

Cost estimation products include a cost estimation summary and cost estimation notes.

The budget for an engineering and technology project represents the actual amount of money authorized for completing a project. The budget establishes the benchmark against which cost performance will be measured throughout the course of a project. One of the main differences between the cost estimate and the budget is that the estimate might contain contingencies while the budget typically does not. Some engineering and technology firms choose not to budget contingencies, but instead hold them in reserve to be used only if necessary. Outputs of the budgeting process are the budget summary and the cost performance baseline.

KEY TERMS AND CONCEPTS

Cost estimate
Labor (personnel)
Materials
Equipment/technology
Services
Facilities
Cost of money (interest rate)
Inflation allowance
Contingency
Overhead
Expert judgment

Analogous estimating
Parametric estimating
Bottom-up estimating
Three-point estimating
Reserve analysis
Vendor bid analysis
Estimating software
Budget
Budget summary
Cost performance baseline

REVIEW QUESTIONS

1. Define the term *cost estimate* as it relates to engineering and technology projects.
2. What is meant by *padding* a cost estimate?
3. What is the formal term for *padding* in a cost estimate?
4. List the direct costs that are most widely applied in developing cost estimates.
5. What are indirect costs (give examples)?
6. How is the scope statement for a project used in developing a cost estimate?
7. How is the schedule for a project used in developing a cost estimate?
8. How is the human resource plan for a project used in developing a cost estimate?
9. How is the risk register for a project used in developing a cost estimate?
10. How are enterprise environmental factors for a project used in developing a cost estimate?

11. Explain the following cost estimation methods:
    a. Expert judgment
    b. Analogous estimating
    c. Parametric estimating
    d. Bottom-up estimating
    e. Three-point estimating
    f. Reserve analysis
    g. Vendor bid analysis
12. What are cost estimation notes and how are they used?
13. Define the term *budget* as it relates to engineering and technology projects.
14. What are two outputs of the budgeting process?
15. What is the most desirable relationship between budgeted costs and actual expenditures?

APPLICATION ACTIVITIES

The following activities may be completed by individual students or by students working in groups:

1. Identify an engineering and technology firm in your area that will cooperate in completing this project. Ask to see the cost estimate for a specific project. Ask what cost estimation method(s) was used to develop the estimate. Ask if contingencies were built into the estimate and on what basis (e.g., percentage of the overall budget, percentage of individual line items).

2. Ask the same firm from Activity 1 to show you the budget for the project in question. How does the budget compare with the estimate? Were contingencies left out of the budget? Was the project completed within budget? How did the project manager monitor cost performance throughout the course of the project?

ENDNOTES

1. Project Management Institute, *A Guide to the Project Management Body of Knowledge,* 4th ed. (Newtown Square, Pennsylvania: Project Management Institute, 2008), 170–171.

2. Ibid.
3. Ibid., 171–173.
4. Ibid., 174.

CHAPTER 6

Project Planning

Human Resource, Communication, Procurement, and Quality Plans

It has already been mentioned in an earlier chapter, but bears repeating here that the various planning processes must be divided into chapters in order to present them in an organized manner that allows students to learn the material systematically. Consequently, this chapter explains how to develop the human resource, communications, procurement, and quality plans for an engineering and technology project. In actual practice, work on developing these plans begins immediately following project initiation and may proceed simultaneously with work on the other planning functions explained in this book in Chapters 3, 4, 5, and 7. The human resource, communication, procurement, and quality plans are all components of the larger project management plan.

THE HUMAN RESOURCE PLAN

In actual practice, the human resource plan is developed early in the planning process so that the project team can be assigned immediately following project initiation or shortly thereafter. Managing the human resources assigned to a project requires project managers to plan, organize, and lead project teams. Planning is explained in this chapter. Organizing and building the team is explained in Chapter Eight and leading the team is explained in Chapter Twelve. This strong emphasis on planning, organizing, building, and leading project teams grows out of the author's experience that has taught an invaluable lesson. The lesson is: Project managers can manage schedules, risk, quality, budgets, and processes, but they have to lead people. In other words, once the people needs of the team have been planned for (this chapter) and once the team has been organized and built (Chapter Eight), it must be effectively led (Chapter Twelve) if projects are to be completed on time, within budget, and according to specifications.

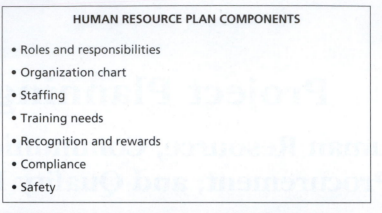

FIGURE 6.1 The human resource plan should contain at least these components.

According to the Project Management Institute, a comprehensive human resource plan will have the following components (Figure 6.1):[1]

- Roles and responsibilities
- Organization chart
- Staffing
- Training needs
- Recognition and rewards
- Compliance
- Safety

Roles and Responsibilities

This component of the human resource plan covers four critical elements: roles, authority, responsibility, and competency. The *role element* describes the various types of positions that are needed for the project in question (e.g., electrical engineer, systems engineer, quality manager, manufacturing specialist). The *authority element* explains the range of authority for all members of the team including the project manager. Who can make what types of decisions, commit project resources, approve expenditures, and so on. The *responsibility element* explains the work that each project member is assigned to perform. The *competency element* explains knowledge, skills, and capacity required of each project team member (i.e., the required qualifications for each role).

An effective way to approach this component of the human resource plan is to develop a special form that can be used for each type of role required in the project team. Figure 6.2 is an example of such a form. In this example, the role in question is for CAD technicians who will be assigned to the project. Notice that the form clearly states that the CAD technicians who are included in the project team will have no budget or decision-making authority, but they will be included in the decision-making process. Their responsibilities and required competencies are also summarized. The human resource plan would contain a form such as the one in Figure 6.2 for all positions in the project team, including the project manager.

ABC ENGINEERING AND TECHNOLOGY, INC.

PROJECT ROLES AND RESPONSIBILITIES

ROLE:
CAD Technician

AUTHORITY:
CAD technicians assigned to this project have no budget or decision-making authority. However, they will be included as stakeholders when team decisions are made.

RESPONSIBILITY:
CAD technicians assigned to this project are responsible for documenting the designs of engineers, applying the appropriate geometric dimensioning/tolerance symbols, and developing materials and parts lists.

COMPETENCY:
CAD technicians assigned to this project must be competent in the use of AutoCAD software and well-versed in the concept of geometric dimensioning and tolerancing. Experience in working on the type of project in question is preferred but not required.

FIGURE 6.2   Each role required in the project would have a form similar to this example.

Another type of form that can be used for summarizing the responsibilities of project team members is the RACI chart. RACI stands for *Responsible, Accountable, Consult,* and *Inform.* Figure 6.3 is an example of a portion of a RACI chart for a project. The chart summarizes what the four team members shown are responsible and accountable for. It also shows who should be consulted and just informed about various aspects of the project. It would not be uncommon for a human resource plan for a project to contain the type of form shown in Figure 6.2 as well as a RACI chart such as the one in Figure 6.3.

RACI Chart Project XYZ — Jones Engineering				
	Team Member			
Activity	**Smith**	**Garcia**	**Do-Vo**	**Ming**
Design	A	I	C	C
Manufacture	C	C	A	C
Quality	I	A	C	R
Test	I	R	I	A

R = Responsible
A = Accountable
C = Consult
I = Inform

FIGURE 6.3   RACI charts are useful for summarizing the responsibilities of team members.

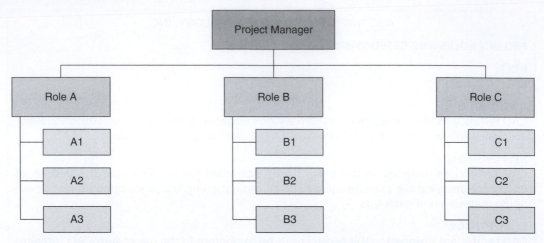

FIGURE 6.4   Template for a hierarchical team organization chart.

Organization Chart

The human resource plan should contain an organizational chart showing all the roles (positions) that make up the project team. The most commonly used type of organization for projects is the hierarchical organization. However, some project managers choose to use a matrix format. Figure 6.4 is a template for a hierarchical organizational chart for a project team. Figure 6.5 is a template for a responsibility assignment matrix.

Staffing

The staffing component of the human resource plan summarizes how and when the project manager will acquire the human resources needed for the project in question. The staffing component of the human resource plan should have at least the following three elements:

- Acquisition of staff
- Time frames
- Release method

Responsibility Assignment Matrix						
	Team Member					
Assignment/Activity	1	2	3	4	5	6

FIGURE 6.5   Team chart in matrix format.

The *acquisition* element explains where the needed staff for the project will come from. Team members might be internal personnel, external hires, contractors, or a combination of these. This element also explains how the project manager will go about acquiring team members. For example, will the project manager need the assistance of the firm's human resource department or, in the case of contractors, the procurement department? Will the project manager have to negotiate with other managers in the firm to secure the services of internal personnel or will the authority to simply select who is needed be provided by higher management? These types of questions are answered in the acquisitions element of the staffing component.

Time frames summarize how much time will be required of each team member and over what period. This element of the staffing component becomes very important when the project manager must draw team members from internal departments and negotiate with the managers of those departments for the people needed. Figure 6.6 is a histogram of the type project managers can use for summarizing and displaying staff time frame needs. In Figure 6.6, the project manager is going to need a lot of time from assembly workers over an 11-month period. The time varies from month to month. It begins with fewer than 40 hours in January but builds to a high of 240 hours in May. The chart does not display how many personnel are needed, but rather the number of hours. This will give the project manager and the manufacturing manager flexibility in assigning human resources to the project.

The *release method* element of the staffing component of the human resource plan summarizes when and how team members will be released from the project. This is important

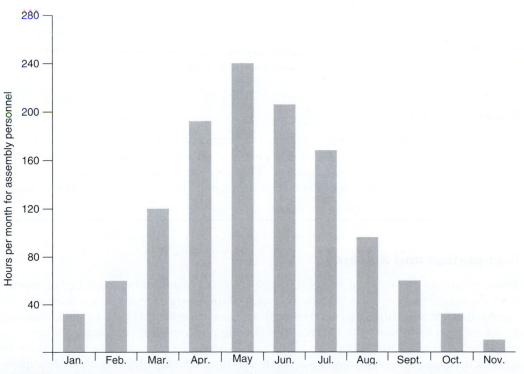

FIGURE 6.6 Time frame histogram for assembly workers.

because it gives the team members and their supervisors a measure of certainty in determining when the team members will be available for other assignments and duties. One of the most important reasons for nailing down how and when team members are released from a project is project cost. As long as team members are assigned to a project, their salary, benefits, and other costs to the firm are charged to the project. Once they are released, these costs accrue against other project or general overhead. Consequently, project managers who are responsible for completing their projects within budget want to be clear on when and how the release point is defined.

Training Needs

It is not uncommon for personnel assigned to project teams to require training. Completing the project might involve a new process, material, technology, or performance standard. The project might require certain personnel to obtain a professional or technical certification or the project manager might simply want to provide some teamwork training. Regardless of why training will be needed, the human resource plan should have a training component that explains what type of training will be provided, for whom, by whom, when, and where.

Project Management Scenario 6.1

This Seems Like a Lot of Planning

Sanchita Gomez had never managed a project team. She had served on numerous project teams, but never as the project manager. The Kal-Tech Project would be her first as a project manager and Gomez was finding the job involving more than she thought. Just this morning she had commented to a colleague: "This sure seems like a lot of planning. For example, take the human resource plan. I thought I would just select good team members and get to work." Gomez's friend—an experienced project manager—just laughed and said: "If you don't plan it, you can't manage it."

Discussion Questions

In this scenario, Sanchita Gomez is surprised to learn that so much planning is required of project managers. Students are sometimes just as surprised as Gomez to learn how much planning there is in project management. What do you think Gomez's colleague meant when he said, "If you don't plan it, you can't manage it." Do you agree with this statement? Why or why not?

Recognition and Rewards

Recognition and rewards can be used to help motivate team members to higher levels of performance. If they are to be used as motivational tools, who decides which team members receive them and according to what criteria? Are they team awards or is commendable performance on a team part of the criteria for selecting personnel for organization-wide recognition and rewards. This component of the human resource plan should explain in specific terms what types of recognition and rewards will be available, who chooses the recipients, and how recipients are chosen (selection criteria).

Compliance

With some projects, compliance becomes a major issue. For example, government contracts typically come with a list of compliance requirements. These requirements might include such things as equal employment opportunity, environmental, safety, health, and small-business set aside compliance standards. Even on private contracts that do not involve federal, state, or local government agencies directly, there are often government regulations that apply (e.g., EPA and OSHA standards). Of course, customers have their own compliance requirements that are included in the specifications for the project. Hence, this component of the human resource plan must list all applicable compliance requirements and explain how the project team will meet the requirements.

Safety

Most engineering and technology firms fall under the jurisdiction of the Occupational Safety and Health Administration or OSHA as well as its state level affiliates. Further, engineering and technology firms have ethics and competitiveness-related obligations to provide a safe and healthy work environment for their personnel. Firms that are lax in this regard are not only behaving in an unethical manner, but they are also undermining their own competitiveness. Highly skilled personnel who are injured cannot help a firm compete. The costs of workers' compensation, medical care, and lawsuits can reduce a firm's profit margin or eliminate it altogether. Hence, this component of the human resource plan explains how the firm will achieve the goal of providing a safe and healthy work environment for its personnel. Typically, this component simply refers readers to the firm's organization-wide safety and health plan.

Once the roles and responsibilities, organization chart, staffing, training, recognition, reward, compliance, and safety components have been developed, the human resource plan is complete. Often the amount of time and effort required to complete a human resource plan can be reduced by using the plan from a previous project and updating it appropriately. In any case, it is wise for firms to maintain a standard template for the human resource plan so that project managers can avoid starting over from scratch with each new project.

THE COMMUNICATION PLAN

The communication plan is developed to ensure effective communication among all project stakeholders throughout the course of the project. A sure way to guarantee problems on an engineering and technology project is to allow poor communication. Different stakeholders need to know different types of information about the project and some need to be informed of problems, proposed solutions, and progress more frequently than others. But all stakeholders in a project need to be kept up-to-date as appropriate, and they need to have convenient and reliable channels for making their views, concerns, and recommendations known.

According to the Project Management Institute, a good communication plan will contain the following information:[2]

- Communication requirements of each stakeholder
- Information to be communicated to each stakeholder, including the frequency, language, format, and level of detail
- Reason for the information being provided to each stakeholder

- Individual responsible for providing the information to each stakeholder
- Individual with authority to approve the release of confidential information
- Individuals or groups allowed to receive confidential information
- Information conveyance methods (e.g., written report, e-mail, press release, formal letter or memorandum)
- Time, budget, and other resources that will be devoted to communication
- Chain of command for dealing with issues that cannot be handled at a given level
- Method for keeping the communication plan updated throughout the course of the project
- Glossary of important terms that will be used in communicating information to stakeholders
- Flowcharts showing information flow, reports, meetings
- Communication constraints that result from legislation, government regulations, organizational policies, or other sources
- Templates and guidelines for meetings, reports, and e-mail
- Information and guidelines for using the project Web site if one is to be used including what information will be made available via the Web site

Devoting the time and effort necessary to ask and answer pertinent questions about communication before executing the project can prevent an untold number of problems. These questions include at least the following: (1) Who are the project's stakeholders? (2) What types of information does each stakeholder need? (3) How often does each stakeholder need the information? (4) In what format does the stakeholder need the information? and (5) Who should be responsible for providing the needed information to each stakeholder? The answers to these questions will form the core of the project's communication plan.

PROCUREMENT PLAN

Every engineering and technology project requires certain resources in order to be completed on time, within budget, and according to specifications. Invariably, some of these resources must come from outside of the organization. In addition to raw materials and supplies, the project team might need to acquire professional services from other individuals or organizations. Subcontracting a portion of the project work to another organization or individual is a common practice that is part of the procurement process.

Developing a procurement plan involves determining what resources are needed, identifying potential suppliers/providers of the resources, deciding which suppliers/providers to contract with, and contracting with the suppliers/providers that can best meet the firm's needs. A comprehensive procurement plan should contain at least the following components:[3]

- Types of contracts to be used in procuring the needed resources
- Evaluation criteria for evaluating bids and quotes from suppliers/providers
- The resources that will be acquired by the firm's procurement department and those that can be acquired directly by the project management team
- Standardized procurement documents
- How multiple suppliers will be managed

- How procurements will be coordinated with other aspects of the project such as scheduling and performance monitoring/reporting
- Constraints that could affect the procurement process
- Assumptions that the procurement process will be based on
- How lead times will be handled to ensure on-time deliveries
- How make-or-buy decisions will be made
- Coordinating delivery dates with the project's schedule
- Identifying risks that must be mitigated through the purchase of performance bonds or insurance contracts
- Establishing the format for the procurement/contracts statements of work
- Identifying any prequalified suppliers/providers
- Summary of the procurement metrics to be used to manage contracts and evaluate suppliers/providers

Contract Types Used in Procuring Project Resources

The legal relationships as well as the risk sharing relationships between the engineering and technology firm and the sources that provide external resources for projects are established by contracts. Consequently, project managers need to be familiar with the types of contracts and variations of these types that are widely used. The three most widely used types of contracts are *fixed-price, cost-reimbursement,* and *time/material* contracts.[4] Of these, the most widely used and preferred type is the fixed-price. The essentials of these two types of contract are as follows:

- *Fixed-price contracts.* With fixed-price contracts, the engineering and technology firm agrees to pay a specified amount for a defined product or service. The price does not vary, although some firms build incentives into the contract for early delivery or other considerations. On the other hand, fixed-price contracts can also have penalties built into them that apply if the product or service does not meet contractual specifications. Comprehensive, accurate specifications are the key to effective fixed-price contracts. There are two types of fixed-price contracts: *firm fixed price* and *incentivized fixed price*. Firm fixed price contracts establish a firm amount that will be paid for a specified product or service. As the name implies, the price does not change unless the engineering firm requests a change of some kind (e.g., change to the specifications). The incentivized fixed price contract sets a firm price for acceptable delivery and performance, but then builds in incentives for exceeding carefully specified expectations.
- *Cost-reimbursement contracts.* Cost-reimbursement contracts are sometimes called *cost-plus* contracts. This is because the supplier/provider bills the engineering and technology firm for actual costs plus a specified fee. With this type of contract, the engineering and technology firm contracts for products or services without really knowing how much they will ultimately cost. Cost-reimbursement contracts can include incentive clauses for exceeding performance and delivery expectations.
- *Time/material contracts.* Time/material contracts are used most often when contracting for professional services when the engineering firm is not sure how much time will be required. They are also used to procure materials when it is not known how

much of a given material will be needed. Hence, they are sometimes referred to as *open-ended contracts*. Engineering and technology firms avoid time/material contracts when possible for obvious reasons. However, they are sometimes necessary. When using this type of contract, engineering and technology firms can build in a clause that specifies the maximum number of hours they will pay for the professional services in question. If the contract is for material, the firm can specify a maximum unit price it will pay.

Statements of Work

An important component of the procurement plan is a statement of work for each product or service that will be contracted for. If the procurement is for professional services, the statement of work specifies precisely what work is needed, the required qualifications of those who propose to provide the services, rates, hours, where the work will be done, timeframes and deadlines, and quality specifications. If the procurement is for materials, the statement of work specifies the type, amount, and quality required as well as delivery dates and locations. Developing comprehensive statements of work is a critical part of procurement planning because these statements can determine if the engineering firm gets the product or service needed or a lesser and possibly unacceptable result.

Make-or-Buy List

When an engineering and technology firm accepts a contract to provide a product and/or service for a customer, decisions have to be made concerning what part of the contract is going to be completed in-house and what part—if any—to procure from external sources. Doing the work internally is called a "make decision." Procuring the work from external sources is called a "buy decision." The make-or-buy list is an annotated list of the work that will be done internally and that which will be done externally. Each entry on the list is annotated with a brief explanation of the rationale for the make-or-buy decision for that entry.

Procurement Documents

When procuring goods and services from external sources, it is important to use standardized forms and other types of paperwork. Standardization protects the engineering and technology firm from charges of treating potential suppliers/providers differently while ensuring that competitive quotes and bids submitted by them can be properly evaluated. The most commonly included procurement documents in the procurement plan are as follows:

- *Request for proposal (RFP).* This document is used to solicit proposals for completing the work as stated in the statement of work. The proposal is required to contain the price, capabilities of the supplier/provider, and approach that will be used to satisfy the statement of work.
- *Request for quote (RFQ).* This document is used to solicit quotes. A quote is less detailed than a proposal and is typically used when the engineering and technology firm knows the capabilities and approach of the potential supplier/provider and just needs to know what the product or service in question will cost.
- *Invitation to bid (ITB).* An invitation to bid is used to solicit bids from a broad base of suppliers/providers with the intent of selecting the lowest qualified bid.

Selection Criteria

When the engineering firm plans to solicit competitive bids or proposals, it must have predetermined criteria for selecting the best bid or proposal. What follows are the types of selection criteria typically included in procurement plans:[5]

- *Responsiveness.* Has the supplier/provider submitted a responsive bid or proposal? A responsive bid or proposal is one that properly and clearly addresses all aspects of the statement of work. A responsive proposal shows that the supplier/provider understands what it is being asked to provide as well as when, how, and according to what specifications.
- *Overall cost.* Has the supplier/provider submitted the lowest overall price? Sometimes the lowest bid is not the least costly when all factors are considered. For this reason, engineering and technology firms typically consider not just the purchase price of the goods or services they solicit, but the overall cost. The overall cost is the purchase price plus operating costs. For example, you are in the market for buying a car and you have narrowed the choices down to two. One car costs less than the other but will cost more to operate because it will use more gasoline and must use only the highest octane brand. Further, the reliability rating of the lower priced car is not good. Hence, upkeep and maintenance will cost more. When considering the overall cost, you would be wise to purchase the car that has the higher purchase price.
- *Capabilities.* Does the supplier/provider have the capabilities required in the statement of work? This criterion becomes especially important when procuring professional services and expertise.
- *Risk.* Does the supplier/provider accept its share of the risk contained in the statement of work? Does the supplier/provider appear to have a plan and the means to minimize risk appropriately?
- *Management processes.* Does the supplier/provider appear to have the management processes in place to properly manage its portion of the project? To ensure that the answer to this question is in the affirmative, some engineering and technology firms require the suppliers/providers they work with to be ISO 9000-compliant or compliant with some other recognized quality management standard.
- *Methods and techniques.* Are the methods and techniques proposed by the supplier/provider appropriate? Will they produce the desired result in a way that meets expectations?
- *Warranty.* Does the supplier/provider warrant its work for the expected period of time? Is the warranty full or limited?
- *Financial capacity.* Does the supplier/provider have the financial capacity to perform what will be expected? If it will need to borrow money, does it have the credit rating to do so?
- *Production capacity and capabilities.* Does the supplier/provider have the production capacity and the production capabilities to perform as expected on the job?
- *Eligibility for government set-asides.* Occasionally, an engineering and technology firm might receive a government contract containing a specification that a percentage of the job be subcontracted to a small business, minority-owned business, or disadvantaged business. When this is the case, have the suppliers/providers claiming a given status submitted documentation of their eligibility for government set-asides?
- *Past performance.* Has the supplier/provider performed well on similar jobs in the past? What is the extent of its experience?

- **References.** Has the supplier/provider included a list of references that can attest to its performance on other contracts? Is the feedback from references positive?
- **Intellectual property or proprietary rights.** Does the supplier/provider claim intellectual property rights or proprietary rights to the goods or services it proposes to supply? If so, how does this affect the selection process?

When the procurement plan has been completed, it will guide the engineering and technology firm's personnel in actually conducting procurements and managing the procurement process. Conducting procurements is part of the execution phase of the project.

THE QUALITY MANAGEMENT PLAN

Engineering and technology firms develop quality policies. Then, they develop quality management plans for implementing those policies. Figure 6.7 is an example of a quality policy. The quality policy sets the tone for the entire organization and, in turn, project teams concerning quality. The quality plan provides more specific information concerning how the quality policy will be carried out on a daily basis. Ideally, the quality management plans developed by project managers and their teams will be subsets of the firm's larger, organization-wide quality plan. In fact, a well-developed quality plan for an engineering and technology company will contain specific information relating to projects. When this is the case, project managers need only refer to the appropriate sections of the larger quality plan when developing one for a specific project. However, in the event that the engineering and technology firm does not have an organization-wide quality plan, the project manager will have to develop one for the specific project in question.

In developing a quality plan, it is important to approach the planning first on the macro level and second on the micro level. The macro level of quality is referred to as *quality management* or "Big Q." The micro level of quality is referred to as *quality control* or "Little Q." Quality management or Big Q involves the continual improvement of the quality of a firm's products, services, people, processes, and environment. The micro level of quality or Little Q is a subset of quality management that focuses on ensuring that the deliverables of specific project meet or exceed project specifications.

ABC ENGINEERING AND TECHNOLOGY, INC.
QUALITY POLICY

ABC Engineering and Technology is committed to providing products and services that exceed customer requirements every time. Our aim is to go beyond achieving customer satisfaction to achieving customer delight. ABC's quality policy encompasses three key principles:

• Maintaining a customer focus at all times.

• Continually improving the performance of our processes, people, and products.

• Maintaining a quality-first corporate culture.

The decisions, behavior, and actions of all personnel in our organization will be guided by this policy.

FIGURE 6.7   Sample quality policy.

Although quality is everybody's responsibility in that everyone in an engineering and technology firm plays a critical role in ensuring quality, most firms have a quality management or quality assurance department. Smaller firms may not have a full-fledged quality management department but they typically have at least a quality professional whose principal responsibility is ensuring effective quality management. Regardless of whether a firm has a quality management department or relies on a single quality management professional, an effective quality management program has the following characteristics:

- Views quality as a corporate strategy for gaining a competitive advantage in the marketplace
- Committed to exceeding customer satisfaction
- Focused on continual improvement of people, processes, products, services, and environments
- Focused on maintaining peak performance
- Dedicated to the application of scientific decision making
- Views training as an ongoing, never-ending prerequisite for continual improvement
- Promotes unity of purpose organization-wide (quality is everybody's responsibility)
- Committed to empowering employees and participatory decision making
- Committed to implementing comprehensive quality systems that have been tried, tested, and shown to be effective (e.g., Six Sigma, Lean, Lean Six Sigma)
- Committed to establishing and deploying effective quality control (Little Q) processes and tools for measuring actual performance against planned performance

Quality control or Little Q is project specific. It involves applying quality tools, such as cause-and-effect diagrams, Pareto charts, histograms, scatter diagrams, run charts, control charts, stratification, and other tools for ensuring that actual performance meets or exceeds planned performance.

Guidance from Quality Pioneers

The two most widely known pioneers of quality management in America are W. Edwards Deming and Joseph M. Juran. Each of these pioneers contributed much to the global quality movement that transformed industry throughout the industrialized world in last two decades of the twentieth century. Their impact is still felt today because of their fundamental and foundational contributions. Deming's *Fourteen Points* and Juran's *Ten Steps to Quality Improvement* still provide the best guidance available for establishing an effective quality program in a firm. Project managers should be familiar with both before participating in the development of quality plans for their projects.

Deming's contributions to the quality movement worldwide would be difficult to overstate. Many consider him the *father of quality*. Deming's philosophy is summarized and operationalized by his Fourteen Points:[6]

- Create constancy of purpose toward the improvement of products and services.
- Adopt the new philosophy organization-wide, beginning with top management.
- Stop depending on inspection to ensure quality—build quality into the product or service from the outset.
- Stop awarding contracts on the basis of low bids.
- Improve continuously and forever the system of production and service, to improve quality and productivity and thus reduce costs.

- Institute training on the job.
- Institute leadership.
- Drive out fear so that everyone may work effectively.
- Break down barriers between departments so that people can work as a team.
- Eliminate numerical goals for the workforce.
- Eliminate slogans, exhortations, and targets for the workforce as they create adversarial relationships.
- Remove barriers that rob employees of their pride of workmanship.
- Institute a vigorous program of education and self-improvement.
- Make the transformation everyone's job and put everyone to work on it.

The contributions to quality of Juran rank near those of Deming. His quality materials have been translated into 14 different languages and the Emperor of Japan awarded him the Order of the Sacred Treasure medal in recognition of his efforts to develop quality in Japanese industry. Juran's ten steps for quality improvement are as follows:[7]

- Build awareness of both the need and the opportunity for improvement
- Set goals for improvement
- Organize to meet the goals that have been set
- Provide training
- Implement projects aimed at solving problems
- Report progress
- Give recognition
- Communicate results
- Keep score
- Maintain momentum by building improvement into the company's regular systems

The guidance provided by Deming and Juran will help project managers play a more effective role in developing and carrying out quality plans for their projects.

Components of the Quality Plan

Although most engineering and technology firms will have a comprehensive quality plan that was developed and is maintained by the quality management department, it is not uncommon for customers to require that a project-specific quality plan be developed. When this is the case, project managers may draw on the firm's existing quality management plan for both guidance and content. For the purposes of this text, it will be assumed that the project manager must develop a quality plan for the project in question. A quality plan for an engineering and technology project should have at least the following components (Figure 6.8):

- ***Quality management approach for the project.*** Project managers must be concerned with the quality of the project's deliverables and the quality of the processes through which the deliverables are produced or provided. This component of the quality management plan explains how the quality of deliverables and processes is managed and ensured. It should contain a list of the deliverables for the project and the corresponding processes for each deliverable. It should also indicate what measures will be used to ensure the quality of each deliverable and process.
- ***Quality-related definitions.*** This component of the quality plan lists the terms most frequently used in the quality plan to ensure that all stakeholders who use the plan

---

**COMPONENTS OF THE
PROJECT QUALITY PLAN**

✓ Quality management approach

✓ Quality-related definitions

✓ Quality objectives for the project

✓ Process quality measures

✓ Product quality measures

✓ Quality-related responsibilities of team members

✓ Quality tools to be employed

✓ Reporting procedures

FIGURE 6.8 The quality plan should contain at least these components.

are speaking the same language and have the same understanding of terminology. The types of terms that are typically found in the definitions component of a project quality plan include the following: continual improvement, Six Sigma, Lean, Lean Six Sigma, Statistical Process Control (SPC), Just-in-Time (JIT), benchmarking, ISO 9000, Pareto charts, cause-and-effect diagrams, check sheets, histograms, run charts, control charts, flowcharts, process boundaries, process configuration, and metrics.

- *Quality objectives for the project.* The broad success criteria for all engineering and technology projects are to complete the project on time, within budget, and according to specifications. This component of the project quality plan translates this criterion into more specific terms. For example, quality objectives relating to the schedule might be: (1) to complete 50 percent of the project by _____ (a specified date) and (2) to complete 100 percent of the project by _____ (a specified date). A quality objective relating to the budget might be: to complete the project within budget with no over-budget expenses. A quality objective relating to specifications might be: to exceed all specifications set forth in the project contract.

- *Process quality measures.* This component of the project quality plan lists the measures that will be used to ensure process quality for all processes that will be engaged on behalf of the project. The process is listed along with the expected level of performance (based on the specifications). Then the actual quality assurance activity is listed along with how frequently it will be applied to the process. For example, a manufacturing process the measure uses might be SPC. The frequency of this measure would be *continual.* For another type of process, the measure might be a quality audit that is conducted weekly for the duration of the project.

- *Product quality measures.* This component of the project quality plan lists the measures that will be used to ensure that all deliverables meet quality specifications before they are shipped. Each deliverable is listed along with its applicable quality expectations. Then, the specific measure that will be used to ensure that the deliverable meets these expectations is listed along with the frequency of its application. For example, if the deliverable is a printed circuit board, the quality measure might be a series of tests that will be applied to each completed board.

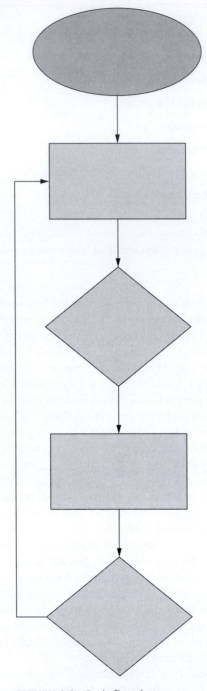

FIGURE 6.9   Basic flowchart.

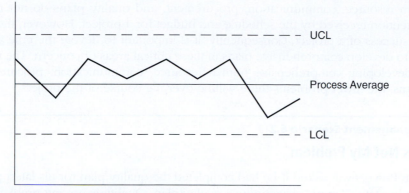

FIGURE 6.10 Basic control chart.

- *Quality-related responsibilities of team members.* In the most competitive engineering and technology companies, quality is everybody's responsibility. Firms that allow their personnel to think that quality is the job of the quality management department only will not be competitive for long. The same is true of project teams. Every member of the team is responsible for doing his or her part to ensure that quality expectations are met or exceeded on their projects. However, some team members have specific responsibilities relating to quality. This component of the project quality plan lists the broad quality-related responsibilities of the team and any specific responsibilities of individual team members.
- *Quality tools to be employed.* There are a variety of tools used to manage and monitor quality in engineering and technology firms. Some of the more widely used are flowcharts, control charts, and cause-and-effect diagrams (see Figures 6.9, 6.10, and 6.11).
- *Reporting procedures.* This component of the project quality plan explains how process and product quality results will be reported, how often, and to whom. Since quality reports are typically done by exception, these reports are typically about problems that have occurred. The project manager assigns a team member to maintain process and product quality logs that are kept up-to-date and used as the basis for developing reports for various stakeholders. The logs should contain at least the following information: (1) name or number of the product/process, (2) date the product or process was reviewed, (3) results/findings of the review (the problem), (4) how the quality problem was resolved, and (5) date the quality problem was solved.

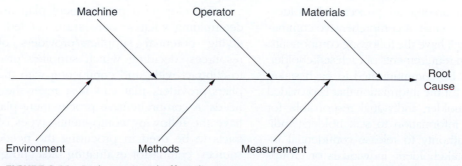

FIGURE 6.11 Basic cause-and-effect diagram.

The human resource, communication, procurement, and quality plans do not receive the level of attention received by the schedule and budget for a project. However, they are critical to the success of a project. Consequently, it is important to devote the time and energy necessary to develop comprehensive plans in these critical areas of concern. The more work put into developing comprehensive human resource, communication, procurement, and quality plans, the fewer problems there will be over the course of the project.

Project Management Scenario 6.2

Quality Is Not My Problem

When John Powers was asked if he had completed the quality plan for his latest project yet, he did not hesitate. "I'm not developing a quality plan. Quality is not my problem. I expect the director of the quality department to handle it." Powers' colleagues were shocked by his answer. After all, their company's motto was "Quality is everybody's job." One of the project managers quoted the company's motto to Powers, but he remained adamant that quality was the job of the quality department.

Discussion Question

In this scenario, John Powers is adamant that quality is not his job or his concern. Discuss his opinion from the perspective of the project manager's responsibility to complete all projects according to specifications. Do you agree with Powers? Why or why not?

SUMMARY

The human resource, communication, procurement, and quality plans are all components of the larger project management plan for an engineering and technology project. A comprehensive human resource plan will have the following components: roles and responsibilities, organization chart, staffing, training needs, recognition and rewards, compliance, and safety. The communication plan is developed to ensure effective communication among all project stakeholders throughout the course. A comprehensive communication plan will have the following components: communication requirements of each stakeholder, information to be communicated to each stakeholder, reason for the information that is provided to each stakeholder, individual responsible for providing the information to stakeholders, individual with authority to release confidential information to stakeholders, individuals or groups allowed to receive confidential information, infor-

mation conveyance methods, resources that will be devoted to communication, chain of command for dealing with issues, method for keeping the communication plan up-to-date, glossary of important terms used in communicating information to stakeholders, flowcharts showing the information flow, communication constraints, templates and guidelines for meetings, and guidelines for using the project Web site.

Developing a procurement plan involves determining what resources are needed, identifying potential suppliers/providers of the resources, deciding which suppliers/providers to contract with, and contracting with the suppliers/providers that can best meet the firm's needs. A comprehensive procurement plan will have the following components: types of contracts to be used in procuring the needed resources, criteria for evaluating bids and quotes, resources that will be acquired by the firm's procurement office and those that will be procured directly by the project team, standardized procurement documents, how multiple suppliers will be managed, how procurements will be coordinated with other aspects of the project, constraints that could affect the procurement process, assumptions the procurement process will be based on, how lead times will be handled to ensure on-time delivery, how make-or-buy decisions will be made, coordination of delivery dates with the project schedule, identification of risks, the format for statements of work, identification of prequalified suppliers/providers, and summary of the procurement metrics that will be used to manage contracts.

Ideally, the quality management plan for a project will be a subset of the firm's larger, organization-wide quality plan. When this is the case, project managers need only refer to the appropriate sections of the larger quality plan when developing one for a given project. However, in the event that an engineering and technology firm does not have an organization-wide quality plan, the project manager will have to develop one for the specific project in question. When this is the case, the project quality plan should have the following components: quality management approach for the project, quality-related definitions, quality objectives for the project, process quality measures, product quality measures, quality-related responsibilities of team members, quality tools to be employed, and reporting procedures.

KEY TERMS AND CONCEPTS

Human resource plan
Roles and responsibilities
Role element
Authority element
Responsibility element
Competency element
Staffing
Training needs
Recognition and rewards
Compliance
Safety
RACI (*Responsible, Accountable, Consult, and Inform*)
Acquisition
Time frames
Release method
Communication plan
Procurement plan
Fixed-price contracts
Firm fixed price
Incentivized fixed price
Cost reimbursement contracts
Cost-plus contracts
Time/material contracts
Open-ended contracts
Statement of work

Make-or-buy list
Procurement documents
Request for proposal (RFP)
Request for quote (RFQ)
Invitation to bid (ITB)
Responsiveness
Overall cost
Capabilities
Risk
Management processes
Methods and techniques
Warranty
Financial capacity
Production capacity and capabilities
Eligibility for government set asides
Past performance
References
Intellectual property or proprietary rights
Quality management plan
Quality management approach
Quality-related definitions
Quality objectives
Process quality measures
Product quality measures
Quality tools
Reporting procedures

REVIEW QUESTIONS

1. What are the components of a comprehensive human resource plan for a project?
2. What is explained in the role element of the roles and responsibilities component of a human resource plan?
3. What is explained in the authority element of the roles and responsibilities component of a human resource plan?
4. Explain the term *RACI*.
5. What is the purpose of the communication plan?
6. What is involved in developing a procurement plan?
7. Explain the concept of the fixed-price contract.
8. Explain the concept of the cost-reimbursement contract.
9. Explain the concept of the time/material contract.
10. What is a statement of work and how is one used in procurement?
11. What is the make-or-buy list in a procurement plan?
12. Compare and contrast the concepts of the RFP and RFQ.
13. What is a responsive proposal?
14. Explain the concept of *overall cost* as it applies to competitive bids or proposals.
15. List and briefly explain the components of a quality plan for a project.

APPLICATION ACTIVITIES

The following activities may be completed by individual students or by students working in groups:

1. Identify an engineering and technology firm that will cooperate in completing this activity. Ask to see a human resource, communication, procurement, and quality plan for a project. Evaluate the plan using the material in this chapter as the criteria. Are the plans comprehensive? Formal or informal? Detailed or broad and general?
2. Assume that you will be asked to serve as the project manager for a class project that is to be developed and submitted at the end of the term. The hypothetical class project will be to develop a seminar on the project management concept of your choice. The project will consist of a comprehensive written report and a PowerPoint presentation made in class. You will be allowed to select the students to serve on your project team. Develop a human resource plan for this project.
3. Using the guidelines provided in this chapter, develop a communication plan for the following hypothetical project: Developing a new educational software package for tutoring students in basic writing skills. The customer for the project is College Tutoring, Inc. Other stakeholders are your team members and higher management in your firm (Software Engineering, LLC).
4. Assume that your company has won a contract to convert a line of military cargo aircraft into civilian aircraft that can be leased to private corporations for transporting executives. Your company plans to solicit competitive proposals to provide the seats for the aircraft. Develop a statement of work that will accompany your company's RFP.
5. Assume that you are a project manager for an engineering and technology firm that has no quality policy. You have been tasked by the firm's CEO to write a quality policy. Develop a draft that could be submitted to the CEO.

ENDNOTES

1. Project Management Institute, *A Guide to the Project Management Body of Knowledge,* 4th ed. (Newtown Square, Pennsylvania: Project Management Institute, 2008), 222–225.
2. Ibid., 257.
3. Ibid., 324–325.
4. Ibid., 322–324.
5. Ibid., 327–328.
6. W. Edwards Deming, *Out of the Crisis* (Cambridge: Massachusetts Institute of Technology Center for Advanced Engineering Study, 1986), 168.
7. Juran Institute, Inc. Retrieved from www.juran.com on November 1, 2011.

CHAPTER 7

Project Planning
The Risk Management Plan

Poorly planned projects are often remembered by what went wrong with them. In fact, even with well-planned projects, it is not uncommon for unforeseen events and unanticipated problems to be what is remembered most about a project. This is why risk management is so important a part of project management. There is always risk involved when an engineering and technology firm undertakes a project of any kind—large or small. Consequently, identifying, prioritizing, and minimizing risk is an important responsibility of the project manager.

Because it is necessary to divide the content for the planning phase of projects into several chapters, risk management is explained in this chapter following chapters on developing schedules and the budget as well as quality, human resource, and communication plans. However, it should be noted that the various planning functions for projects— scheduling, budgeting, and risk management as well as developing quality, human resource, and communication plans—are typically undertaken simultaneously or nearly so. At the very least, there is overlap among the various planning functions for projects.

The point has been stressed throughout this book that regardless of other applicable success criteria, there are three that apply to all projects. These criteria are time, cost, and quality meaning projects are to be completed on time, within budget, and according to specifications. Satisfying these three criteria is the responsibility of the project manager in every project. Unfortunately, engineering and technology projects are like everything else in life. Things do not always go according to plan. This fact is the origin of the maxim that claims *anything that can go wrong will go wrong,* sometimes referred to as *Murphy's Law.*

Although the anonymous Mr. Murphy might have overstated his case somewhat, an engineering and technology project is a complex undertaking with a lot of moving parts that must come together at the right time and in the right way if the project is to be completed on time, within budget, and according to specifications. There is much room for error and problems in projects. Consequently, identifying, assessing, and minimizing risk is part of the project planning process. In larger firms, there may be risk management professionals or even a risk management department to assist the project manager with these tasks. In smaller firms, project managers may have to undertake the risk management process on their own.

RISK DEFINED

Risk is simply the level of probability that things will not go as planned and that unplanned events will occur and have a detrimental effect on a project's success. What is always at risk with projects is their successful completion. This, in turn, creates other risks. For example, engineering firms run the risk of damaging their reputations, financial integrity, and, in turn, competitiveness when they fail to successfully complete projects. Several factors can increase the level of risk in projects (Figure 7.1), which are as follows:

- **Duration.** The longer it takes to complete a project the more likely it is that something will go wrong. This is like saying the longer one is behind the wheel of a car, the higher the probability of having an accident. Increasing the duration of a project correspondingingly increases the project's exposure to events and factors that can undermine plans.
- **Lapse time.** A lapse between being awarded a contract and beginning work on a project increases the likelihood of something going wrong. The longer the lapse, the greater the risk. For example, material prices might increase, labor unions might go on strike, material shortages might occur, qualified subcontractors might get tied up with other projects, and the economy might go into a tailspin. This is why it is important to keep lapse time to a minimum in projects.
- **Inexperience.** Generally speaking, the more experience an engineering and technology firm has with a given type of project the less risk associated with the project. Inexperience can lead to errors and introduce problems into a project. Of course, there has to be a first time for every type of project, but there is a definite correlation between

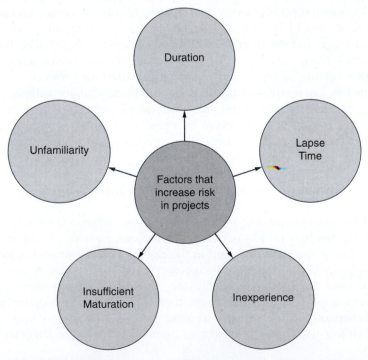

FIGURE 7.1   These factors can multiply the level of risk for a project.

risk and the level of experience of a project team (i.e., less experience equates to greater risk).

• ***Insufficient maturation.*** As processes and technology mature, they become more reliable. This is because over time personnel get better at operating processes and technologies. Further, bugs and problems in the processes and technologies are identified and corrected. Projects in which new technologies and/or new processes are introduced have a higher level of risk than those that use proven technologies and processes.

• ***Unfamiliarity.*** Engineering and technology firms sometimes use subcontractors to produce one or more deliverables for a project. Larger firms often prequalify subcontractors to ensure they can do the jobs called for in the project. However, it is not uncommon for firms to have to work with subcontractors they are not familiar with. Unfamiliarity in any form increases the level of risk on a project.

RISK MANAGEMENT DEFINED

Project managers must be good risk managers. Helping minimize risk on projects is an important responsibility of project managers. From a project manager's perspective, risk management is defined as follows:

> Identifying risks that might negatively affect the successful completion of projects, assessing their potential impact, developing mitigation plans, and implementing the plans in ways that minimize the risk.

The term "minimize" in this definition is important because that is the best project managers can do. Rarely can risk be eliminated. There are just too many factors outside of the project manager's control that can introduce risk into a project. Many of these are identified when developing the list of enterprise environmental factors. Positive items on the list of enterprise environmental factors can be used to help minimize risk. Negative factors increase risk. Hence, their potential impact must be minimized. To minimize and manage risk, project managers should do the following (Figure 7.2):

• ***Be aware that risks exist on all projects.*** Project managers should assume that risks are present in any project no matter how large or small. Ignoring risk as an issue when planning projects is a mistake that can be costly. The first question to ask in the initiation

---

Risk Management Steps

• Be aware that risks exist on all projects.

• Identify project-specific risks.

• Assess the potential consequences of the risks.

• Communicate the risks and their potential consequences to stakeholders.

• Develop and implement risk mitigation plans.

• Monitor the effectiveness of risk mitigation strategies.

FIGURE 7.2   Project managers must perform all of the steps in the risk management process.

stage of a project is: What are the risks? The answer to this question is essential to making an informed go/no-go decision concerning whether to pursue the project.

- *Identify project-specific risks.* Although some risk factors are generic in that they apply to most projects, it is important to go beyond the *usual* and look for project-specific risks. The same type of project undertaken by a different project team can have different levels of risk. Consequently, it is important for project managers to identify the specific risks that apply to every individual project they manage.

- *Assess the potential consequences of the risks.* One type of risk might be easily dealt with in even its worst-case scenario. However, some risks have the potential to completely undermine a project. Consequently, it is important for project managers to carefully assess all risks that apply and determine what might actually happen—from best-case to worst-case scenario—should the risk actually come into play. The level of risk associated with a project can have an effect on how the project is approached. Of course, it can also be the determining factor in deciding to accept or reject a given project. But since this determination is made during project initiation, if the project is in the planning phase the firm has already decided to pursue it. Consequently, assessing the potential consequences of all applicable risks is a critical planning function.

- *Communicate the risks and their potential consequences to stakeholders.* Once all applicable risks have been identified—at least as best as they can be—the risks and their potential consequences should be communicated to all project stakeholders. It is especially important for the project manager and all members of the project team to understand the risks that apply to their projects.

- *Develop and implement risk mitigation plans.* Since risk cannot be eliminated altogether, the next best course of action is to develop strategies for minimizing it. These mitigation strategies, taken together, become the risk mitigation plan for the project. All projects should have a risk mitigation plan. For example, if there is a risk that a hard-to-obtain type of material will not be delivered on time what can be done to work around this situation or, in other words, to minimize the risk associated with it.

- *Monitor the effectiveness of risk mitigation strategies.* Once the risk mitigation plan is implemented, the project manager should monitor the effectiveness of the mitigation strategies for the duration of the project. In addition to monitoring the risks that were identified earlier in the process, project managers should be vigilant in looking for others that might arise during the course of the project. It is not always possible to identify all project risks on the front end. Consequently, it is not uncommon for other risks to unexpectedly crop up during the course of the project. When this happens, mitigation steps should be taken immediately and the mitigation plan should be updated accordingly.

CLASSIFICATIONS OF RISK FACTORS

The risks associated with engineering and technology projects may be classified in different ways. One of the more effective ways to classify risk is according to where control lies:[1]

- *External—Unpredictable.* These types of risks arise from third parties, acts of nature, and other factors over which the engineering and technology firm has no control. They are completely unpredictable. For example, natural disasters such as floods, hurricanes, tornados, and earthquakes can have devastating effects on firms and their projects, but they cannot be predicted.

- ***External—Predictable but Uncertain.*** It can be predicted that these types of risk will occur but not the extent to which they will occur. For example, it can be predicted that the customer might ask for late-in-the-project changes, but not the type or extent of the changes. Market changes and regulatory issues can also be predicted as potential problem causers, but not the extent of the problems they might cause.
- ***Internal—Technical.*** These risks arise primarily from technologies that are used in any phase of a project's life cycle. Risk increases when new and untried technologies are used in construction projects since technologies often have bugs that must be worked out over time. Another factor that increases risk when using new technologies is the ability of people to operate the technologies since they are on the low end of the learning curve at the beginning of the project. Consequently, internal-technical risks are typically tied to such issues as performance, quality, complexity, and specifications.
- ***Internal—Nontechnical.*** These risk factors grow out of human and organizational issues. For example, an individual who has always performed well on past projects may not perform well on the current project for a variety of reasons. Nontechnical internal risks are typically associated with problems in such areas as teamwork, funding, higher management support, and communication.
- ***Legal/Ethical—Civil and Criminal.*** These risks grow out of an organization's ethical and legal obligations to project stakeholders. The risk is that the firm, project team, or an individual member of the team might commit some type of ethical, civil, or criminal violation relating to the project and that the violation will lead to mediation, arbitration, or adjudication. For example, a negligent act that is perpetrated by or tacitly approved by the project manager or higher management that results in a death on the job could lead to criminal charges. These charges and the tragedy and controversy surrounding them could affect the firm's ability to perform as expected on the job.[2]

Figure 7.3 is a graphic representation of an RBS or Risk Breakdown Structure.[3] Project managers should develop or at least participate in the development of an RBS for each project they are involved with and share it with all members of the team. It can be an excellent tool for triggering the thinking of team members as they try to identify project risks.

RISK AS IT RELATES TO PROJECT SUCCESS CRITERIA

Throughout this book, the point is made over and over that three basic success criteria always apply to engineering and technology projects: time, cost, and quality. Meeting these criteria means completing projects on time, within budget, and according to specifications. When considering risk, it is a good idea to add two criteria: environmental and occupational safety. These additional criteria mean that projects should also be completed: (1) in a manner that is environmentally friendly and (2) without safety and health violations. One can argue that the latter two criteria are really subsets of the first three. They can be treated in this way—as subsets—or they can be broken out as separate criteria.

The rational for making environmental and safety/health concerns success criteria is simple. A project that is completed on time, within budget, and according to specifications at the expense of the environment or the safety/health of employees or the general public will be a short-term success but a long-term failure. Fines, negative publicity, and other expenses levied by government regulatory agencies can eventually turn an ostensibly successful project into a failure. In fact, otherwise competent and successful engineering and

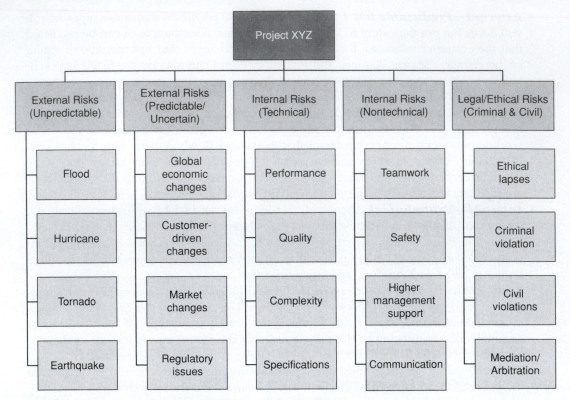

FIGURE 7.3 Template for a Risk Breakdown Structure (RBS) for projects.

technology firms have been driven into bankruptcy by the litigation and fines associated with regulatory violations.

Hence, it is advisable to look at the risk associated with projects as it relates to the following five broad success criteria: (1) time, (2) cost, (3) quality, (4) environment, and (5) safety/health. There are specific risk factors associated with each of these five criteria. These risk factors are as follows:[4]

- **Time-related risk factors.** There are a number of factors that can put completing a project on time at risk. These factors include: too tight a schedule, poorly planned schedule, change orders, government bureaucracy, material delivery problems, work slowdowns or strikes, and natural disasters.
- **Cost-related factors.** There are a number of factors that can put completing a project within budget at risk. These factors include: too tight a schedule, poorly planned schedule, poor cost estimating, disputes among stakeholders, poor employee performance, equipment/technology problems, government bureaucracy, work slowdowns and strikes, and natural disasters.
- **Specification/quality-related factors.** There are a number of factors that can put completing a project according to specifications at risk. These factors include: rush the work because the schedule is too tight, disorganization caused by poor planning, poor attention to specifications when preparing the cost estimate, poor workmanship,

lack of sufficiently skilled personnel, and poor coordination between and among stakeholders.

- **Environment-related factors.** There are a number of factors that can introduce environmental risks into a project. These factors include: questionable practices induced by too tight a schedule, frustration over working with an impenetrable government bureaucracy, poor management, poor supervision, and lack of knowledge of environmental regulations.
- **Safety/health-related factors.** There are a number of factors that can introduce safety and health related risks into a project. These factors include: unsafe practices induced by rushing the work, poor supervision, insufficient support of safety/health measures from higher management, inexperienced workers, use of potentially toxic materials, working at heights, worker reluctance to wear appropriate personal protective gear, lack of knowledge concerning appropriate safe and healthy work practices, and poor supervision.

Project Management Scenario 7.1

There are risks in every endeavor

Professor Jones is introducing her project management class to the concept of risk. Before getting into risk as it relates specifically to engineering and technology projects, she wants to make the point that there are risks in every endeavor. She believes that seeing how a common activity unrelated to the class has risks will help her students better understand the concept. Consequently, Professor Jones gives her class the following assignment: Assume that you and several of your friends plan to take a trip to the beach for Spring Break. The nearest beach is 600 miles away and you plan to drive. Your goal is to spend several days relaxing on the beach and having fun. Identify all of the risks inherent in this trip that might inhibit your fun.

Discussion Question

In this scenario, Professor Jones wants her students to understand that there is risk in every endeavor. Assume that you are in this class. Make a list of all risks you can identify that might prevent you from achieving the goal of having fun at the beach.

RISK IDENTIFICATION PROCESS

The first step in identifying the risk associated with a project is to get organized. This means the project manager should do the following: (1) form the risk identification team (some larger organizations have risk management departments), (2) distribute the Risk Breakdown Structure template, (3) select the risk identification methods to be used, and (4) decide what the output of the risk identification process will be (see Figure 7.4).

Risk Identification Team

If the engineering and technology firm is small, the risk management team might consist of only the project manager and one or two personnel. However, in most cases—especially with

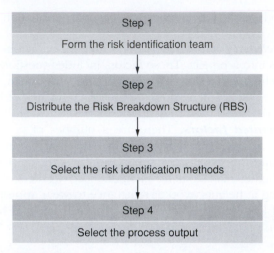

FIGURE 7.4 Risk identification process.

larger firms—the risk management team will consist of the project manager, risk management experts, and members of the project team. Although the risk management team should be kept to a manageable size, anyone from the firm who can contribute to accurately identifying project risks and planning appropriate responses may be included.

Risk Breakdown Structure Template

Figure 7.3 shown earlier in this chapter is a template for a Risk Breakdown Structure for engineering and technology projects. The RBS template is used to trigger the thinking of the risk management team in identifying potential project risks. The team should never be limited by the RBS template since there might be project-specific risks that have no corresponding category on the template. Before beginning discussions and brainstorming activities and even during these activities, members of the risk management team can add categories of risk to the top of the template or specific areas of risk to the existing categories. However, in most cases the team will find that the template covers most of the predictable areas of risk commonly encountered in engineering and technology projects.

Risk Identification Methods

There are several different methods that can be used to identify risks for a specific project. The most common of these include the following (Figure 7.5):

- Review of the project documents
- Brainstorming sessions
- SWOT analysis
- Experience review
- Review of professional literature
- Survey of experts/Delphi technique
- Expert judgment

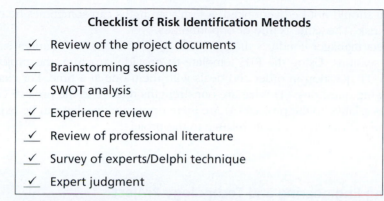

FIGURE 7.5 Project managers can use these methods to identify risks.

REVIEW OF PROJECT DOCUMENTS. A thorough review of the project documents by all members of the risk management team is an effective way to identify project-specific risks. The members of the risk management team have their own areas of expertise and corresponding perspectives. Having people from a variety of backgrounds and perspectives examine the project documents can reveal risk factors that even the most experienced engineering and technology professional might not see.

A review of the project documents might reveal the following potential problems: (1) the project will take longer to complete than originally thought, (2) there is going to be too much lapse time between signing the contract and beginning the work, (3) the engineering and technology has no experience with the type of project in question, (4) technologies not previously used will be required to complete the project, (5) it will be necessary to use unfamiliar subcontractors on part of the project, (6) the specifications are not as well-defined as they should be, (7) the cost estimate leaves very little room for error, (8) regulatory issues might be a problem, and (9) stringent government regulations apply.

BRAINSTORMING SESSIONS. Leading the risk management team through a comprehensive brainstorming session or several sessions is an excellent way to identify project-specific risks. The brainstorming sessions are given structure by the RBS and are facilitated by the project manager. For large projects, the project manager might conduct one brainstorming session for each of the major risk categories: external/unpredictable, external/predictable but uncertain, internal/technical, internal/nontechnical, and legal/ethical—criminal/civil. For smaller projects, all of the risk categories might be covered in just one or two brainstorming sessions. Regardless of the number of sessions, the goal is to have people of different areas of expertise and different perspectives think creatively about the issue of risk, guided but not limited by the RBS.

SWOT ANALYSIS. A comprehensive SWOT analysis (strengths, weaknesses, opportunities, threats) is an effective way to identify project-specific risks. The strengths, weaknesses, opportunities, and threats referred to in this section are those relating to the engineering and technology firm's ability to complete the project in question on time, within budget, according to specifications, and in a safe and environmentally friendly manner. The tendency is to think of risk only in terms of the negative (weaknesses and threats), but there is a positive side to risk that should not be overlooked. A firm's strengths can sometimes be exploited to help minimize risk. The same is true of opportunities.

The project manager conducts the SWOT analysis sessions in a manner similar to the brainstorming sessions. Using the RBS template to provide structure, the project manager poses each SWOT question in order and deals with them one at a time. For example, consider the following questions: (1) What are our strengths relative to the project? (2) What are our weaknesses relative to the project? (3) Are there opportunities that can be exploited with this project? and (4) Are there threats relative to the project?

Challenging Engineering and Technology Project

ENGINEERING PERSONALIZED MEDICINES

Different people respond to the same medicines differently. This is one of the problems facing healthcare professionals everyday with every patient treated. When it comes to medicine, one size does not fit all. Unfortunately, medical professionals do not currently have a reliable means of making the types of diagnoses necessary to tailor medicines to individuals. Human DNA contains more than 20,000 genes, is a unit of chemical code. The overall genetic blueprint of every human being is basically the same. The variants that account for individual differences are found in only about one percent of the human DNA. But that one percent is enough to make every human being an individual when it comes to physical traits. One of those traits that makes people difference is their tolerance of and reaction to medicines. Developing a reliable way to sort out individual differences in tolerance of and reaction to medicines is a challenging project for engineers and technologists.

One of the main problems that must be solved before medicines can be personalized is the development of better, more reliable systems for quickly assessing an individual's genetic profile. Another problem that must be solved is how to collect and manage massive amounts of data on individuals. Yet another problem that must be solved is the development of inexpensive but fast diagnostic tools such as gene chips and sensors that are capable of detecting minute amounts of specific chemicals in the blood. Developing the tools needed for diagnosis, assessment, and sensing is the engineering and technology challenge relating to the personalization of medicines.

Meeting this challenge will require creative thinking, innovation, and persistence on the part of engineering and technology professionals. It will also require effective project management. *Before proceeding with this chapter, stop here and consider how managing the development of diagnostic, assessment, and sensing tools for the personalization of medicines will require the various process and people skills of project managers.*

Source: Based on *National Academy of Engineering.* www.engineeringchallenges.org/cms/8996/9129.aspx?printThis=1

The engineering and technology firm's successful track record in completing similar projects could be a strength. The need to work with unfamiliar subcontractors on a portion of the project could be a weakness. An opportunity might be the chance to earn performance bonuses built into the contract by reaching certain milestones ahead of time. A threat might be looming labor unrest that could result in a strike or a work slowdown.

The advantage of using the SWOT analysis as a risk management method is that it looks at both sides of the risk equation. By so doing it sometimes reveals positives that can be exploited in ways that help the firm overcome or, at least, minimize the negatives. For example, a weakness might be inexperience on the project team. An offsetting strength might be a strong project manager who is especially good at monitoring and mentoring inexperienced personnel.

EXPERIENCE REVIEW.   The experience review is a method available to firms that have experience in completing the type of project in question. Experience is an excellent teacher provided, of course, that students are paying attention. The value of the experience review is increased if it is conducted prior to brainstorming or SWOT analysis sessions. When done in this way, the brainstorming and SWOT analysis sessions can be informed by the experience of the risk identification team members.

When conducting an experience review, the project manager will occasionally ask experienced personnel to *testify* before the risk management team so that the team gains the benefit of their experience. Another way to conduct the experience review is to assign each member of the risk management team to interview colleagues who have worked on similar projects to determine lessons they have learned and can pass along. An important responsibility of the risk management team is to ensure that the engineering and technology firm benefits from its own experience.

REVIEW OF PROFESSIONAL LITERATURE.   An effective but often overlooked risk identification method is the review of professional literature. This is another of those methods that should be undertaken before conducting brainstorming or SWOT analysis sessions so that the sessions can be informed by its findings. An effective way to conduct the review of professional literature is to give each member of the risk management team the assignment of reviewing his or her professional literature to locate articles and other references that explain problems that occurred with other similar projects. These problems, particularly if they surface more than once or even repeatedly, point to potential risks for the current project.

SURVEY OF EXPERTS/DELPHI TECHNIQUE.   Surveying experts anonymously for their input can be an effective way to identify project-specific risks. The experts can be internal or they can be professional colleagues or a combination of both. Engineering and technology professionals often join professional organizations. Through these associations they meet colleagues with expertise in different aspects of the profession. These colleagues may be willing to respond to a survey on the basis of problems they have encountered when working on projects similar to the one in question. The survey instrument solicits their input about project risks. This input is summarized, and the summary is circulated among the experts again for additional consideration. The process is repeated until the list of risks has been pared down to those that are most likely to be associated with the project in question. This is another of those risk identification methods that should be undertaken prior to conducting brainstorming or SWOT analysis sessions so that the sessions can be informed by its findings.

EXPERT JUDGMENT.   One of the reasons each member of the risk management team was selected to participate in the first place is the expertise he or she brings to the table. Each member of the team has a certain amount of experience and training relating to the project in question. Consequently, one of the most important assets each team member brings to

---

Summary of Risk Statements XYZ Project

External Risks—Unpredictable

- If hurricane season is unusually active, the project may not finish on time.
- If hurricane hits during the project, the work in progress might be severely damaged.

Internal Risks—Technical

- If the new material control software does not live up to expectations, the material delivery, unloading and storage process will breakdown resulting in delays.

Internal Risks—Nontechnical

- Because of the firm's OSHA violations or its last job, if safety procedures are not followed meticulously, heavy fines might be assessed.
- The team is inexperienced in working with this type of project.

FIGURE 7.6   Sample risk statements.

the table is expert judgment. Expert judgment comes into play in all of the other risk identification methods, but it is especially important when reviewing the project documents.

To a layperson, the project documents might appear to be just an attractive set of drawings, an impressive set of specifications, and an explicitly written contract. But to a professional with expertise in some aspect of the project in question, the documents can be invaluable sources of information for identifying risk, especially when reviewed within the structure of the RBS. Of course, as is always the case when using the RBS it should be an enabling tool rather than a limiting device. Members of the risk management team should begin by thinking inside the box defined by the RBS and end by thinking outside the box.

Process Output

The output of the risk management process is a list of potential risks organized under the broad categories specified in the RBS along with any new categories added to the RBS during the risk identification process. Each entry in the list should be stated in a standard format that will inform the risk response process. A helpful format for risk statements is the cause-and-effect format (e.g., If a certain *cause* occurs, it will have the following *effect*.). Figure 7.6 contains examples of risk statements written in this format and tied to their corresponding risk category from the RBS. Each of the risk statements shown in Figure 7.6 is written in the cause-and-effect format. When risk statements are written in the cause-and-effect format, it is easier to develop risk response strategies.

QUALITATIVE RISK ANALYSIS

Once the risks for a project have been identified, they must be analyzed. Risk analysis is the process used to assess the qualitative or quantitative value of risk as it relates to specific risk factors. Qualitative risk analysis assesses the *probability* that a given risk factor will have an *impact* on the project and the extent of the impact. Quantitative risk analysis assesses the *probability* that a given loss will occur and the *magnitude* of the loss.

Both types of risk analysis are complex undertakings requiring specialized expertise. Hence, both types of risk analysis can require the participation of risk management, especially on larger more complex projects. Of the two—qualitative and quantitative—the more widely used is qualitative risk analysis. Qualitative risk analysis is an important step that should be completed prior to developing risk responses because it can inform the risk response process. Risk probability assessment involves estimating the likelihood that a given risk will actually occur. Risk impact assessment involves estimating the potential effect each risk identified might have on the success of the project in question.

The risk management team is responsible for establishing risk probability definitions and corresponding impacts. Probability is a number between 0 and 1. A risk probability of 0.0 means the risk will never happen. A risk probability of 1.0 means it is guaranteed to happen all the time. Impact can also be assigned numerical values between 0 and 1. For example, 0.0 impact means the risk factor will have no effect, but an impact of 1.0 means it will have a devastating effect. By graphing probability on one axis and impact on the other, the risk management team can determine which of the following levels of risk applies (Figure 7.7):

- Low risk
- Moderate risk
- High risk

Different engineering and technology firms will have different levels of risk tolerance. However, generally speaking risk management teams become concerned when the risk factor falls into the moderate range or higher.

Another approach is to transform the graph in Figure 7.7 into a probability and impact matrix such as the one shown in Figure 7.8. With this approach, the probability and impact scales of the graph are placed along each respective axis to form a grid. The risk value of each intersection on the grid is computed (probability × impact = risk value). The risk management

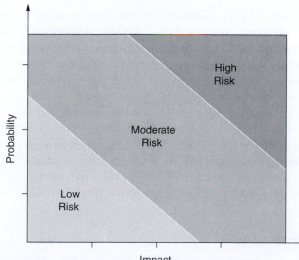

FIGURE 7.7 Plotting risk.

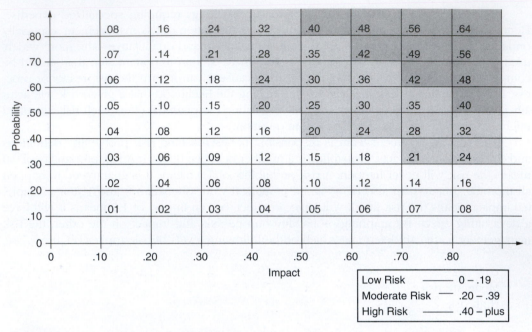

FIGURE 7.8   Computing and plotting risk values.

team must decide what it considers the cutoff points for low, moderate, and high risk. In Figure 7.8 the following cutoff points were established:

- 0 − .19 = Low risk
- .20 − .39 = Moderate risk
- .40 plus = High risk

The probability value chosen by the risk management team is multiplied times the estimated impact value for each risk factor. The resultant risk value determines if the risk of the factor in question is low, moderate, or high. The darker shaded areas in Figure 7.8 represent the high risk zone. The lighter shaded areas represent the moderate risk zone. Any risk factors for the project in question that fall into either of these two zones will receive special attention when the risk management team develops its risk responses.

One of the weaknesses of qualitative risk analysis comes into play at the point where the risk management team plots the estimated probability and impact values on a graph such as the one in Figure 7.7 or 7.8. Expert judgment is required in selecting the probability and impact values. Consequently, the old information technology maxim that says *garbage in/garbage out* applies when performing any form of risk analysis. Hence, ultimately qualitative risk analysis is only as good as the judgments made by the members of the risk management team.

Risk Response Strategies

In developing appropriate responses to risk, project management teams have three broad options: (1) elimination, (2) transfer, and (3) minimization.

ELIMINATION. With this option, the risk management team attempts to completely eliminate the risk in question. For example, assume that one of the risks identified relates to an important material needed for the project. The risk is that the material might not be delivered on time. One way to eliminate this risk is to eliminate this new material and substitute another material that can be delivered on time.

TRANSFER. With this option, the risk management team recommends paying a third party to assume some or all of the risk in question. The most commonly used risk transfer method is the purchase of insurance. For example, assume that a risk factor identified is that the facility where the project in question will be completed is in a hurricane prone region of the country. A hurricane could cause the workforce to be evacuated or it even damage or destroy the facility in question. This risk could be transferred by purchasing special insurance for the project.

MINIMIZATION. With this option, steps are taken to mitigate the impact of the risk in ways that minimize the firm's exposure. There are two types of minimization strategies: (1) minimize the chance that the risk will occur and (2) minimize the amount of damage should the risk occur. For example, assume that one of the risks identified is the potential for heavy OSHA fines if accidents or injuries occur during the course of the project. The firm could minimize the chance that violations will occur by providing mandatory safety training for all personnel who will work on the project. The firm could minimize the amounts of the fines should violations occur by asking OSHA to provide assistance in establishing a safe job site. An experienced risk management team would recommend doing both.

Once the risk management team has determined which of the risk statements represent moderate to high risks, the statements are listed in order of priority beginning with the highest risk. Then risk response strategies—elimination, transfer, or minimization—are planned for each.

Figure 7.6 contains five risk statements that were developed earlier in the risk identification phase of the process. Assume that one of these statements had risk values that put them in the moderate to high risk range. The risk statement along with a corresponding risk response strategy is as follows:

- *Internal Risk—Nontechnical:* Because of the firm's OSHA violations on its last job, if safety procedures are not followed meticulously, heavy fines might be assessed. To minimize the risk associated with this situation, the following strategies will be applied: (1) Train all supervisors on their responsibilities for ensuring that their personnel work safely, (2) Require personnel who will work on the project to complete a two-day safety seminar before beginning their work, and (3) Request consulting assistance from OSHA in establishing a comprehensive safety program for the firm.

QUANTITATIVE RISK ANALYSIS

Quantitative risk analysis is another approach used by risk management experts to assess the risks incumbent in a project. Quantitative analysis can be a complex undertaking—hence the need for risk analysis experts to assist. Quantitative risk analysis is used to address the following three broad questions:

- How long will the project take to complete (Will it go past the deadline)?
- How much will it cost to complete the project on time (Will it go over budget)?
- Will the project team be able to meet all specifications (Will quality standards be met)?

These questions are just another way to get at the three success criteria that apply to all projects (i.e., time, cost, and quality). Of these three criteria, quantitative analysis is typically applied to determinations of time and cost.

QUANTITATIVE ANALYSIS TOOLS

Quantitative analysis experts use a variety of tools to perform an analysis on a given risk factor. One of the most commonly used tools is the distribution curve. Figure 7.9 contains examples of three of the most frequently used distribution curves in quantitative analysis: Normal Distribution or "Bell Curve" (so named because it is shaped like a bell), Triangular Distribution, and Beta Distribution. Another widely used quantitative analysis tool is the Decision Tree.

Distribution Tools in Quantitative Analysis

A normal distribution curve such as the one shown in Figure 7.9 has a probability density function and a cumulative distribution function. The probability density function of the normal

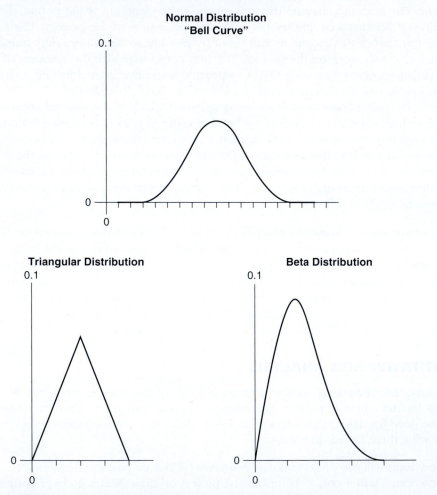

FIGURE 7.9 Widely used distributions in quantitative analysis.

Cumulative Distribution Function

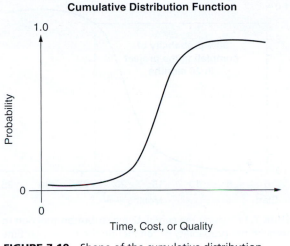

FIGURE 7.10   Shape of the cumulative distribution function.

curve is shaped like a bell as shown in Figure 7.9. Probability values are plotted on the vertical axis and time, cost, or quality values are plotted on the horizontal. The curve for the cumulative distribution function takes on the shape of an elongated "S" as shown in Figure 7.10. As with the probability density function of the normal curve, the cumulative distribution function has the probability values plotted on the vertical axis and the time, cost, or quality values plotted on the horizontal axis.

The shapes of the curves for the probability density function and the cumulative distribution functions are mathematically prescribed and can be computed by applying standard equations. Changing the variables in the mathematical equations changes their shapes. In the days when these equations had to be computed manually, using distribution curves was a labor-intensive task. However, now that they are embedded in computer software and preloaded in electronic calculators, risk analysis personnel need only supply the variables.

Since this type of assistance is readily available for both types of functions—probability density and cumulative distribution—project managers are well-advised to focus on using the distribution curves for analyzing time, cost, and quality risks. For example, Figure 7.11 shows the cumulative distribution function applied to determine the probability that a project can be completed in 24 months as specified in the bid package. The plot in Figure 7.11 shows that there is a 90 percent probability that the 24-month goal can be met. Now assume that there is a substantial incentive bonus for completing the project six months early—in 18 months. The dotted line on the graph shows that the probability of receiving the incentive bonus is approximately 47 percent.

Decision Tree in Quantitative Analysis

The decision tree is a quantitative analysis tool used to analyze cost options in projects. For example, assume that an engineering and technology firm has received several bids from suppliers to provide certain materials for a project. Only two of the bids were completed in accordance with instructions. Consequently, the firm must choose between these two bids

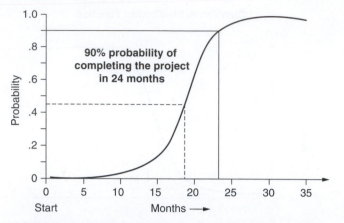

FIGURE 7.11   Using the cumulative distribution function to determine the probability of completing the project on time.

(Supplier A and Supplier B). The risk management team decides to use a decision tree as a tool in making the decision.

Suppliers are instructed to provide three prices: a *base price* for delivering the material on time, an *incentive* price for delivering the material early, and a *penalty* price for delivering the material late. Figure 7.12 is a decision tree that was developed by the risk management team. The tree has two main branches—one for each supplier—and three branches off of each main branch. The smaller branches represent the on-time, early, and late prices submitted by the suppliers. The probabilities that were determined for each scenario—on time, early, late— for each supplier are shown on the respective branches of the decision tree. For example, the probability that Supplier A will complete the project early is .2, on-time is .5, and late is .3.

With the decision tree presented as shown in Figure 7.12, the computations for de- termining the likely price for the project can be completed. In this step, the price on each branch of the tree is multiplied times its probability value as follows:

Supplier A:

.2 × \$200,000 = \$40,000
.5 × \$150,000 = \$75,000
.3 × \$100,000 = \$30,000
Total = \$145,000

Supplier B:

.2 × \$90,000 = \$18,000
.5 × \$150,000 = \$75,000
.3 × \$225,000 = \$67,500
Total = \$160,500

These computations show that the firm can expect to pay \$145,000 with a probability of .7 or a 70 percent chance of having the material delivered on time or early if it chooses

Decision Tree

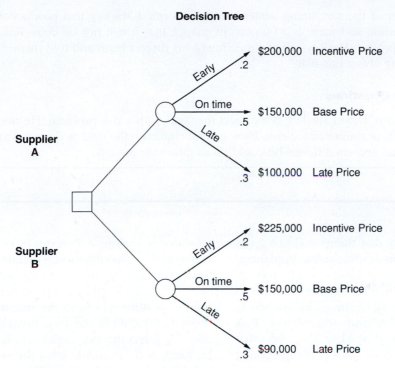

FIGURE 7.12   Decision trees are helpful tools for decision making.

Supplier A. The firm can expect to pay $160,500 with a probability of .8 or an 80 percent change if it chooses Supplier B. This clarifies the decision of the risk management team. It can expect to pay $15,500 less with a 70 percent chance of completing the project on time or early or $15,500 more with an 80 percent chance. The determination would be made by deciding which is more important—an 80 percent chance of finishing on time or early or saving $15,500.

There are risks inherent in all projects. Those risks relate to completing the job on time, within budget, and according to specifications as well as safely and in an environmentally friendly manner. Consequently, risk must be managed. Risk management is an imperfect process, but the process and methods described in this chapter will allow project managers to manage risks as well as they can be managed on projects of any size. The better that project managers and others involved in risk management become at this important process the more likely it is that their projects will be profitable.

Project Management Scenario 7.2

We need to do something about this risk

David Mackey is concerned about one risk in particular. His company has ordered a new five-axis machining center. If the machine works as well as the one Mackey saw demonstrated, his team should be able to complete the current project well ahead of schedule and earn a substantial incentive bonus. This is the good news. The bad news is that the manufacturer of the machining center has just informed Mackey that production has fallen behind schedule and there is a 60 percent chance that it will not be delivered on time for the current project. Mackey called a meeting of his project team and told them: "We need to do something about this risk."

Discussion Questions

In this scenario, David Mackey is a project manager with a risk problem. He needs to eliminate, transfer, or minimize the risk. How can he eliminate the risk? Is there a way to transfer the risk? What are some things he could do to minimize the risk?

SUMMARY

Risk is the possibility that things will not go as planned and that counterproductive unplanned events could arise during the course of a project. Several factors can increase the level of risk in a project: duration, lapse time, inexperience, insufficient maturation, and unfamiliarity. Risk management is defined as identifying risks that might negatively affect the proper completion of projects, assessing their potential impact, developing mitigation plans, and implementing the plans in ways that minimize the risk.

To minimize risks, project managers should: (1) be aware that risks exist on all projects, (2) identify project-specific risks, (3) assess the potential consequences of the risks, (4) communicate the risks and their potential consequences to stakeholders, (5) develop and implement risk mitigation plans, and (6) monitor the effectiveness of risk mitigation strategies. Risk factors can be classified as external—unpredictable, external—predictable, internal—technical, internal—nontechnical, and legal/

ethical—civil/criminal. Risk can be evaluated as it relates to the following success criteria for projects: time, cost, quality, safety, and the environment.

The risk identification process includes the following steps: (1) form the risk management team, (2) distribute the Risk Breakdown Structure, (3) select the risk identification methods to be used, and (4) decide what the output of the risk identification process will be. Risk identification methods include the following: review of the project documents, brainstorming sessions, SWOT analysis, experience review, review of professional literature, survey of experts/Delphi technique, and expert judgment.

Qualitative risk analysis assesses the probability that a given risk factor will have an impact on the project as well as the extent of the impact. Quantitative risk analysis assesses the probability that a given loss will occur and the magnitude of the loss. Risk response strategies fall into three broad categories: (1) elimination, (2) transfer, and (3) minimization.

KEY TERMS AND CONCEPTS

Murphy's Law
Duration
Lapse time
Inexperience
Insufficient maturation
Unfamiliarity
External—unpredictable

External—predictable
Internal—technical
Internal—nontechnical
Legal/ethical—civil and criminal
Time-related risk factors
Cost-related risk factors
Specification/quality-related risk factors

Environment-related risk factors
Safety/health-related risk factors
Risk management team
Risk Breakdown Structure template
Review of project documents
Brainstorming sessions
SWOT analysis
Experience review
Review of professional literature
Survey of experts/Delphi technique

Expert judgment
Qualitative risk analysis
Risk response strategies
Elimination
Transfer
Minimization
Quantitative risk analysis
Distribution tools
Decision tree

REVIEW QUESTIONS

1. Explain the concept of Murphy's Law and how it applies to projects.
2. Define the term *risk* as it relates to projects.
3. How can the following factors affect risk in a project: duration, lapse time inexperience, insufficient maturation, and unfamiliarity?
4. Define the concept of risk management.
5. What should project managers do to minimize risk?
6. List and briefly explain four classifications of risk factors.
7. Name a time-related risk factor.

8. Name a cost-related risk factor.
9. Name a quality-related risk factor.
10. Name an environment-related risk factor.
11. Name a safety-related risk factor.
12. List the steps in the risk identification process.
13. List and briefly explain the most commonly used risk identification methods.
14. What is assessed in qualitative risk analysis?
15. What is assessed in quantitative risk analysis?
16. List and explain the three broad options of risk response.

APPLICATION ACTIVITIES

1. Contact an engineering and technology firm in your community that will cooperate with you in completing this activity. Ask the firm's representative to explain the most common risk factors the firm has to deal with on its projects. Make a list of these risks and discuss them with classmates.
2. Use the Risk Breakdown Structure template in Figure 7.3 to complete this activity. Using the RBS template as a starting point, identify as many risks as you can that might apply if developing a new software package for tutoring students in Calculus.
3. Using the probability/impact matrix in Figure 7.8, assign probability and impact values for each of your risk statements from the previous activity (Number 2) and plot them to determine if the risk is low, moderate, or high.
4. Use the cumulative distribution graph in Figure 7.11 to complete this activity. What is the probability that a certain project can be completed in 10 months? 20 months? 30 months?

5. Use the decision tree in Figure 7.12 to determine which supplier's bid to accept. Two suppliers have submitted bids to provide all of the materials for your project. Each has submitted an on-time, early, and late-bid price. Your risk management team has determined that the probability of on-time, early, and late delivery for each supplier and the corresponding bid amounts are as follows:

Supplier A

On time (.4): $1,525,000

Early (.1): $2,212,000

Late (.3): $998, 785

Supplier B

On time (.2): $1,058,000

Early (.1): $1,678,906

Late (.4): $1,000,000

Determine the likely price for each supplier and decide which supplier will receive the contract to supply materials for your project. Explain your decision.

ENDNOTES

1. International Marine Contractors Association. "Identifying and Assessing Risk in Construction Contracts: An IMCA Discussion Document," 5. Retrieved from www.imca-int.com on February 1, 2012.

2. Ibid.

3. Project Management Institute, *A Guide to the Project Management Body of Knowledge*, 4th ed. (Newtown Square, Pennsylvania: Project Management Institute, 2008), 280.

4. Patrick X., W. Zou, Guomin Zhang, and Jia-Yuan Wang, *"Identifying Key Risks in Projects: Life Cycle and Stakeholder Perspectives."* Retrieved from www.pres.net/Papers/Zou_risks_in_constructionprojects.pdf on January 30, 2012.

Project Execution
Build the Project Team

One of the most important responsibilities of project managers is teambuilding. An important aspect of teambuilding is conflict management. Typically, a team is assigned to the project manager at the same time as the project charter. The earlier the team members are assigned to a project, the more help they can give the project manager on the project. But having a team and having an effective, productive team are two different propositions. Effective, productive project teams whose members are committed to the team's mission and will strive to perform at peak levels to achieve that mission rarely just happen. Rather, they must be built. Teambuilding should begin as soon as team members are assigned to a project and it should continue through the duration of the project.

TEAMBUILDING DEFINED

A team is a group of people working together to accomplish a common mission. In the context of project management, a team's common mission is the successful completion of the project in question. Regardless of the type or size of the project, the common mission of a project team's members is to complete the project on time, within budget, and according to specifications. Project teams are formed to accomplish this specific purpose. The better the members of the project team work together and cooperate in achieving the project's mission, the more effective the project team will be. Molding a disparate group of individuals with different backgrounds, talents, motivation levels, and perspectives into mutually supportive, cooperative team members is the responsibility of the project manager. The concept is known as *teambuilding*.

COMMON MISSION: THE BASIS OF EFFECTIVE TEAMWORK

Having a common mission is essential to effective teamwork. Commitment to the team's mission is the foundation on which effective teamwork is built and conflicts within the team are resolved. Having a common mission is what brings project team members together and gives them a reason to be mutually supportive and cooperative in spite of their differences in background, talent, motivation level, and perspective. Getting team members to commit to a common mission is the project manager's biggest challenge in building the project team.

People in engineering and technology firms have their own agendas, egos, personalities, and perspectives. These agendas, egos, personalities, and perspectives will be reflected in the members of project teams. The project manager's challenge is to take a disparate group of individuals with their own agendas, egos, personalities, and perspectives and help them embrace a common mission: the successful completion of the project they are assigned to.

Of course, there is more to effective teamwork than just committing to a common mission, but without a common mission the project is like a house without a foundation. However, once all members of a team have committed to the common mission of completing the project on time, within budget, and according to specifications, the other ingredients for success will be easier to accomplish.

People, as a rule, are individualistic, and this is especially true of Americans. Further, individualism is reinforced by American society. It is the rights of the individual that are spelled out in the U.S. Constitution and, more specifically, the Bill of Rights—not the rights of the group or team. Respect for the individual is deeply ingrained in the American psyche. America's heroes tend to be rugged individualists who started with nothing and succeeded by triumphing over great odds. Consequently, convincing members of project teams to put their individualistic tendencies aside for the good of the team can be a challenge. This is one of the reasons that teambuilding never really ends or, at least, not until the project in question is completed. Throughout the course of a project, the project manager must constantly work on teambuilding.

Project Management Scenario 8.1

Mack is not a team player

Why don't you want Mack on your project team? He is one of the best engineers in this company. "I know all about Mack's skills. In fact, normally Mack would be my go-to guy. But this project is going to require especially good team players, and Mack is not a team player. He is a my-way-or-the-highway kind of guy. Consider what I am facing on this project. The customer is demanding and stubborn. Because of the accident that occurred last month, the regional OSHA representative is going to practically live with us. Add Mack's sometimes coarse personality into the mix and I will have a potentially volatile situation on my hands. I will have my hands full just trying to keep the customer and the OSHA representative from drawing their swords. I don't need Mack adding fuel to the fire."

Discussion Question

In this scenario, the project manager does not want Mack, a talented engineer, on the project team because he is not a good team player. Is this a legitimate concern? Have you ever worked, either in school or on a job, with someone who did a good job individually but did not work well as part of a team? If so, describe the situation. As a project manager, would you avoid a potential team member because she is not a good team player?

BUILDING THE PROJECT TEAM

Part of building an effective team is choosing team members wisely. Unfortunately, project managers do not always get to choose their team members—certainly not all of them. Project managers may choose or at least have a voice in choosing some members of the project team—provided the individuals they want are not already committed to another project. However, often project managers do not have the freedom choose all of their team members. This fact just makes teambuilding an even more challenging undertaking.

Regardless of how project managers get their team members for projects, the concept of *fit* should be considered whenever possible. For example, in *Project Management Scenario 8.1* the project manager did not want Mack on his team in spite of Mack's talent as an engineer. The problem was fit. Because of his personality and approach to the job, Mack would not have been a good fit for the project in question. The project manager thought Mack would clash with the customer and the regional OSHA representative. On another project, Mack might have worked out better but on the project in question he was a poor fit.

The message in this scenario is that, to the extent possible, it is important to consider fit when assigning individuals to teams. Even the most talented individual can undermine the work of the team if she does not fit in as a good team player. During the selection process it is important to consider how well prospective team members will fit in with the rest of the team and if they will be good team players.

Developing a Mission Statement for the Team

Because the team's mission seems obvious to them, some project managers make the mistake of thinking it will be obvious to team members. This is a bad assumption. The mission of every project team should be defined in writing for team members—even those who have served with the project manager on numerous teams. The team's purpose, when put in writing, becomes its mission statement.

Part of the mission of any project team in is to complete the project in question on time, within budget, and according to specifications. This purpose constitutes the core of the team's mission. The mission, in turn, should be made clear from the outset and reiterated frequently throughout the course of the project. This is a critical step in building the project team. When a project team first comes together, the members might be complete strangers to each other and the project manager. In such cases, the mission statement can be an invaluable tool for helping project managers mold a group of disparate individuals into a team and for keeping team members focused on why the team exists in the first place. The mission statement can be an equally useful tool even in situations where the project team members and project manager already know each other.

A project team's mission statement should state the name of the project along with a brief description followed by this statement: *The mission of this team is to complete the project on time, within budget, and according to specifications.* Some firms prefer to add safety and the environment to the mission statement while others believe that these two factors are covered by the time and budget criteria (i.e., failing to provide a safe and environmentally friendly workplace can result in fines, medical costs, legal expenses, and schedule delays). What follows are examples of mission statements for project teams:

- Project XXX is a new software release for a comprehensive computer-aided design software package. The purpose of the project team is to complete the project on time, within budget, and according to specifications.

- Project YYY consists of the development, design, prototyping, testing, and manufacturing of a low-voltage power supply system for the F-92 commercial jetliner. The purpose of the project team is to complete the project on time, within budget, and according to specifications.
- Project ZZZ is the establishment of a company-wide safety program to reduce workers' compensation, medical, insurance, and legal costs. The purpose of the project team is to complete the project on time, within budget, and according to specifications.

Each of these sample mission statements contains three distinct elements: (1) the name of the project, (2) a brief description of the project, and (3) purpose of the project team that is responsible for the project. Together, these three elements make up the mission statement for the team that will complete the project.

These examples are intentionally simple, easy to read, and easy to understand. This is how the mission statements for project teams should be written. Project managers should keep simplicity and understandability in mind when developing their teams' mission statements. A mission statement is a tool for communicating the team's purpose with team members and other stakeholders, some of whom may be laypeople. Consequently, a good rule of thumb is to make the team's mission statement understandable to people who might know little about engineering and technology.

Understanding the Team-Related Responsibilities of Project Managers

Becoming a project manager can be an important step up the career ladder to higher-level management positions in an engineering and technology firm. In fact, building and leading project teams is excellent training for higher levels of management. Consequently, when an opportunity arises to build and lead a project team, it is important to get it right. What follows are typical responsibilities of project managers who are called on to build and lead project teams:

- Serve as the team's connection with higher management in the firm.
- Serve as the leader of the project team, the person with ultimate responsibility for completing the project on time, within budget, and according to specifications.
- Serve as the official record keeper for the team. Records include minutes of meetings, correspondence, agendas, and a variety of different kinds of reports. Typically, the project manager will appoint a recorder to take minutes during meetings. However, even if recordkeeping is delegated it is still the responsibility of the project manager to ensure that it is done and done properly. This means the project manager must ensure that minutes are completed in a prompt manner, accurate, distributed to all who should receive them (preferably within 48 hours of the meeting), and kept on file for future reference.
- Participate in team discussions and debates, but take care to avoid dominating.
- Implement team recommendations that fall within the realm of the project manager's authority—provided they are worthy recommendations—and work with higher management to implement worthy recommendation that fall outside of it.
- Prevent counterproductive conflict in the team that might undermine its performance and resolve counterproductive conflict that cannot be prevented.

Developing Positive Working Relationships in Teams

Project teams work most effectively when individual team members form positive, mutually supportive relationships. Positive working relationships among team members can be the difference between having a high-performing team and having one that is just mediocre or, worse, a failure. Learning how to help team members establish, develop, and nurture positive working relationships is important for project managers. To build positive working relationships in their teams, project managers should do the following:

- Help team members understand the importance of being honest and reliable. Effective teamwork is not possible in an environment of mistrust.
- Help team members develop an attitude of mutual support. Stress that team members are supposed to help each other as they work together to accomplish the team's mission.
- Help team members develop a supportive attitude (i.e., "We are in this together") toward each other as they struggle with the challenges of completing the project.
- Help team members understand that they need to help each other deal with the pressure and corresponding stress involved in getting the project completed on time, within budget, and according to specifications.

These are the basics. Competence, trust, communication, reliability, and mutual support are the foundation on which effective teamwork is built. Any amount of time devoted to improving these factors is a good investment for project managers.

EXPLAINING THE ROLES OF TEAM MEMBERS

It is important that individual team members understand their respective roles in the team. Consequently, project managers must be prepared to explain the roles team members will play in carrying out the work required to complete the project. Specifically, team members need to know: (1) their primary role in the team, (2) their supporting role(s) in the team, and (3) their range of authority (i.e., what they are able to do unilaterally and what requires approval from the project manager).

Knowing these things will equip team members to do their primary jobs in the team while also mutually supporting other team members. Knowing these things will also eliminate the risk of conflict that can arise when: (1) a team member thinks a teammate is encroaching on his territory (primary area of responsibility) or (2) a team member thinks a teammate is reluctant to pitch in and provide mutual support. Finally, knowing these things will ensure that team members have the latitude to make decisions about their work within clearly defined boundaries without running the risk of overstepping their authority.

EXPLAIN HOW THE TEAM IS SUPPOSED TO OPERATE

Team members need to know how decisions are made and what role they play in the process, how conflict is resolved, how members are to communicate with each other and the project manager, how problems are to be solved, and how the daily work of the team is supposed to be done. Knowing the parameters and how to operate within them will enable team members to give their best to accomplishing the mission of the team without the distraction of wondering if they are overstepping their authority.

Expectancy is an important aspect of human motivation. People like to know what to expect. Team members will want to know what the project manager expects of them, what their fellow team members expect them, what the customer expects, and what higher management expects. Not knowing these things can be a demotivator. Consequently, time spent by the project manager in explaining what is expected and how the team is supposed to work will be time well invested.

FOUR-STEP MODEL FOR BUILDING EFFECTIVE TEAMS

Good teams do not just happen—they must be built. Building effective project teams is best accomplished using a systematic four-step model:

1. Assess
2. Plan
3. Execute
4. Evaluate

This four-step model is applied as follows: (1) assess the team to identify weaknesses that must be improved and strengths that can be exploited, (2) plan teambuilding activities based on the results of the assessment, (3) execute the planned teambuilding activities, and (4) evaluate results (see Figure 8.1). The four-step model for teambuilding presented herein is effective for project teams in engineering and technology settings. It allows project managers to approach teambuilding in a systematic manner that bases all actions and activities on factual information rather than guess work. It also allows project managers to tailor their teambuilding efforts specifically to the team in question rather than applying a one-size-fits-all approach.

Assessing a Project Team's Strengths and Weaknesses

John is the new coach of a basketball team about which he has only limited information. He knows the team has a mediocre record, but he knows very little about the strengths and weaknesses of individual team members. Consequently, the first thing he wants to do is conduct a comprehensive assessment of the team's abilities. He wants to identify specific strengths and weaknesses. With an assessment completed, John will have a better idea of

---

FOUR STEPS TO EFFECTIVE TEAMBUILDING

- **ASSESS** the team to identify weaknesses that must be improved and strengths that can be exploited.

- **PLAN** teambuilding activities based on the results of the assessment.

- **EXECUTE** the planned teambuilding activities.

- **EVALUATE** to determine if the teambuilding activities were effective and repeat the cycle continually throughout the project.

FIGURE 8.1 Building effective project teams is a systematic process.

what he needs to do to turn the team into a winner. This same approach can be used by project managers each time they are to lead a new team.

A mistake commonly made by project managers is beginning teambuilding efforts before determining where things stand with the team. This mistake can lead to wasted time and resources. Project managers who begin teambuilding activities without first assessing strengths and weaknesses are operating in the dark. Accurately assessing the strengths and weaknesses of teams is an important and necessary first step in teambuilding. An effective assessment will help the project manager avoid wasting time and other resources strengthening what is already strong while failing to work on weak areas.

For project teams to be effective and productive, several factors must be present. At a minimum these factors include the following:

- Clear direction that is understood by all team members (e.g., mission, goals, ground rules).
- Team players on the team (e.g., team first—me second).
- Accountability measures that are understood by all (e.g., evaluation of performance).

What follows are specific criteria in each of these three broad areas that can be used for conducting an assessment of a project team's strengths and weaknesses.

DIRECTION AND UNDERSTANDING.   Members of teams need to be clear concerning what the team is supposed to accomplish. They need to have clear direction that is fully understood. The following criteria can be used to determine if a project team has direction and if it understands that direction:

- Does the team have a clearly stated mission?
- Do all team members understand the mission?
- Does the team have a set of goals that translate its mission into more specific terms?
- Do all team members understand the goals?
- Does the team have a schedule for completing the project?
- Do all members of the team know the schedule and intermediate deadlines within it?

CHARACTERISTICS OF TEAM MEMBERS.   Effective teamwork requires that members of teams be good team players. While it is necessary for members of project teams to think independently and critically, once decisions are made they must come together as a team to implement them. Further, throughout the course of a project, team members must interact in mutually supportive ways. This requires team members to be team players who are willing to sacrifice their individual agendas for the good of the team. The following criteria can be used to determine if a project team's members are good team players:

- Are all team members open and honest with each other all the time?
- Do all team members trust each other?
- Do all team members put the team's mission and goals ahead of their personal agendas all the time?
- Do all team members know they can depend on each other?
- Are all team members committed to accomplishing the team's goals?
- Are all team members willing to cooperate with each other to accomplish the team's mission?

- Do all team members take the initiative to ensure that the project is completed on time, within budget, and according to specifications?
- Are all team members patient with each other when working on solutions to problems?
- Are all team members resourceful in finding ways to get the job done in spite of obstacles?
- Are all team members punctual for work, meetings, assignments, and in meeting deadlines?
- Are team members tolerant of individual differences among members of the team (i.e., intellectual, racial, cultural, gender, and political differences)?
- Are all team members willing to persevere when the work becomes difficult?
- Are all team members mutually supportive of each other?
- Are all team members comfortable stating their opinions, pointing out problems, and offering constructive criticism in team meetings?
- Do all team members support team decisions once they are made?

ACCOUNTABILITY. People in teams need to know how the team's performance will be evaluated and what accountability measures will be used. The following criteria can be used to determine if team members understand the accountability aspects of their work:

- Do all team members know how team progress/performance will be measured?
- Do all team members understand how success is defined for the team?
- Do all team members understand how team decisions are made?
- Do all team members know their respective responsibilities?
- Do all team members know the responsibilities of all other team members?
- Do all team members understand their authority within the team?

CONDUCTING A TEAM ASSESSMENT. Once a project team is formed the project manager will begin to learn where things stand with each team member. But rather than rely on informal observations, it is important for project managers to formalize their assessment of the team and conduct the assessment in a structured, systematic way. An effective way to accomplish this goal is to use the criteria listed in the previous sections to build an assessment instrument that can, in turn, be used to record observations of team members and their respective strengths and weaknesses.

Turning the criteria into an assessment instrument can be accomplished by adding a rating scale. A rating scale like the following example that assigns numerical values to each possible response is recommended:

Completely true	6
Somewhat true	4
Somewhat false	2
Completely false	0

Once project managers have completed the assessment, they use the scores to determine the team's strengths and weaknesses. Any criterion that receives a score of less than "4" or "somewhat true" represents a weakness that should be singled out for improvement. Any criterion that receives a score that is higher than "4" represents a strength that can be exploited to maximize the team's performance. Figure 8.2 is an example of an assessment instrument that can be used to systematically identify the strengths and weaknesses of a project team.

Instructions

To the left of each item is a blank for recording your perception regarding that item. For each item, record your perception of how well it describes your team. Is the statement *Completely True (CT), Somewhat True (ST), Somewhat False (SF),* or *Completely False (CF)*? Use the following *numbers* to record *your* perception.

> CT = 6
> ST = 4
> SF = 2
> CF = 0

Direction and Understanding

_____ 1. The team has a clearly stated mission.

_____ 2. All team members understand the mission.

_____ 3. The team has a set of goals that translate the mission into more specific terms.

_____ 4. All team members understand the goals.

_____ 5. The team has a schedule for completing the project.

_____ 6. All team members know the schedule and intermediate goals within it.

Characteristics of Team Members

_____ 7. All team members are open and honest with each other.

_____ 8. All team members trust each other.

_____ 9. All team members put the team's mission and goals ahead of their personal agendas all of the time.

_____ 10. All team members are committed to accomplishing the team's mission.

_____ 11. All team members are willing to cooperate to accomplish the team's missions.

_____ 12. All team members will take the initiative to ensure the project is completed on time, within budget, and according to specifications.

_____ 13. All team members are patient with each other when working on solutions to problems.

_____ 14. All team members are resourceful in finding ways to accomplish the team's mission in spite of obstacles.

_____ 15. All team members are punctual for meetings, assignments, and deadlines.

_____ 16. All team members are tolerant of the individual differences of team members.

_____ 17. All team members are willing to persevere when the work becomes difficult.

_____ 18. All team members are mutually supportive.

_____ 19. All team members are comfortable expressing opinions, pointing out problems, and offering constructive criticism.

_____ 20. All team members support team decisions once they are made.

FIGURE 8.2 Sample assessment instrument.

Accountability

_____ 21. All team members know how team progress/performance will be measured.

_____ 22. All team members understand how team success is defined.

_____ 23. All team members understand how team decisions are made.

_____ 24. All team members know their respective responsibilities.

_____ 25. All team members know the responsibilities of all the other team members.

_____ 26. All team members understand their authority within the team.

FIGURE 8-2 (*continued*)

Developing a Teambuilding Plan

Teambuilding activities should be planned on the basis of the results from the assessment of strengths and weaknesses. For example, assume the assessment shows that the team does not understand the mission. Clearly, an important teambuilding activity will be to either rewrite the existing mission statement or do a more thorough job of explaining it or both. If the results of the assessment show that team members do not understand the schedule, an important teambuilding activity will be to clarify the schedule including milestones and intermediate deadlines. Regardless of what the assessment reveals about the team, the results should be used as the basis for planning activities to correct weaknesses and exploit strengths.

Executing Teambuilding Activities

Teambuilding is an ongoing, never-ending process. The idea is to make a team better and better as time goes by. The project manager may need to work with selected team members individually. For example, if it becomes clear that members of the team do not trust one of their teammates the project manager may need to pull that member of the team aside and work quietly to help correct the situation. In other instances, the entire team might be involved in a teambuilding activity. For example, when the team has just been formed and is having its first organizational meeting, presenting the members with a mission statement, goals, and a schedule can be a teambuilding activity because it can serve to get everyone on the same page from the outset.

The key is to base all teambuilding activities on the results of a systematic assessment. Project managers should never make the mistake of assuming that they know the strengths and weakness of team members—unless, of course, they have worked with them on other projects. A good rule of thumb for project managers is assess first and then base decisions concerning what developmental activities to provide on the results of the assessment.

Evaluating Teambuilding Activities

If teambuilding activities are effective, weak areas pointed out by the assessment will be strengthened. A simple way to evaluate growth and improvement in the team is for the project manager to periodically review the current behavior of team members against the original assessment. If these ongoing observations show that progress is being made, nothing more is required for the time being. If not, the project manager has additional work to do. If a given teambuilding activity appears to have been ineffective, the project manager should try something else. Involving team members in identifying ways to strengthen the team and improve its performance can itself be an effective teambuilding activity.

INITIATING THE TEAM'S WORK

Before beginning work on a project, it is a good idea for project managers to hold a team initiation meeting. This meeting is different than the kickoff or project initiation meeting held during the initiation phase of the project. The kickoff meeting is about the project. The team initiation meeting is about the team. Some project managers combine the two meetings, but the recommended approach is to hold two separate meetings because the purpose and emphasis of each meeting is different.

The purpose of the team initiation meeting is to give the team a good start at working together to complete the project successfully. This can be achieved by: (1) discussing the team's mission; (2) confirming that all team members are committed to completing the project on time, within budget, and according to specifications; (3) reviewing what the project manager expects concerning members being good team players; and (4) reviewing the team's accountability measures. The team initiation meeting allows the project manager to set the tone for how the team is to approach its work and how its members are to interact with each other and the project manager.

TEAMS SHOULD BE COACHED

Project managers soon learn that if team members are going to work together, they have to be coached. This is especially the case when project managers have no line authority over team members. In other words, coaching is even more important when project managers are not the "boss" in the traditional sense of the concept—which is often the case. Even when project managers have line authority, team members typically respond better to coaching than to bossing.

Project managers need to understand the difference between bossing and coaching. Bossing, in the traditional sense, involves giving orders and evaluating performance. Bosses approach the job from a perspective that says, "I'm in charge—do what I say." Coaches, on the other hand, approach the job from a *follow-me* perspective. Coaching is about leading rather than ordering. An important goal of a coach is to lead the project team to peak performance so as to achieve the best possible project outcome. To this end, coaches work with team members to influence, encourage, equip and empower them to do their part in making the project a success. Project managers can become effective coaches by doing the following:

- Developing a team charter. A team charter contains the team's mission, goals—which are project milestones—and ground rules.
- Continuing team development and teambuilding activities throughout the course of the project.
- Mentoring individual team members or providing mentors for them as appropriate.
- Promoting mutual respect between and among team members.
- Working to make human diversity within the team an asset.
- Setting a positive example of everything that is expected of team members.

Team Charter

It is not hard to imagine a coach from the world of sports calling his team together and saying, "This year we have one overriding purpose—to win the championship." In one simple statement, this coach has clearly and succinctly defined the team's mission—the first component of a team charter. From this statement, team members should immediately realize that everything they do will be aimed at winning the championship. The coach's statement of the mission was brief, to the point, specific, and easily understood. Project managers should be just as specific in explaining their teams' mission, ground rules, and goals/milestones to team members. How to write a mission statement for a team was explained earlier in this chapter.

DEVELOPING THE TEAM CHARTER. A team charter is a document consisting of three major components: (1) mission statement for the team, (2) ground rules for team members, and (3) team goals/milestones (see Figure 8.3). The team charter is used to provide direction for team members and to make sure they understand their responsibilities in helping complete the project in question. The mission statement is developed by the project manager. The goals/milestones come from the project schedule. The ground rules may be developed by the project manager and presented to the team or with the assistance of team members. The latter approach is recommended.

DEVELOPING THE MISSION STATEMENT. Developing a team mission statement was covered earlier in this chapter. This is a reminder that a well-written team mission statement is brief but comprehensive and easy to understand, even by laypeople. The example in Figure 8.3 satisfies these criteria.

DEVELOPING THE GROUND RULES. A project team's ground rules answer the following question: As we work together to complete this project, how are we to interact with each other? There are two ways to develop the ground rules, which are as follows: (1) the project manager develops them on the basis of past experience and his or her observations of team members or (2) the project manager provides team members with a list of potential ground rules and asks them to choose the ones they think are most important. With this approach the project manager typically asks team members to choose their top 10 ground rules from a list of 15 or more.

Developing the ground rules for a team presents the project manager with an interesting challenge. On the one hand, in order to achieve buy-in, it is important that all team members accept the ground rules as their own. On the other hand, project managers need to maintain a modicum of control over the process of developing the ground rules to ensure that important ground rules are included in the final list. Some project managers develop the ground rules for their teams and present them to team members as a fait accompli. Others choose to involve team members in developing the ground rules. The approach used should be the one the project manager intuitively feels will work best with the specific team members in question. On one project, it might be best to simply provide the ground rules. On another project, it might be important to involve the team members in developing the ground rules. The challenge for the project manager is to accurately determine which approach is best for the team in question.

Team Charter
XYZ PROJECT TEAM

Mission Statement

Project XYZ is a new release of a flight simulation software package. The mission of this team is to complete the project on time, within budget, and according to specifications.

Ground Rules

As we work together to accomplish the team's mission, members will be guided by the following ground rules:

- **Punctuality.** Team members will be punctual in arriving at work and in meeting deadlines.

- **Honesty.** Team members will be open, honest, and frank with each other at all times.

- **Dependability.** Team members will conduct themselves in ways that show they can be depended on.

- **Responsibility.** Team members will take responsibility, individually and as a group, for accomplishing the team's mission.

- **Mutual support.** Team members will work cooperatively in helping each other complete the project successfully.

- **Initiative.** Team members will take the initiative in finding ways to solve problems, overcome roadblocks, and keep the work of the team flowing.

- **Conflict management.** Team members will do what is necessary to prevent counterproductive conflict from undermining the team's work. Team members who disagree over ideas and recommendations will do so without becoming disagreeable.

Project Goals/Milestones

1. Complete preliminary design by January 15.
2. Complete final design by February 7.
3. Complete coding by March 11.
4. Complete preliminary tests by March 30.
5. Complete debugging process by April 20.
6. Complete field tests by May 30.
7. Complete user manual by June 15.
8. Closeout project by June 30.

FIGURE 8.3 Sample team charter.

What follows is a list of possible ground rules for team charters. The following list can be used as a guide in developing ground rules for any project team:

- *Honesty.* Team members will be open and honest with each other at all times.
- *No personal agendas.* Team members will put the team's needs ahead of their personal agendas in all cases.
- *Dependability.* Team members will conduct themselves in ways that show they can be depended on.

- *Commitment.* Team members will be committed to completing the project on time, within budget, and according to specifications and to meeting all other pertinent expectations of management and the customer.
- *Responsibility.* Team members will take responsibility for their individual performance as well as the team's performance.
- *Mutual support.* Team members will be mutually supportive in carrying out their responsibilities in the team.
- *Initiative.* Team members will take the initiative in helping the team accomplish its mission.
- *Patience.* Team members will be patient with each other as they work together to solve the problems that can occur during projects.
- *Resourcefulness.* Team members will be resourceful, innovative, and creative in finding ways to get the project completed on time, within budget, and according to specifications in spite of obstacles, inhibitors, and problems.
- *Punctuality.* Team members will be punctual for team meetings and other project activities.
- *Tolerance.* Team members will be tolerant of individual differences in team members.
- *Perseverance.* Team members will persevere in getting the job done when difficulties arise and during times of adversity.
- *Conflict management.* Team members will express their opinions, make recommendations, point out problems, and offer constructive criticism in a tactful manner. When team members disagree, they will do so without being disagreeable. In addition, team members will solve differences among themselves in a responsible, professional manner that contributes to team morale and performance.
- *Decisions.* Team members will participate in the decision-making process by offering input before decisions are made. Once a team decision has been made, all members will support it fully and do their best to carry it out, even if they do not agree with it (unless, of course, there are ethical problems with the decision).

FINALIZING THE TEAM CHARTER.   Once the team's mission statement has been finalized and the team's ground rules selected, the team charter can be finalized. The final step involves adding team goals which are simply statements of project milestones taken from the project schedule. The final team charter will contain the team's mission, ground rules, and goals/milestones. Although these components should be reviewed from time to time, typically they will not change during the course of the project. However, there may be occasions when contract changes will cause corresponding changes in the team's charter. In these cases, the team charter should be revised accordingly so that it always accords with the latest version of project documents.

Coaching and Team Development

Project teams should be handled in a manner similar to sports teams when it comes to team development. Coaches of sports teams work constantly on developing the skills of individual team members and the team as a whole. Project managers should do the same. Team development activities should be continual, and they should go on throughout the course of the project. Developing the skills of individual team members and building the team as a whole should be viewed by project managers a normal part of the job.

Project managers should learn to take the long view when it comes to developing team members. In most organizations, project managers can count on having the team members they lead today serve with them on future projects. The time and effort invested in developing a member of a current team might pay added dividends when that individual serves on future teams.

Coaching and Mentoring

Project managers need to be good coaches for several reasons. One of these reasons is that good coaches are good mentors. This means they establish nurturing, developmental relationships with team members. Developing the technical skills of team members, improving the contributions individuals make to the team, and helping team members become better team players are all mentoring activities. Project managers who want to be effective mentors can help team members develop by:

- Helping them develop teamwork skills
- Making sure they understand all components of the team's charter
- Helping them learn to be good team players
- Helping them understand other people and their points of view
- Teaching them how to behave in unfamiliar settings or circumstances
- Giving them insight into differences among people

Coaching and Mutual Respect

It is important for team members to respect their coach, for the coach to respect team members, and for team members to respect each other. The following strategies will help project managers earn the respect of their team members and team members earn the respect of each other:

- **Trust made tangible.** Trust is established by: (1) setting a positive example of being trustworthy, (2) honestly and openly sharing information, (3) explaining personal motives, (4) refusing to play favorites, (5) giving sincere recognition for a job well done, and (6) being consistent in applying discipline. Doing these things will build trust and mutual respect. The project manager must set the example of doing what is necessary to build trust, but may also have to conduct one-on-one conversations with team members to help them see the importance of trust and how to build it. Project managers can make instruction on how to build trust part of team meetings and can even conduct meetings devoted specifically to teambuilding.
- **Appreciation of people as assets.** This strategy is especially important for building a strong team. It will also help project managers gain the support and cooperation of team members. Even in this age of advanced technology, people are still a team's most valuable asset. To perform at their best, team members must be treated like assets that can appreciate in value if properly developed over time. Appreciation for people is shown by: (1) respecting their thoughts, feelings, values, and fears, (2) respecting their individual strengths and differences, (3) respecting their desire to be involved and to participate, (4) respecting their need to be winners, and (5) respecting their need to learn, grow, and develop. Organizations often claim that people are their most valuable asset, but too few actually follow through and treat people like valuable assets.

Words are not enough. Consequently, project managers must work with team members in ways that continually increase their value to the team and then let the team members know they are valued.

- *Communication that is clear and candid.* Communication can be made clear and candid if project managers do the following: (1) open their eyes and ears—observe and listen, (2) be tactfully candid, (3) give continual feedback and encourage team members to do the same, and (4) confront conflict in the team directly and immediately before it can fester and blow up. People in teams want to be informed and they want to know that the information they are given is the truth.
- *Unequivocal ethical standards.* Ethical standards can be made unequivocal by: (1) adopting the organization's code of ethics at the team level or, if the organization does not have a code of ethics, working with the team to develop its own code, (2) identifying ethical conflicts or potential conflicts as early as possible and acting to resolve them, (3) recognizing and rewarding ethical behavior, (4) correcting unethical behavior immediately—never ignoring it, and (5) making all members of the team aware of the team's code of ethics. In addition to these strategies, project managers must set a consistent example of living up to the highest ethical standards themselves.

Coaching and Human Diversity

America is the most diverse country in the world and this diversity is reflected in the composition of project teams. Diversity in all of its forms—racial, gender, cultural, political, religious, and intellectual—can be an asset to project teams or it can be a liability. The difference comes in how project managers and team members handle diversity. For this reason, it is important that project managers invest the time and effort necessary to make diversity an asset in their project teams.

Most of the future growth in the labor force in the United States will consist of women, minorities, and immigrants. People in these groups will bring new ideas and new perspectives to their jobs, precisely what engineering and technology firms need if they are going to stay fresh, current, and competitive. Consequently, project managers will need to be effective in handling diversity in ways that make it an asset for their teams. This can be a challenge.

In spite of the progress that has been achieved in making the American workplace both diverse and harmonious, some people—consciously or unconsciously—still erect barriers between themselves and those they view as being different. These barriers can quickly undermine the trust and cohesiveness on which teamwork is built, especially when the team has a diverse membership. To keep this from happening, project managers can apply the coaching strategies listed in Figure 8.4.

Identify the Specific Needs of Different Groups

Project managers should ask women, ethnic minorities, and older workers in their teams to describe any unique inhibitors they face in trying to do their jobs—things they see but the project manager might not. Then make sure that all team members understand these barriers, and are willing to work together as a team to mitigate, eliminate, or accommodate them. The differences in people that occur in project teams are much less likely to cause problems if team members feel comfortable enough with each other to discuss them openly, honestly, and frankly.

COACHING STRATEGIES FOR HANDLING HUMAN DIVERSITY IN PROJECT TEAMS

- Identify the needs of different groups within the team (e.g., racial, gender, age, cultural, national origin).

- Confront cultural clashes directly and immediately—do not allow them to simmer below the surface until they eventually blow up.

- Eliminate cases of institutionalized bias (e.g., too few female restrooms in an organization that once had a predominantly male workforce).

- Help people find common ground.

FIGURE 8.4   These strategies can be used to prevent diversity-related problems from undermining the effectiveness of project teams.

Confront Cultural Clashes Directly and Immediately

When diversity-based conflict occurs in teams, project managers should confront it directly and immediately. This approach is particularly important when the conflict is based on such issues as race, culture, ethnicity, age, and/or gender. Because these issues are so deeply personal to individuals, they are potentially more volatile than everyday disagreements over work-related matters. Consequently, conflict that is based on or aggravated by human diversity should be dealt with quickly and effectively.

Few things will polarize a team faster than diversity-related disagreements that are allowed to fester and grow. When diversity-related disagreements are allowed to persist, more of the team's productive energy will be devoted to conflict than to completing the project in question. People who harbor resentment over cultural clashes cannot focus as intently as necessary on doing the jobs the project manager needs them to do.

Identify and Eliminate Institutionalized Bias

Engineering and technology firms that have done things a certain way for a long time are susceptible to a concept known as *institutionalized bias*. Institutionalized bias is bias that—although not necessarily intended—occurs when society changes but the organization does not. Consider the following example. An engineering and technology firm that has historically had a predominantly male workforce now has one in which there are a growing number of women. However, the organization's facility still has 10 rest rooms for men and only one for women, a circumstance left over from how things used to be. This is an example of institutionalized bias. Although there is no bias intended, it exists none the less because societal changes have not been matched by corresponding organizational changes.

Engineering and technology firms and, in turn, their project teams can find themselves unintentionally slighting members simply out of habit, tradition, or unwitting circumstances. When the demographics of an organization and its teams change but its habits, traditions, procedures, and work environment fail to accommodate the change, the result can be unintended discrimination. Eliminating institutionalized bias is important because failing to do so will undermine a team's morale and performance. If the bias applies more broadly than to just one team, it will ultimately undermine organizational excellence.

An effective way to eliminate institutionalized bias is to circulate a blank notebook and ask team members to record—without attribution—instances and examples of it they have encountered. After the initial circulation, repeat the process periodically. The input collected will be helpful in identifying institutionalized bias that can then be eliminated. By collecting input directly from team members and acting on it promptly, project managers can ensure that discrimination that results from organizational inertia is not creating or perpetuating resentment among personnel who need to be focused on completing the projects they are assigned to.

Help People Find Common Ground

People are surprisingly similar—regardless of outward differences such as race, ethnicity, gender, and native language. The hopes, fears, desires, and ambitions of people of every race, gender, national origin, and culture are not that different. In fact, they are very similar. For example, when the author graduated from Marine Corps boot camp at Parris Island, South Carolina, many years ago, the grandstands were filled with proud parents who represented a collage of American culture.

There were Hispanic parents of young men from the inner cities of America's northeast. There were African American parents of young men from the rural south. There were Caucasian parents from northern and southern suburbs. Some parents were well-off economically and some were obviously poor. Others were in the economic middle class. On a normal day, these parents would probably have had no contact with each other and nothing in common—at least outwardly. But on the day their sons became United States Marines, they had a shared pride that transcended differences of language, race, and economic status. Cultural differences did not matter to these proud parents because they had something in common that was more important than their differences.

Project managers can use this concept of having something important in common to strengthen their teams by helping members find common ground and then helping them see that this common ground makes them more alike than different. The one thing they all have in common is the desire to complete the project they are assigned to on time, within budget, and according to specifications. They have different reasons for wanting the project to be a success, but the fact that they share this mission gives them common ground nonetheless.

HANDLING CONFLICT IN TEAMS

In even the best teams there will be conflict. Even when all team members agree on a goal, they can still disagree on how best to accomplish it. Conflict is an ever-present reality in teams, a reality that can undermine effective teamwork unless it is confronted promptly and resolved in a positive manner. Members of project teams have sufficient energy to do their work or to engage in conflict, but not both. Conflict can sap the energy of a team, cause it to lose its focus on the project, and undermine its performance.

Preventing Problems That Can Cause Conflict

The best way to handle conflict is to prevent it. Project managers will never completely eliminate conflict, but they can limit the amount of it they have to deal with by preventing the problems that cause it. The following strategies will help project managers prevent conflict:

- **Clarify assignments.** It is important for people in teams to understand what work is assigned specifically to them. Clarity of assignments is important to project teams and team members. If it is not clear who does what, two situations—both of them bad—can arise: (1) team members can get into turf battles and (2) team members can be confused concerning who is responsible for completing a given assignment. These two situations almost always lead to conflict.

- **Clarify roles.** Although it is important for members of teams to be mutually supportive in doing their work, it is also important for them to understand their specific and primary roles in the team. For example, on a baseball team each player has a primary role: pitcher, catcher, first base, second base, third base, shortstop, right field, center field, and left field. Although the players in these positions are responsible for backing each other up and mutually supporting each other, each player has primary responsibility for his specific position. This is how teams in engineering and technology firms are supposed to work. To prevent confusion and conflict, project managers must clarify the roles of all team members so that conflict does not arise over the issue of territoriality. This means that project managers must be prepared to take team members aside and have one-on-one discussions about roles. The author once had to deal with a team member who had been a project manager but was moved out of that role for poor performance. This former project manager, thinking he knew more than he did, got into the habit of going around the project manager (the author) directly to team members. He would countermand assignments made by the project manager and micromanage team members who were trying to do their jobs. It took several one-on-one conversations and documentation of problems he had caused to convince this team member to play his role in the team and let others play theirs.

- **Encourage team members to talk through their differences.** People in teams will have differences of opinion and perspective. This is normal and natural. In fact, it is good as long as it is not allowed to become personal and counterproductive. However, differences of opinion and perspective can easily develop into conflict. To prevent this from happening, project managers should encourage their team members to talk through their differences. The better they communicate, the less likely it is that differences will escalate into conflict. Occasionally, project managers will need to serve as mediators and facilitators to help team members talk through their differences. This is an appropriate role for project managers and one they have to learn to play well.

- **Encourage team members to learn how to disagree without being disagreeable.** Frank and open discussions and even debates concerning project-related issues are good things provided team members can disagree without being disagreeable. Consequently, project managers should help team members learn that there is nothing wrong with disagreeing with each other as long as they do not become disagreeable in the process. Disagreement can be a good thing if handled well. The opinion of one team member might sharpen the opinion of another. The opinion of one team member might change the opinion of another. But these things will happen only if team members can disagree without taking things personally and becoming disagreeable.

- **Handle conflict promptly.** When project managers see conflict brewing between team members, they should deal with it right away. Putting off dealing with simmering conflict is a sure way to guarantee it will escalate and eventually blow up. The time to intercede and deal with conflict between and among team members is the moment it is noticed. Conflict that is ignored is usually conflict that just gets worse.

Potential Human Responses to Conflict in Teams

Conflict in project teams is, if not inevitable, highly likely. Consequently, it is important for project managers to understand the various ways in which people respond to conflict and how to ensure that their responses are team-positive and resolution-oriented. The various responses to conflict fall into one of the following categories:

- ***Escape responses.*** Escape responses represent one extreme of the continuum of possible responses to conflict. Escape responses are negative because they hurt the team and the person who does the escaping. Escape responses include denial, flight, and suicide. While it is not likely that project managers will have to deal with suicides on the part of team members, one should not assume. Consequently, it is best for project managers to be aware of the three escape responses to conflict. People who respond to conflict by trying to escape it in any of these ways are so averse to conflict that they will go to extremes to avoid it. All escape responses are harmful. Consequently, project managers should be vigilant in observing team members and in acting quickly if they notice any who might take an escape response to conflict.

- ***Attack responses.*** Attack responses represent the other extreme of the response continuum. They are negative in that they hurt the team, the victims, and the person who perpetrates the attack. Attack responses include litigation/formal grievances, assault, and murder. Murder and assault are not common in project teams, of course, but litigation and formal grievances are. One team member filing a lawsuit or, more likely, a formal grievance against another can tear a team apart. Only assault or murder could have been more detrimental to team cohesion. When any of the attack responses to conflict happen, team members take sides or go into defensive mode. When team members take sides, the conflict spreads and becomes more intense. When they go into defensive mode, they focus only on the conflict and its potential outcomes and not on the project. Consequently, project managers must stay closely connected to team members so they can anticipate the potential for an attack response and move quickly to prevent it.

- ***Resolution responses.*** Resolution responses are positive in that they can lead to a resolution that is good for all stakeholders. Although all of the responses explained in this paragraph are resolution responses, it is important to understand that they are not all equally positive. Consequently, the various resolution responses are listed in order of preference: overlook, reconcile, negotiate, mediate, and arbitrate. Often the best way to resolve conflict is for those involved to simply *overlook* what brought it on in the first place and move on. This occurs when both parties realize that fighting over a certain issue is going to get them nowhere, so they agree to forget it and move on. When this does not work, *reconciliation* is the next option. Reconciliation is the forgive-and-forget response. This means that those involved shake hands, apologize, and agree to refocus on the project's mission. When this option does not work the parties in question can *negotiate* some type of mutually agreeable resolution. In this case, neither party gets everything desired, but each gives enough that both parties are willing to shake hands and move on. *Mediation* is called for when the other responses have not worked and a mediator must be brought in to facilitate discussions and an agreement. A mediator—the project manager—serves as a referee and tries to guide the conflicting parties to a resolution that is in the best interests of the team and the project. When this does not work, the only remaining positive response is *arbitration*. With this option, the arbitrator serves as a decision maker. In the current context, arbitration means the project manager listens to both sides and makes a decision that the team members in conflict must accept.

Project Management Scenario 8.2

All this diversity is causing conflict

Melinda Morris has been a project manager for less than a month and things are not going well. The team members for the XYZ Project are the best the firm has to offer in their respective positions. Consequently, one would think the project team for the XYZ Project would be effective and smooth-running. It's not. In fact, diversity-related disagreements have turned the team into a group of warring cliques that devote more time to disagreements than to the team's work. Morris is at a loss concerning what to do. When a colleague asked her how things were going with her team, Morris responded: "Not well. I have a very diverse team, and all of this diversity is causing conflict. I don't know what to do. If you have any suggestions, I would like to hear them."

Discussion Question

In this scenario, Melinda Morris is a new project manager with a big problem. Her team has fallen into diversity-related squabbling and she does not know what to do about it. Have you ever been a member of a team in any setting where members did not get along because of diversity-related differences? If so, what problems did the disagreements in your team cause? If Melinda Morris asked for your advice concerning how to resolve the conflict in her team and how to handle the diversity issues so that problems do not recur, what would you tell her?

SUMMARY

Having a common mission is essential to effective teamwork because the foundation on which teamwork is built is a common mission. Ensuring that all members of a team buy into its collective mission is one of the most important duties and sometimes difficult challenges of the project manager. The first step in building a project team is developing its mission statement. The core of a project team's mission is always to complete the project on time, within budget, and according to specifications. A complete mission statement contains three distinct elements: name of the project, description of the project, and statement of purpose.

Fit is an important consideration when selecting the members of a team. People who do not work well with others, no matter how talented they may be in their respective professions, do not make good team members. Having team members who do not fit in can disrupt the team's work. Members who lack the characteristics of good team players can undermine the effectiveness of the whole team.

Project managers are responsible for serving as: the project team's connection with higher management, the official record keeper for the team, participants in team discussions (without dominating the discussions), implementer of team recommendations, and facilitator of positive working relationships among team members. Teambuilding proceeds in four steps: assess, plan, implement, and evaluate. For teams to be effective they must have a clear direction, members who are good team players, and accountability.

An assessment is conducted to identify strengths and weaknesses in a project team. The results of the assessment are used as the basis for developing a teambuilding plan. The plan is implemented and then given time to take effect. Finally, the team is evaluated to determine if the teambuilding activities resulted in improvements. Teams should be coached rather than bossed. Project managers can become effective coaches by: developing a team charter for the project team (mission, ground rules, and goals/

milestones), making teambuilding and team development ongoing activities, mentoring individual team members or providing mentors for them, promoting mutual respect among team members, working to make diversity an asset in the team, and setting a positive example of everything that is expected of team members.

Conflict will occur in even the best teams. Consequently, project managers must be prepared to deal with it. There are three categories of human responses to conflict: escape, attack, and resolution responses. Escape responses are denial, flight, and suicide. Attack responses are litigation/formal grievances, assault, and murder. Resolution responses are to overlook, reconcile, negotiate, mediate, or arbitrate. Project managers should work with team members to encourage resolution responses.

KEY TERMS AND CONCEPTS

Fit
Assess
Plan
Implement
Evaluate
Direction and understanding
Characteristics of team members
Accountability
Team assessment
Teambuilding plan
Executing teambuilding activities
Evaluating teambuilding activities

Team charter
Mission statement
Ground rules
Coaching and team development
Coaching and mentoring
Coaching and mutual respect
Coaching and human diversity
Institutionalized bias
Conflict in teams
Escape responses
Attack responses
Resolution responses

REVIEW QUESTIONS

1. What is the basis of effective teamwork?
2. What is the one consistent purpose of project teams?
3. Explain the concept of fit and why it is important when choosing the members of teams.
4. How can project managers develop positive working relationship in their teams?
5. Summarize the four steps for building effective teams.
6. What is meant by giving a team direction and understanding?
7. List five characteristics that are important for members of teams.
8. What is meant by accountability as it applies to teams?
9. Explain briefly how to assess the strengths and weaknesses of a team.

10. Explain briefly how to develop a teambuilding plan.
11. What is the difference between "bossing" and "coaching" a team?
12. What is a team charter?
13. Develop an example of a ground rule that might be part of a team charter.
14. Why is it important for project managers to confront cultural clashes in their teams directly and immediately?
15. What is institutionalized bias? Give an example.
16. List five techniques for preventing problems in teams that might lead to conflict.
17. List and explain the three categories of human responses to conflict.
18. List the various reconciliation responses to conflict.
19. Explain the difference between mediation and arbitration from the project manager's perspective.

APPLICATION ACTIVITIES

The following activities may be completed by individual students or by students working in groups:

1. Use the assessment instrument in Figure 8.2 to complete this activity. Choose two criteria from each section of the assessment instrument (e.g., two from "Direction and Understanding," two from "Characteristics of Team Members," and two from "Accountability"). Assume that these six criteria were rated very low during the team assessment. Develop a teambuilding plan for overcoming the team's weaknesses in these six areas.

2. Assume that you are a new project manager about to lead your first team. None of the team members report directly to you. Develop a plan for: (1) holding your team members accountable for completing the project on time, within budget, and according to specifications and (2) preventing conflict within the team.

CHAPTER 9

Project Execution:

Procurements

Because this book is intended primarily for students who are preparing to work as project managers in engineering and technology-related fields, this chapter approaches the concept of procurement from the perspective of engineering and technology firms that serve as prime contractors for projects. What is contained herein is the information that a project manager needs to know to be a positive participant in or even a leader of the procurement process for an engineering and technology project.

Before work can begin on a project, the engineering and technology firm must identify, locate, and obtain the necessary materials and equipment, as well as any subcontractors who will be needed for any aspect of the project. The process is known as *procurement*. Procurement, like so much of project management, is a process. Hence it has inputs, procedures/methodologies, and outputs. The most common inputs are the project documents, requests for proposals (RFP), and requests for quotes (RFQ).

The most common procedures/methodologies include: (1) accepting responses submitted on a low-bid, best-value, micro, small, or sole-source basis as specified; (2) evaluating the bids and quotes provided by subcontractors and materials suppliers; (3) determining if a given subcontractor or materials supplier is a responsible provider; and (4) determining if a given subcontractor or materials supplier is a responsive bidder. The most important outputs of the procurement process are the contracts awarded to responsible and responsive subcontractors and materials providers.

This chapter will help prospective and practicing project managers as well as other engineering- and technology-related personnel develop the knowledge and skills they will need to be active and positive participants in the procurement process. As has been explained throughout this book, in smaller firms the project manager might actually have to wear multiple hats—one of which is procurement manager. In larger firms, there will be a procurement professional or even a procurement department. Regardless of the project manager's level of involvement in the procurement process, he or she will need to know the basics of procurement presented in this chapter.

PROCUREMENT METHODS

There are a number of different methods used to procure the materials, equipment, and subcontractors needed in projects. The method chosen depends on such factors as the magnitude of the expected price of the materials, equipment, and/or service, the philosophy of

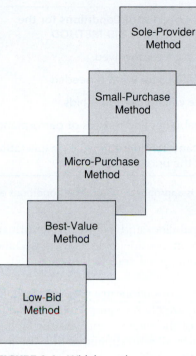

FIGURE 9.1 Widely used procurement methods.

the engineering and technology firm making the procurements, and the documented record of those who propose to provide the materials and/or services. The most widely used procurement methods in engineering and technology firms are as follows (Figure 9.1):

- Low-bid method
- Best-value method
- Micro-purchase method
- Small-purchase method
- Sole-provider method

Project managers are likely to use all of these at some point in their careers. Consequently, it is important to be familiar with all of these methods.

Low-Bid Method

The low-bid method is also called the low-price or best price method. This method is most appropriate for use when the following conditions exist (Figure 9.2):

- A substantial amount of money is involved.
- There are precise specifications for what is needed.
- At least two providers are willing to submit bids.

**Checklist of Conditions for the
LOW-BID METHOD**

✓ A substantial amount of money is involved

✓ There are precise specifications for what is needed

✓ At least two providers are willing to submit bids

✓ Providers have equal capabilities and records of performance

✓ Price is the most important consideration (i.e., expectations of other considerations may be lowered to meet the price criterion)

FIGURE 9.2 The low-bid method is appropriate when these conditions exist.

- Providers have equal capabilities and records of performance.
- Price is the most important consideration (i.e., expectations of other considerations may be lowered to meet the price criterion).

With this method, bidding instructions are prepared and made available to subcontractors or materials suppliers. Bids are often required to be submitted sealed. When sealed bids are called for representatives of the engineering and technology firm open them during a bid-opening conference to which all who submitted bids are invited. This approach ensures that all who submitted bids can see that everything about the process is above board and open. Their bids have not been tampered with.

When using the low-bid method, it is important to provide precise instructions to bidders as well as comprehensive and complete specifications. Precisely worded instructions to bidders will help prevent procedural challenges when the bids are opened and the low bidder is named. Comprehensive specifications will help ensure more accurate bids. It is also important to provide sufficient time for bidders to develop their proposals. Rushing bidders is likely to result in less accurate bids since the bid price submitted is the result of hurried guesswork rather than thoughtful deliberations. Bids should be due no later than a time and date certain that is specified in the instructions to bidders. Bids that come in after the specified deadline should be returned unopened to the bidder.

With the low-bid method, the bid is awarded to the firm that meets the following criteria: (1) submits a *responsive* bid (i.e., one that complies with all bidding instructions and meets all specifications), (2) is declared to be a *responsible* bidder (i.e., a bidder that can do the work or provide the materials in strict accordance with the provisions of the contract), (3) submits the lowest price for the products/services needed, and (4) acknowledges that the low-bid price submitted is fixed and final unless altered on the basis of mutually agreed-to-change orders. It is important to understand that all four of these criteria apply when choosing the *low-bid* method. If the other three criteria are not applied along with the *low-bid criterion*, subcontractors and materials suppliers can simply win the bid by providing unrealistically low bids they cannot possibly honor.

Best-Value Method

Best-value procurement is a method that seeks to award contracts to subcontractors and materials suppliers on the basis of their ability to provide the best overall value. Best-value is a

Checklist of Conditions for the
BEST-VALUE METHOD

✓ A substantial amount of money is involved

✓ The best production approach or material for the job is not known

✓ More than two providers are willing to submit bids

✓ Overall value is more important than the lowest possible price

✓ The opportunity to weigh quality, price, service, and past performance according to internal priorities is important

---

FIGURE 9.3 The best-value method is appropriate when these conditions exist.

combination of best quality, best price, best service, and best record of performance. When all of these factors are considered in combination, the bidder that can provide the best result is considered the best-value provider. It is not uncommon for a best value bid to be higher than other competing bids. This is because when considering quality, price, service, and performance record the price might be higher, for example, to achieve the level of quality required. The best-value method is appropriate for use when the following conditions exist (Figure 9.3):

- A substantial amount of money is involved.
- The best production approach or material for the job is not known.
- More than two providers are willing to submit bids.
- Overall value is more important than the lowest possible price.
- The opportunity to weigh quality, price, service, and past performance according to internally established priorities is important.

With the best-price method, the contract is awarded to the responsive and responsible bidder that proposes to provide the best value for the price. As with the low-bid method, when using the best-price method it is important to provide comprehensive bidding instructions, accurate specifications, and sufficient response time to ensure thorough, accurate bids.

Micro-Purchase Method

The micro-purchase method is used when the cost of the material or service is negligible. The engineering and technology firm must decide what level of price is negligible and the definition will vary from firm to firm. However, the cutoff point is typically less than $5,000. A micro-purchase is typically a direct purchase without quotes. When using the micro-purchase method, firms need to have a fairly accurate idea of what the material or service should cost so there is assurance that the price paid is *fair* and *reasonable*. To know if a price is fair and reasonable, it is necessary to have an accurate price expectation before deciding to use the micro-purchase method.

Like the small-purchase method, there is always a temptation to break up a purchase into component pieces so that the micro-purchase method can be used instead of the other methods. This is rarely justified and should, therefore, be rarely done. Circumventing the appropriate procurement method to compensate for poor planning or other nonemergency factors is unwise. Engineering and technology firms that reinforce inefficiency will get more inefficiency.

Small-Purchase Method

The small-purchase method is used when it would cost more to go through a formal bidding process than the service or material needed is likely to cost. Engineering and technology firms typically establish procurement levels that allow the formal bidding process to be circumvented for the sake of time and efficiency. The cutoff point will vary from firm to firm depending on the size of the firm and the philosophy of its higher-management team. However, it is not uncommon for a firm to apply the small-purchase method for materials or services that should cost between $75,000 and $150,000.

When using the small-purchase method, the firm asks for quotes. The process is informal and may actually be done over the telephone. In fact, it often is transacted by telephone, especially when dealing with a known and trusted provider. At least two and occasionally more quotes are solicited to make sure that the laws of competition apply. When using this method, acceptance of the winning quote can be done over the telephone with the contract to follow later.

There is an important point to understand about using the small-purchase method. There will be times when procurement personnel might be tempted to break up a contract into component pieces that do not exceed the established price level for the small-purchase option. This is sometimes done to avoid having to go through the bidding process. When this happens, the culprit is usually time. For example, if due to an emergency there is no time to go through the bidding process. To ensure that the service or material is provided on time, procurement personnel break up the contract into small enough amounts that each piece of the contract can be obtained using the small-purchase method. There may be times—hopefully few—when this approach has to be used. If so, they should be few and far between. Further, this *workaround* should not be used to compensate for poor planning or to ensure that a given provider gets the contract without having to bid on it.

Sole-Provider Method

The sole-source method of procurement is noncompetitive in that it seeks to obtain the needed service or material from just one source. This method is used only when the other procurement methods are not appropriate for some legitimate reason. Using the sole-source method to procure materials and services is appropriate when one or more of the following conditions exist:

- The needed material or service is available from only one source.
- There is not sufficient time to use a competitive procurement method such as low-bid or best-value.
- There are factors that make the sole-source method legitimate (e.g., the owner specifies that a given provider must be used).
- Only one provider is able to meet the quality criteria in the specifications.
- Special circumstances that could not be anticipated make this method the most appropriate method.

Just as there are conditions that can make the sole-source method a legitimate option, there are also the following conditions that should not be used to justify using this procurement method:

- To compensate for poor planning
- To ensure that the contract in question goes to a favored provider
- To garner favors or even illicit compensation from a given provider
- To gain a competitive advantage based on factors other than performance

---

Project Management Scenario 9.1

I think we should sole source the bid and hire ABC, Inc.

There is disagreement among members of the procurement team of Consolidated Manufacturing Processes, Inc. (CMP), concerning what procurement method to use for the new X-Tech project. CMP is the prime contractor for producing a new generation of unmanned aircraft for the Air Force. Most of the structural components for the aircraft will be a new composite material. CMP has worked with ABC, Inc., a materials science firm, on several other projects in which their performance was superb. Consequently, several members of the procurement team want to sole source the structural elements of the project to ABC, Inc. The project manager, Sarah Ferguson, is the leading voice for the sole-source option. She has already stated emphatically: "I think we should sole source the bid and hire ABC, Inc. If we do we can be guaranteed that it will be done right and on time without any problems, excuses, or change orders." CMP's procurement director disagrees. He thinks the bid should be let on a competitive basis using the low-bid method.

Discussion Question:

In this scenario, the project manager and the procurement director disagree concerning which procurement method to use in selecting a structural material subcontractor. Join one side or the other and explain why you think that option is best in this situation.

Challenging Engineering Project

MAKING SOLAR ENERGY ECONOMICAL

As the demand for fossil fuels continues to grow and their reserves continue to dwindle, the need for alternative fuel sources becomes increasingly critical. One of those alternative sources is solar energy. The good news is that solar energy has enormous potential. It is readily available, transmitted free of charge from its source, and environmentally friendly. The bad news is that cost and other factors currently limit its share of energy consumption to only one percent, while fossil fuels still account for 85 percent.

The big challenge facing engineering and technology professionals now is how to make solar energy economical. In order to become a major source of energy, the following challenges will have to be met and conquered:

- Improve the efficiency of solar cells. Commercial solar cells now convert sunlight into electricity at an efficiency rate of only 10 to 20 percent.
- Develop new materials to cut down on the fabrication costs of solar cells.
- Develop new materials that will enhance the effectiveness of capacitors, superconducting magnets, and flywheels to solve the problems associated with storing solar energy.

Meeting these challenges will require creative thinking, innovation, and persistence on the part of EQT professionals. It will also require effective project management. *Before proceeding with this chapter, stop here and consider how making solar power economical will require all of the process and people skills of project management.*

Source: Based on *National Academy of Engineering.* http://www.engineeringchallenges.org

PREPARING AN RFP OR RFQ PACKAGE

The procurement methods used most frequently by engineering and technology firms—low-bid and best-value—require that project managers or project managers working in conjunction with procurement professionals to prepare an RFP or an RFQ package for bidders. A comprehensive bid package will contain the following information (Figure 9.4):

- Invitation to bid
- Announcement of the pre-bid conference
- Bidding instructions
- Bidding form

These components together make up the bid package or RFP/RFQ that is provided to subcontractors and materials suppliers that might wish to submit a bid. The provider selected is then awarded a contract. Contracts are covered later in this chapter.

Invitation to Bid

The invitation to bid alerts subcontractors and materials suppliers that a project is being planned that they might want to participate in. The invitation to bid provides potential participants with enough information to know if they would like to submit a bid and if they are capable of providing the service or materials being requested. This latter concern is

---

Contents of a Comprehensive
BID PACKAGE

- Invitation to bid

- Announcement of the pre-bid conference

- Bidding instructions

- Bidding form

FIGURE 9.4 Bid package for an RFP or RFQ.

important because when the bids are evaluated, one of the concerns of the engineering and technology firm that is the prime contractor is finding *responsible* bidders.

Recall that a responsible bidder is one that can perform the service or provide the materials in strict accordance with all specifications, requirements, and expectations. To help bidders understand if they can perform as required the invitation to bid contains the following information:

- **Project description.** This is a brief description of the type and size of the project. Is the project large, mid-sized, or small? Does the project fit within the bidder's core competencies? Does the bidder currently have the capacity to take on such a project? The project description component of the invitation to bid answers these questions. All of these questions have a bearing on whether the potential bidder should invest the time and effort to submit a bid. For example, a small machining firm that specializes in nonprecision work may not want to bid on a large project that requires extremely tight tolerances. Correspondingly, a large firm that specializes in large-volume projects would probably not bid on a project to manufacture a small number of parts. By providing a comprehensive description of the project as the first component of the invitation to bid, engineering and technology firms can save themselves time and effort they would otherwise have to spend evaluating bids submitted by unqualified bidders. Correspondingly, the project description can save potential bidders the time and effort of preparing a bid on a job they are not likely to be able to complete according to specifications.

- **Project location.** This component describes the precise location of the project. This information is important because subcontractors typically have a range of operation within which they work. Some bid only on local jobs, while others are willing to bid on jobs that are regional, national, or even international. Location becomes even more important than usual during difficult economic times. Generally speaking subcontractors and materials suppliers want to work on projects that are as close to their primary facilities as possible. However, a rule of thumb for engineering and technology firms is that the worse the economy becomes the farther materials suppliers are willing to ship and the farther away subcontractors are willing to be. Of course, the farther a subcontractor has to ship its work or a material supplier has to ship its goods, the more their costs will be. These factors will become part of their deliberations as they develop their bids. When economic times are particularly bad, subcontractors and materials suppliers are often willing to cut their profit margins sharply just to have enough work to stay in business. However, during normal economic times, this is rarely the case.

- **Project start and completion dates.** This component lets bidders know the time-frame within which they will have to provide the service or materials. With materials there will be a more specific date since when materials arrive where they are needed is critical. Too early and they have to be stored and secured. Too late and they throw the project off the schedule. For subcontractors, the start and completion dates define how long their personnel and other resources will have to be committed to the project in question. This is an important consideration for firms that take on parts of projects because they are often working on more than one job at a time. If a subcontractor submits a bid containing the proper assurances that its personnel and other resources will be available for the project, it is making a commitment that will become part of the contract. Violating a contractual obligation can be costly for a subcontractor.

- **Bonds.** This component provides information that automatically eliminates some potential bidders. Consequently, this component is one of the first things subcontractors and materials suppliers will look for in the invitation to bid. The bonding requirement eliminates at the outset any subcontractors or materials suppliers that are unwilling or unable to undertake the expense of getting bonded. A bond is a financial assurance of performance. Bonds are explained in greater detail later in this chapter.
- **Project documents.** This component explains how the documents that subcontractors need to prepare their bids will be made available. These documents include the specifications, drawings, and any other documents that might apply to a given job. In times past, the documents were made available at specified times on specified days at a central location. Subcontractors would make an appointment to review the documents and come to the central location to do so or they would receive them in the mail. In recent times, the ability to provide documents electronically has simplified this aspect of bid preparation.
- **Legal considerations.** This component explains how certain potentially contentious issues will be handled. This component is intended to preclude lawsuits that might arise out of disagreements over the issues covered. For example, this section often contains specific criteria for rejecting a bid or for allowing a subcontractor or materials provider to withdraw a bid with no monetary penalty.
- **Bid deadline.** This component provides a specific date and time by which bids must be received. It should also contain a statement to the effect that bids received after the deadline will not be considered. Often, this will be the first part of the invitation read by potential bidders. Preparing a bid response is a time-consuming process. If the deadline looms too close, some subcontractors and materials suppliers may decide that they simply do not have time to prepare a responsive bid and decline the invitation. Others may decide the work is important enough to work around the clock to respond. This dilemma illustrates why it is so important for engineering and technology firms to give subcontractors and materials suppliers sufficient time to prepare responsive bids. Providing less than sufficient time can result in three things happening—all of them bad. First, responsible subcontractors may choose not to submit a bid, thereby reducing the quality of the bidding pool. Second, subcontractors might rush to complete a bid on time and miss important aspects of the specifications or drawings, thereby submitting unresponsive bids. This also reduces the quality of the bid pool. Finally, a subcontractor might suspect that the quick turn around on bids was a ploy by the engineering and technology firm to rule out all but a few favorite bidders who were informed of the project prior to the invitation. This situation can lead to challenges to the bidding process and even litigation.

Announcement of the Pre-Bid Conference

This component provides the date, time, and location of the bid conference. The bid conference is a meeting involving representative of the engineering and technology firm and potential bidders in which bidders are allowed to ask questions to ensure they understand all aspects of the process, specifications, and instructions. Engineering and technology firms that serve as prime contractors do not always hold bid conferences, but many do. Responding to the questions and concerns of bidders at the outset can prevent problems such as unresponsive bids, misunderstandings concerning specifications, and even legal challenges later in the process.

Bidding Instructions

Bidding instructions tell potential bidders specifically what they need to do submit a responsive bid. Some of the material covered in the bidding instructions may also be found in the invitation to bid, a circumstance that is encouraged since redundancy promotes a better understanding on the part of bidders. Information typically contained in the bidding instructions includes the following:

- Date and time bids are due
- Location where bids must be received
- Instructions for completing the bid form
- Unit prices for work/materials to be provided
- How and where to indicate any additional fees that will be charged
- How and when the winning bid will be announced
- How and when the contract will be awarded
- Special instructions (i.e., instructions concerning any aspect of the project that is out of the ordinary)

Bidding Form

To ensure that bids from various subcontractors can be readily and accurately compared, engineering and technology firms typically provide a bid form such as the one shown in Figure 9.5. Providing a standardized bid form helps ensure that all stakeholders are on the same page when it comes to submitting and evaluating bids. Notice on the form in Figure 9.5 that bidders must acknowledge that they have reviewed all of the project documents, that they are bonded and in what amount and by whom, any part of the required work they will not perform (exclusions), the period of time for which their bids are valid, their base bids plus additional amounts to be added should the engineering and technology firm decide to proceed with alternates, and, finally, remarks.

Requiring all bidders to submit their final bids on the same form allows the engineering and technology firm to make more accurate comparisons when deciding which subcontractor wins the bid and receives the contract. Without this uniformity, bid openings can quickly degenerate into confusion and recriminations. Requiring subcontractors to provide bids on project alternates is also important. A situation that often occurs is this. The engineering and technology firm might want to provide certain additional amenities or a higher quality material or piece of equipment as part of the job. However, its ability to do this will depend on price. In cases such as this, the base bid represents the minimum the engineering and technology firm wants done.

The engineering and technology firm has estimated and budgeted a certain amount for the basic work to be done. If the base bid is below this amount, the firm serving as the prime contractor might decide to include one or more alternates in the contract. These alternate might require the subcontractor to use a higher material or hold to a tighter tolerance or include something else that increases the value of the finished product or service. However, if the base bids will require all that is budgeted for the component of the project in question, the alternate(s) will likely be discarded. A situation that often arises with alternates is this. A subcontractor's base bid is higher than a competitor, but its base bid plus alternates is lower. In this case, the subcontractor with the lowest total for the base bid and any alternates the engineering and technology firm decides to proceed with receives the contract.

BID FORM FOR SUBCONTRACTORS

Project: _____

Date: _____

Time: _____

Location: _____

Bidder: _____

Address: _____

Telephone: _____ FAX: _____

Email: _____

URL: http:// _____

Addenda:

We acknowledge reviewing all

project documents _____

(Initial)

Bonding:

a. Bonded amount: _____

b. Bonding agent: _____

We propose to complete the following work on this project, as shown on the plans and described in the specifications and addenda as follows:

EXCLUSIONS FROM THIS BID:

PROJECTED SCHEDULE:

BASE BID: $

(including taxes)

ALTERNATE 1: $

ALTERNATE 2: $

BIDDER'S REMARKS:

Signature: _____

Printed Name: _____

Title: _____

Bid price valid until _____

(date)

ALTERNATE 3: $

ALTERNATE 4: $

Date: _____

FIGURE 9.5   A standard bid form simplifies the bid review process.

BONDS, ADDENDA, AND ALTERNATES

In the previous sections, the concepts of bonds, addenda, and alternates were part of the narrative. These concepts are important enough in the procurement phase of project management to warrant their own explanations. The following paragraphs explain the basics of bonds, addenda, and alternates as they relate to the job of the project manager:

- ***Bonds for subcontractors.*** Bonding is a way for an engineering and technology firm that is serving as the general contractor for a project to manage the risk that a subcontractor might not perform as expected. A bond is a financial guarantee that the subcontractor will perform or forfeit a specified amount of money. There are different types of bonds, although all of them are in essence the same: a financial guarantee of performance. *Bid bonds* guarantee that the subcontractor will follow through and enter into a contract if it is selected during the bidding process. *Performance bonds* guarantee that the subcontractor will do the work contracted for in accordance with all applicable specifications and other requirements and expectations. A *payment bond* guarantees that the subcontractor will follow through and pay all of its bills relating to the project (e.g., materials and labor). This is important because an engineering and technology firm serving as the prime contractor on a project can be held liable for the unpaid bills of subcontractors in some cases. Consequently, many engineering and technology firms will not accept bids from unbounded subcontractors. Subcontractors purchase bonds in the same way that insurance policies are purchased. Then, if they do not perform, the bonding agent must pay the specified amount of money.

- **Addenda to project documents.** Addenda are written changes to the project documents in response to design changes, errors, or changes of other kinds. Changes to projects are sometimes made between when the project documents are completed and when invitations to bid are sent out. There are times when changes are made even after invitations to bid have gone out but bids have not yet been received or closed out. In both cases, rather than completely rewrite a certain project document the engineering and technology firm will add an addendum to it. If this happens before the invitations to bid are sent out, the bid form should have a line in it to ensure that subcontractors read the addendum and factored it into their estimate (see Figure 9.5). If it happens after the invitation to bid has gone out but before the due date for bids, all subcontractors should be notified of the change in writing. Again, bidders are required to indicate on the bid form that they reviewed all applicable addenda and factored them into their bids.

- **Bid alternates.** It is not uncommon for an engineering and technology firm that is serving as the prime contractor on a project to ask bidders to submit a price for a *base bid* and for *alternates*. The base bid covers the work/materials for the mandatory aspects of the project. The alternates represent discretionary aspects of the project (i.e., things that will be included if the price allows it). Alternates can represent an additional feature or they can represent escalating levels of quality. For example, an engineering and technology firm might specify a quality of material be used. Then it might add alternates for upgrading the materials. Upgrade Level 1 would be alternate one—one level of quality above the base bid. Level 2 would be alternate two—two levels of quality above the base bid. The engineering and technology firm will select either the base bid or one of the alternates depending on price. Regardless of what the alternates in a project actually represent, subcontractors should price them like separate projects and provide a separate price in the bid for each alternate.

CONTRACTS FOR SUBCONTRACTORS AND MATERIALS SUPPLIERS

Once bids have been received and evaluated, the subcontractors and materials suppliers who are selected receive contracts. The contract is an important component in the package of project documents. There are different forms of contracts, but they all tend to contain the same or similar elements. The elements typically contained in a contract between an engineering and technology firm that is serving as the prime contractor for a project and its subcontractors are as follows:

- **Participants.** This clause of the contract contains the contract participants—the parties to the contract. The contract is between two parties: the engineering and technology firm serving as the general contractor and the subcontractor or material provider. Although the participants may seem obvious, there can be no assumptions when entering into a contract. The full and DBA (doing business as) names of both parties are stated. It is not uncommon for engineering and technology firms and subcontractors to have formal corporate names that are different than their DBA names.

- **Description of the work to be provided.** This clause of the contract describes in detail the work the subcontractor will provide or the material the vendor will supply. The more comprehensive and accurate this section is the better. If the description is too long to fit into the actual contract document, it can be attached to it as an appendix. Ideally, this element will be written in a way that leaves no room for disagreement or creative misinterpretation by either party. If disagreements occur during the course of the project over change orders that are requested, this is the first place the project manager will look to resolve the conflict.

- **Starting date.** This clause of the contract contains the starting date of the subcontractor's portion of the project. Depending on the project schedule, the start date might have some slack in it. For example, it might read as follows: *the subcontractor may begin work on January 5 or as late as—but no later than—January 9.* Starting dates for subcontractors are important for several reasons. First, to complete their work on time subcontractors must start on time. Second, subsequent work on the project might depend on the subcontractor completing its work. Finally, when subcontractors submit a bid, it is valid for only a specified period of time. If the subcontractor's starting date is after this valid period, the bid no longer applies and may have to be renegotiated or even rebid.

- **Completion date.** This clause of the contract contains the completion date—the date by which the subcontractor agrees to have all of its work completed. This date is especially critical since subsequent work in the project may depend on the subcontractor completing its work. When one subcontractor falls behind schedule on a project, it can start a chain reaction that throws the rest of the project off the schedule.

- **Contract amount.** This clause of the contract contains the amount the subcontractor submitted to win the bid for the work in question. It is a total of the base bid and any alternates that were accepted during the bid process. This amount will not change unless the subcontractor negotiates a change order with the engineering and technology firm that is serving as the prime contractor.

- **Progress payments.** This clause of the contract explains the milestones that will trigger progress payments to the subcontractor. Not all subcontractor contracts allow for progress payments. Whether or not progress payments are to be made depends on the amount of work to be done and the length of time the work will take to complete. Progress payments provide the subcontractor with cash flow so that it does not have to borrow money to make payroll and pay for materials and equipment it uses in the project. If the subcontractor is borrowing for these purposes, the progress payments allow it to pay the loan off promptly, thereby avoiding interest and other fees.

- **Liquidated damages.** This clause of the contract contains an explanation of what failures will trigger liquidated damages and in what amounts. Sometimes referred to as "late fees," liquidated damages are assessed against the subcontractor for any damages suffered by the prime contractor as a result of failures on the part of the subcontractor. For example, if the subcontractor does not complete the agreed upon work by the agreed upon date and the project schedule is affected, the prime contractor can assess liquidated damages as specified in this element of the contract. For example, assume that the printed circuit board subcontractor on an engineering and technology project does not complete its work on time. This, in turn, means that the assembly team for the firm cannot begin its work on time. The late completion of the printed circuit boards cascades through the remainder of the project costing the prime contractor to be late in delivering the finished product. In a case such as this, the prime contractor would probably elect to invoke the liquidated damages clause of the contract.

- *Retained funds or holdbacks.* This clause in the contract explains how much—usually a percentage of the overall amount—the engineering and technology firm serving as the prime contractor will hold back from the subcontractor as assurance that the work will be completed properly and on time. Subcontractors do not receive their retained funds until the prime contractor is satisfied that the work provided is complete and properly done. By retaining funds, prime contractors build in a certain amount of protection against a subcontractor completing most but not all of the work, then taking its money, and leaving. The amount of retained funds is typically 10 percent, although this amount can vary depending on the level of confidence the prime contractor has in the subcontractor. A subcontractor whose work record is questionable might have to agree to a higher percentage of retained funds than one that has a strong work record and a history of completing its work on time, within budget, and according to specifications.
- *Final payment.* This clause of the contract states the conditions under which the prime contractor will release any retained funds to the subcontractor. This is an important clause because by releasing any and all retained funds the prime contractor is also releasing the subcontractor from further responsibility on the project, other than any responsibilities contained in the subcontractor's warranty—if applicable—and any that have been put in writing and signed by both parties to the contract.
- *List of project documents.* This clause of the contract lists all documents—drawings, specifications, addenda, and so on—that are part of the project documents for the project in question.
- *General conditions.* This clause of the contract contains what is often referred to as *boilerplate* language. Boilerplate language summarizes the standard legal language that should be contained in any contract. The language contained in the general conditions clause of the contract is typically taken from a third-party source that provides wording that has been tested in the courts and proven to be legally valid.
- *Special conditions.* This clause of the contract contains an explanation of any additional conditions of a special nature that apply to the project in question and are not adequately covered by the general conditions clause. Occasionally, there will be special requirements that apply to a project that grow out of the exigencies of that project alone. For example, a certain type of material may be required to accommodate the conditions the finished product will be exposed to.
- *Bonds.* Bonds were explained earlier in this chapter. This clause in the contract explains what types of bonds are required of the subcontractor and in what amounts.
- *Insurance.* This clause of the contract specifies the types of insurance the subcontractor must carry in order to work on the project in question. For example, prime contractors typically require their subcontractors to carry workers' compensation and liability insurance. It is important that this clause and its requirements be included in the contract because the prime contractor can be held liable for the accidents, injuries, and damage of a subcontractor that is allowed to work without the appropriate insurance.

EVALUATING BIDDERS AND BIDS

When an engineering and technology firm that serves as the general contractor on a project solicits bids from subcontractors and materials suppliers, it needs to know the following two things before accepting a bid:

1. Is the bidder responsible—can it actually perform as specified, required, and expected?
2. Is the bid responsive—does it comply in all aspects with the instructions in the RFP or RFQ?

The answers to these questions are even more important than the price quoted by the bidder. This is because the answers to these questions determine whether or not the price submitted by a bidder is actually valid and whether or not the work proposed can actually be performed. If the bidder submits an unresponsive bid, the price quoted cannot be trusted. If the bidder is not responsible, it may not be able to complete the work as specified, regardless of its bid price. Consequently, project managers need to learn how to evaluate bids and bidders.

Evaluating Bidders for Responsibility

A responsible bidder is a subcontractor that can comply with all requirements and expectations in the RFP/RFQ while performing the work in question according to specifications. There are actually two considerations embedded in the responsibility question: (1) *Can* the bidder perform as required? and (2) *Will* the bidder perform as required? Just because a subcontractor can perform as required does not mean it will. Consequently, it is important to answer both aspects of the responsibility question.

The "can" part of the question is answered by evaluating the capabilities of subcontractors from the perspectives of financial resources, organizational infrastructure, skills/experience, and facility/equipment resources. The "will" part of the question is answered by evaluating the subcontractor's record from two perspectives: performance over time and ethics.

To perform as required on a project, a subcontractor must be financially stable. It must have or be able to borrow the money needed to perform on the project. To perform as required, a subcontractor must have the organizational infrastructure required to support its personnel (e.g., operational policies, management/supervisory procedures, an accounting/payroll system). To perform as required a subcontractor must have the personnel available who have the skills and experience to do the required work in the required way. To perform as required a subcontractor must have the equipment called for in the project and appropriate facilities to support its personnel and equipment. Without the appropriate support systems in place, a subcontractor may not be able to perform as required on the project.

The "will" part of the question is answered by examining a subcontractor's performance record and ethics record. Can the subcontractor show that it has performed as expected over a period of time? Can the subcontractor provide the names of other prime contractors it has worked with? What do these references say about the subcontractor? Does the subcontractor appear to be ethical? Are there ethical problems in its background? Do references say the subcontractor has integrity or do there appear to be doubts along these lines? A perfectly capable subcontractor can still fail to perform on a project. Consequently, its records of performance and ethics are important considerations during the evaluation process.

Evaluating Bid for Responsiveness

The bids that should be considered by prime contractors are those submitted by subcontractors who appear to be responsible. Once the responsibility question has been answered, the responsiveness of bids can be evaluated. A responsive bid is one that meets the following criteria:

- Received on time
- Complies with all requirements in the RFP/RFQ
- Prices quoted are fair and reasonable

Responsive bids received from responsible subcontractors are the only ones that should be considered for price comparisons. Once the determinations of responsibility and responsiveness have been made, comparing prices is not difficult. Figure 9.6 is an example of a bid comparison form used by prime contractors for comparing the prices submitted by subcontractors. Notice that these forms compare not just the overall price, but the individual components of that price (tasks) as well as alternates. Using this approach helps the prime contractor make an easy determination of whether subcontractors included all of the work called for in their bids.

ETHICS IN PROCUREMENT

No process relating to engineering and technology projects has more opportunities for ethical lapses than the procurement process. Unethical players in the process have myriad ways to circumvent the process and make deals that are good for them but bad for the project, the engineering and technology firm serving as the prime contractor, and the subcontractor/material supplier. Consider the following situations that have occurred numerous times on engineering and technology projects. A subcontractor submits an unresponsive bid but is awarded the contract anyway after paying "kickback" to a decision maker in the engineering and technology firm that solicited the bids. A material provider wins the bid on a large project by knowingly

BID COMPARISON FORM				
Subcontractors	**1**	**2**	**3**	**4**
Base Price				
Task 1				
Task 2				
Task 3				
Task 4				
Task 5				
ALT 1				
ALT 2				
ALT 3				
ALT 4				
Total Price				

FIGURE 9.6 Sample form for comparing the bids of subcontractors.

bidding substandard materials that do not meet the specifications while claiming that they do. A subcontractor bribes a quality inspector to ignore certain deficiencies in its work.

Ethical Procurement Defined

Ethical procurements are the result of a procurement process that is characterized by integrity, transparency, openness, suitability, and fair competition. When the procurement process operates within an ethical framework, contracts are awarded to subcontractors and materials suppliers on the basis of their ability and commitment to meet all requirements in the RFP/RFQ rather than on factors, such as favoritism, familiarity, illicit compensation, politics, and intimidation.

Competition and Ethical Procurement

One of the most effective ways for an engineering and technology firm to ensure an ethical procurement process is to base the process on fair and open competition. Fair and open competition means that when an engineering and technology firm that is serving as the prime contractor for a project invites subcontractors to submit bids: (1) the net is cast broadly so as to include as many potential bidders as possible, (2) all subcontractors/materials providers are provided with the same documents on which to base their bids, (3) all bidders and bids are evaluated on the basis of the same criteria, (4) all potential bidders have an equal opportunity to attend bid conferences, (5) all bidders have equal access to project documents including addenda, (6) all bidders receive the same communications from the prime contractor during the bid preparation process, (7) all bidders have an opportunity to be present when bids are opened, and (8) all bids are opened at the same time in the same location.

All eight strategies listed for making ethical procurements through the use of competition can be satisfied by a comprehensive system based on RFPs in which the engineering and technology firm casts a broad net, gives recipients of the RFP sufficient time to prepare and submit their bids, and then fairly and objectively evaluates the bids. The centerpiece of the process is the RFP itself. Figure 9.7 is an example of the cover page for a well-designed RFP. Figure 9.8 is the contents page for the same RFP. Notice on the cover page in Figure 9.7

REQUEST FOR PROPOSAL
RFP NO. 1602

Computer Hard Drives

ISSUED:_____(date)_____

CLOSING TIME: _____(date and time)_____

CLOSING LOCATION: FWB Eng-Tech, Inc.
100 Industrial Drive
Fort Walton Beach, FL 32547

CONTACT PERSON: Mark James
Tel: (850) 942-1869
Fax: (850) 942-1870

FIGURE 9.7   Sample RFP.

RFP 1602 Contents

FIGURE 9.8 Contents of a comprehensive RFP.

that recipients of the RFP will be given a date, time, and place for submitting their bids and a contact person to call should they have questions.

Figure 9.8 shows the level of detail RFPs often contain. Notice that the contents page for RFP 1602 explains in detail the proposal process, proposal preparation, and evaluation criteria. The engineering and technology firm that prepared this RFP even includes a receipt confirmation form and a checklist to help bidders ensure that they have completed all of the requirements before submitting their bids. Figure 9.9 is an example of a receipt confirmation for RFP, and Figure 9.10 is an example of a bidder's checklist. Although this may seem like a lot of detail, it is this level of detail that keeps the process fair, objective, and ethical. Every bidder must respond to the same criteria in the same way and during the same time period. Further, all will be evaluated using the same criteria

---

RFP NO. 1602
RECEIPT CONFIRMATION

From: _____   (Name of Bidder)

To: FWB Eng-Tech, Inc., 100 Industrial Drive, Fort Walton Beach, FL 32547
Attention: Mark James
By Email: mark.james@fet.com
By Fax: (850) 942-1870

Re: Request for Proposals, RFP No. 1602 – Computer Hard Drives
We confirm receipt of the RFP and confirm that we will submit a Proposal. Our contact person (one person only) and that person's delivery and mailing addresses and telephone and fax numbers are as follows. All communications (including addenda) regarding the RFP should be directed to our contact person.

Contact Person:

Name: _____

Title: _____

Delivery Address: _____

_____

_____

Mailing Address: _____

_____

Telephone No: _____

_____

Telephone No: _____

Fax No: _____

Email: _____

Signature:_____ Date: _____

Title: _____

FIGURE 9.9   Receipt confirmation for RFP.

BIDDER'S CHECKLIST

This checklist is provided for the convenience of bids. The accuracy or completeness of this checklist is neither warranted by FWB Eng-Tech nor is the checklist necessarily comprehensive. Its use is not mandatory and it does not have to be returned with the Proposal. It is provided as an optional tool.

✓   The bidder has read and understands the requirements of the RFP.

✓   The Receipt Confirmation Form has been submitted.

✓   The Proposal meets all mandatory requirements.

✓   The Proposal addresses all elements of the RFP.

✓   The Proposal clearly identifies the bidder, the project, and RFP number.

✓   The appropriate number of copies of the Proposal have been included in the submittal package.

✓   The Proposal will be delivered to the Closing Location before the deadline.

FIGURE 9.10   This checklist will help ensure a complete bid package.

Other Issues in Ethical Procurement

In addition to having a procurement process that is fair, open, and competitive, engineering and technology firms must be cognizant of the ethical behavior of subcontractors and materials suppliers. For example, an engineering and technology firm that knowingly awards a contract to a subcontractor who employs illegal aliens, underage workers, drug abusers, or wanted felons is committing an ethical breach by aiding and abetting unethical behavior. A firm that knowingly awards a contract to a materials supplier that obtains its materials illegally or abuses its workers is also aiding and abetting unethical behavior. A firm that awards a contract to a subcontractor that has a record of serious safety violations is aiding and abetting unethical behavior. Consequently, it is important for engineering and technology firms that solicit bids to consider the ethics of their subcontractors and materials suppliers.

Project Management Scenario 9.2

I don't think we should accept their bid

Don Markham is the project manager for the Department of Defense project his company just received a contract for. As such he is one of the members of the procurement team that will evaluate the bids from subcontractors and materials suppliers. Before opening the bids that have been submitted, the procurement team is evaluating the bidders to ensure that they are responsible. A problem has arisen in this regard. The most important subcontractor on this job will be the one that provides low-voltage power supplies. Questions have come up about the ethics of one of the bidders: LMN Power Supplies. LMN has a long record of completing its projects on time and within budget. However, it also has a reputation for employing illegal aliens and for its shoddy safety practices. Feedback on the company suggests that there are more accidents and injuries than are actually reported. Apparently, when an illegal alien is involved in an accident the company gets away with not reporting it because the illegal worker—for fear of deportation—refuses to take any action.

Some members of the team want to qualify LMN as being a responsible subcontractor on the basis of the firm's record for completing its projects. Some members want to disqualify LMN on the basis of ethical considerations. The most vocal of the latter group is the project manager—Don Markham. Markham began the debate on LMN when he said: "I don't think we should accept their bid." The debate has gone on for more than an hour with nothing resolved.

Discussion Question:

In this scenario, the procurement team is evaluating bidders to ensure that they are responsible. Several members think that proven performance is the only criterion they should apply in making the determination. Several other members think that ethics should be an added criterion. Assume that you are a member of this procurement team. Which side of the debate would you take and why?

SUMMARY

Before work can begin on a project, the engineering and technology firm serving as the general contractor must identify, locate, and obtain the necessary materials and the subcontractors who will do the hands-on work. The process is known as procurement. The most commonly used procurement methods are as follows: low-bid, best-value, micro-purchase, small-purchase, and sole-provider. A comprehensive bid package (RFP or RFQ) will contain an invitation to bid, announcement of the pre-bid conference, bidding instructions, and bidding form.

The invitation to bid should contain the following information: project description, project location, start and completion dates, bonds, project documents, legal considerations, and the bid deadline. Bidding instructions should contain the following information: date and time bids are due, location where bids must be received, instructions for completing the bid form, unit prices for work and labor, additional fees, how and when the winning bid will be announced, how and when the contract will be awarded, and special instructions.

A bond is a financial guarantee that a subcontractor will either complete the work as specified or forfeit a specified amount of money. There

are three types of bonds that are widely used: bid, performance, and payment bonds. Addenda are written changes to the project documents in response to design changes, errors, or changes of other kinds. Alternates are discretionary add-ons to the basic work being bid—extras that will be added to the contract if funding allows.

A contract for a subcontractor should contain the following information: participants or parties to the contract, a description of the work, starting date, completion date, contract amount, progress payments, liquidated damages, retained funds, final payment, a list of project documents, general conditions, special conditions, bonds, and insurance.

Bidders should be responsible, meaning they should be able and willing to complete the work in question as specified and in accordance with the project documents. Bids should be responsive, meaning they should comply with all requirements in the RFP/RFQ; be submitted on time; and contain prices that are fair and reasonable. Ethical procurement is procurement that is characterized by integrity, transparency, openness, suitability, and fair competition. Fair competition is an effective way to ensure the integrity of the procurement process.

KEY TERMS AND CONCEPTS

Procurement	Participants
Low-bid method	Description of the work to be provided
Best-value method	Starting date
Micro-purchase method	Completion date
Small-purchase method	Contract amount
Sole-provider method	Progress payments
Invitation to bid	Liquidated damages
Pre-bid conference	Retained funds
Bidding instructions	Final payment
Bidding form	List of project documents
Bonds	General conditions
Addenda	Special conditions
Alternates	Insurance
Bid bonds	Responsible bidder
Performance bonds	Responsive bid
Payment bonds	Ethical procurement

REVIEW QUESTIONS

1. Define the term *procurement* as it relates to engineering and technology projects.
2. List and briefly explain the most commonly used procurement methods.
3. List and briefly explain the contents of a comprehensive bid package.
4. What should be contained in the invitation to bid?
5. What should be contained in the bidding instructions?
6. What is the purpose of a bond?
7. List and briefly explain the types of bonds that are widely used.
8. What are addenda?
9. What is a bid alternate? Give an example.
10. List the information that should be contained in a contract given to a subcontractor.
11. What is meant by the term *responsible bidder*?
12. How can a project manager determine if a bidder is responsible?
13. What is meant by the term *responsive bid*?
14. How can a project manager determine if a bid is responsive?
15. Define the term *ethical procurement*.

APPLICATION ACTIVITIES

The following activities can be completed by individual students or by students working in groups:

1. Identify an engineering and technology firm in your region that will cooperate with you in completing this project. Ask the procurement director to walk you through how the company solicits bids, evaluates bids, and awards contracts to subcontractors. Ask what the biggest challenges of the procurement process are. Ask to see an invitation to bid, bids that were submitted by subcontractors, and a contract. Determine which procurement method the company uses most often. Report the findings of your research to the class.
2. Do the research necessary to identify at least one case in which there were ethical failures in the procurement process of an engineering and technology firm. Write a report on your findings and report the results of your research to the class.

CHAPTER 10

Project Monitoring and Control

Once the various elements of the project management plan have been developed and execution is underway, the work of the project must be monitored and controlled. Monitoring and controlling project work involves observing and measuring and, then, making appropriate adjustments and taking appropriate action based on the observations and measurements. Project managers and their team members monitor and control all aspects of a project during execution. Most of their time and effort is spent monitoring and controlling the following project elements (Figure 10.1):

- Scope
- Schedule
- Costs
- Quality
- Risk

MONITORING AND CONTROLLING SCOPE

One of the challenges project managers always face is *scope creep*. Scope creep occurs when the project scope is not properly monitored and controlled so that changes are allowed to be made that exceed the scope of the project as set forth in the scope statement, contract, and other project documents. In most engineering and technology projects, changes will be requested by the customer after the contract has been signed. This is why having an efficient change-order process is critical to effective project management.

When changes to the original scope of a project are requested, the changes must be processed as change orders so that the appropriate adjustments can be made to the project scope and corresponding elements of the project including the budget and the schedule. If the change-order process is not properly designed and operated, changes requested by customers after a contract has been signed will result the engineering and technology firm doing work it is not being paid to do. For example, assume that a fellow student offers you

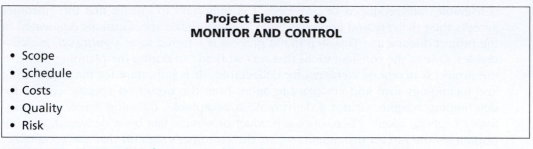

**Project Elements to
MONITOR AND CONTROL**

- Scope
- Schedule
- Costs
- Quality
- Risk

FIGURE 10.1 These project elements must be carefully monitored and controlled.

$50 to tutor him for two hours in a given subject. You agree and show up at the appointed time. After about an hour and 50 minutes of tutoring the student says: "This is going so well I would like you to tutor me in another subject for a couple of hours."

The student's request for additional tutoring exceeds the scope of your agreement with him in two ways: (1) it adds two additional hours beyond what was originally agreed to and (2) it adds a new subject to the mix. This is scope creep. The only way to prevent this situation from resulting in scope creep is to prepare a change order. The change order will summarize the original scope of the project as well as the work that is being added to it. A price for the new work will be negotiated and agreed to by both parties. Once this is done, the original scope, budget, and schedule are revised to accommodate the addition of the new work.

In monitoring and controlling a project's scope, project managers and their teams do three things (Figure 10.2):

- ***Ensure acceptance of deliverables.*** Every project consists of specific deliverables—products and/or services—that are to be provided for the customer. As these

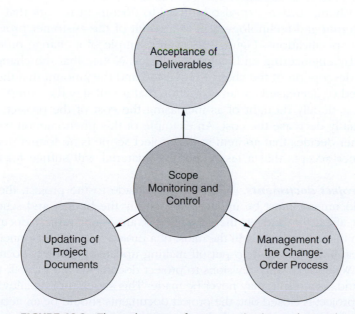

FIGURE 10.2 Three elements of scope monitoring and control.

deliverables are produced or provided, it is important to ensure that the customer accepts their delivery and approves them as meeting the specifications delineated in the project documents. This step in the process is referred to as *signing off* on deliverables. One of the considerations that was worked out during the planning phase of the project is *acceptance criteria* for deliverables. It is important for the engineering and technology firm and customer to agree from the outset on specific criteria for determining whether or not a deliverable is acceptable, meaning it meets the customer's specifications. Then, once a product or service has been delivered, it is important for the project manager to secure the customer's sign off that it is acceptable. Products that do not meet specifications—are not acceptable—require rework that is paid for by the engineering and technology firm. But products that are deemed acceptable and are signed off on by the customer cannot be returned for additional work unless a change order is initiated and a new price and schedule are agreed to. Any number of things could happen to a product after it has been delivered that might damage it or even render it unacceptable. For example, a forklift driver for the customer might drop a product and damage it while transporting it for storage in a warehouse. The engineering and technology firm should not be held responsible for damage that occurs in this way or any other way after a product has been delivered. The best way to ensure that this does not happen is to secure a sign-off on the deliverable from the customer.

• ***Manage the change-order process.*** Managing the change-order process is one of the most important responsibilities project managers and their teams have. Without an effectively managed change-order process it will be difficult if not impossible to satisfy the three basic success criteria that apply to all projects: budget, schedule, and quality. Change orders are used to document requests for work that is outside the scope of a project so that the customer can be charged for the work and so that the project documents—budget, schedule, and scope—can be updated accordingly. Change orders can also be used to document rework that is required of the engineering and technology firm as a result of the customer providing poor or insufficient specifications. Figure 10.3 is an example of a change order form of the type used by engineering and technology firms. Notice that the change order form requires a description of the change requested and the amount that the contract will be increased or decreased (or an indication that it will stay the same). While change orders are generally thought of as increasing the cost of the project, some change orders actually decrease the cost. An example of this phenomenon would be when the customer decides that an item in the project scope is no longer needed or when the customer accepts that a less expensive material will suffice for the project in question.

• ***Update project documents.*** As changes are made to the project, the project management documents must be updated—especially the budget and schedule. Change orders that affect the budget, the schedule, or any other project document must be reflected in those documents. In the midst of a project with all personnel working hard to meet deadlines, it is easy to put off making updates to project documents. This is a mistake. When necessary revisions to project documents are put off, they are easily forgotten and, as a result, may never be made. This should not be allowed to happen. When the project is closed out, the project documents should be an accurate record of everything about the project, including how much work was done, how much it cost,

CHANGE ORDER FORM

Project Number: Date:

Project Name:

Change Order Number:

Description of the Change:

_____

Not valid until signed by the Customer and the Contractor

_____

The Contract Sum prior to this Change Order was...$

The Contract Sum will be (increased) (decreased)
(unchanged) by this Change Order in the amount of ....$

The new Contract Sum including this Change Order will be ....$

The Contract Time will be (increased) (decreased) (unchanged) by............................

The Date of Completion as of the date of this
Change Order is now ....

_____   _____

Customer   Project Manager   Date

_____

By   Date

FIGURE 10.3 Comprehensive change order forms can prevent oversights.

and how long it took. Otherwise, the closeout process is invalid. During the lessons-learned element of the closeout process, if project managers and their team members are working from inaccurate project documents, they will learn the wrong lessons or, at best, inaccurate lessons.

Project Management Scenario 10.1

We don't need a change order—just do it

"I am getting tired of processing all of these change orders. The next time the customer asks for a change, just do it—we don't need a change order." John Marshall, project manager for the Alpha-Beta Project, was clearly frustrated when he had made this statement last week. But now that the customer had requested another change, Marshall's words took on more significance for his team members. Should they "just do it" or should they process the paperwork for a change order? Unfortunately, John Marshall was out of town for a week and nobody wanted to disturb his vacation to ask for a clarification. In his absence, the assistant project manager made the decision to process a change order.

Discussion Question

In this scenario, the project team is in the uncomfortable position of needing to make a decision about whether or not to process a change order after their absent supervisor, the project manager, said that no additional change orders would be processed.

The assistant project manager takes responsibility and makes the decision to process a change order. Did the assistant project manager make the right decision? If you were the assistant project manager, how would you justify processing a change order in this case or any other case?

MONITORING AND CONTROLLING THE SCHEDULE

Monitoring and controlling the schedule is one of the most important responsibilities of project managers. The schedule affects every other aspect of the project. Of course, completing projects on schedule is one of the fundamental success criteria for all projects. If a project falls behind schedule, the project manager is forced to respond by choosing one or more of the following options (Figure 10.4):

- *Increase staff.* If a decision is made to increase staff, that decision will affect the project's human resource plan as well as the budget. Unless a contingency was built into the budget to cover increasing staff, the increased personnel costs will have to come out of the engineering and technology firm's profit line item.
- *Approve overtime.* If a decision is made to approve overtime work for existing project personnel, the budget will be affected. Unless a contingency was built into the budget to accommodate overtime for project personnel, the cost of the overtime will have to come out of the firm's profit line item.
- *Subcontract out some of the work.* If a decision is made to subcontract out some of the work that is behind schedule, the decision will affect the budget unless a

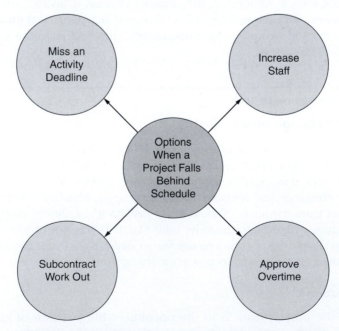

FIGURE 10.4   Project managers have limited options when a project falls behind schedule.

contingency was built into the budget to cover the cost of the subcontracted work. Without a contingency to cover it, the subcontracting costs will have to come out of the firm's profit line item.

- **Miss an activity deadline.** Sometimes project managers will consciously decide to miss the deadline for a specific activity in hopes of making it up within the overall timeframe for the project. When this is done, every other deadline in the project is subject to being adjusted, especially those on the project's critical path. It is sometimes possible to miss the deadline on one project activity without affecting the rest of the schedule, but this is not the typical case. More often than not, missing one deadline causes the project manager to either adjust all other subsequent deadlines or miss the overall project deadline. If the schedule is already tight—which is often the case—making adjustment within the schedule may not be possible. On the other hand, missing the overall project deadline can result in financial penalties, damaged customer relations, and the loss of future business. Figure 10.5 is a portion of a schedule for an engineering and technology project. Observe the projected starting and finish dates for the activities under item 79 on this schedule. If these intermediate deadlines are not met, how will it affect the remaining activities in the rest of the schedule? The affect of missing an activity deadline can create a problem that cascades through the rest of the project schedule.

It should be clear at this point that when monitoring and controlling projects, there is going to be tension between the budget and the schedule. Anything that affects one typically affects the other, either positively or negatively. Falling behind schedule typically adds to the cost of the project and reduces profits. Completing requirements ahead of schedule typically reduce costs and increases profits.

MONITORING AND CONTROLLING COSTS

Completing projects within budget is one of the fundamental success criteria for all projects. Consequently, monitoring and controlling costs is essential to effective project management. Project managers *monitor* costs using a cost baseline, as was explained in Chapter Five. Figure 10.6 is an example of a cost baseline. The cost baseline is used to monitor actual expenditures against budgeted costs throughout the course of a project. Any expenditures that are over or under the budgeted amount for a given line item will be reflected on the cost baseline. Corresponding adjustments should be made to the budget.

It is important for engineering and technology firms to have an effective process for examining, approving, and disapproving adjustments to the budget. Project managers do not just *monitor* actual expenditures against budgeted costs and record the difference, plus or minus. They actually do what is necessary to control costs and keep them within budget. This means they must influence the actions of others as they relate to and affect the budget. Project managers do not just monitor the budget and make adjustments when actual expenditures stray from the cost baseline. Rather, adjustments to the budget should not be made until they are subjected to a thorough examination to determine their cause and until they have gone through an approval process. Of course, the change-order process serves this purpose, but change orders do not always account for all required budget adjustments.

One of the more common causes of budget adjustments is *performance value*. Performance value is the value of the work being performed on a project. When the cost

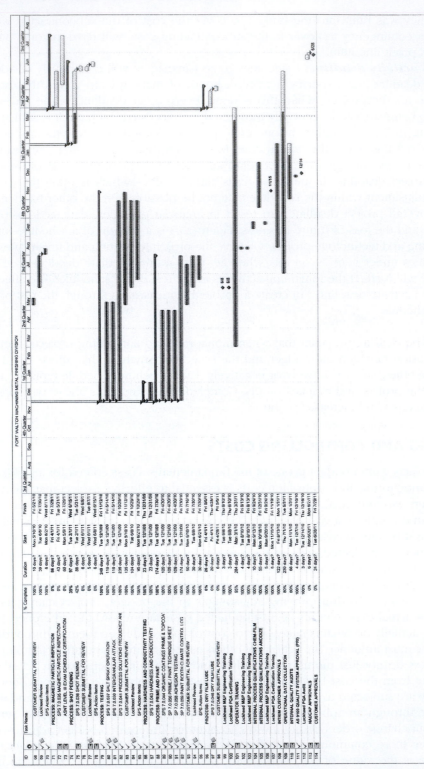

FIGURE 10.5   Missing the deadline for one activity can affect other activities.

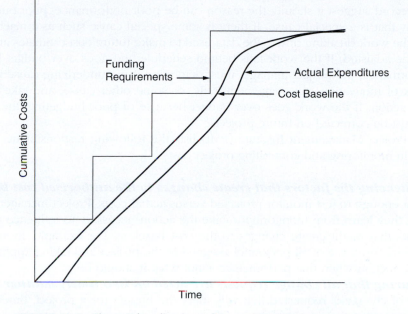

FIGURE 10.6   The cost baseline is used to monitor actual expenditures throughout the project.

estimate and then the budget are developed for a project, the work tasks that must be performed have all been assigned certain values which are represented by the amount of money dedicated to them in the budget. If a certain work package or activity in the project has been budgeted at $1,500 and should require 15 labor hours, that is its projected value to the project. Hence, the work must be completed properly in 15 hours or less for actual performance to match the budget. If the work requires more than 15 hours to complete it will exceed its budgeted amount and actual performance will not equal its projected value.

For this reason, cost monitoring and control involve more than just monitoring actual expenditures against budgeted costs. The process also involves monitoring, influencing, and controlling the performance of those doing the work on the project. Work performance versus budgeted costs is an important aspect of cost monitoring and control. This aspect answers the critical question, "Are we getting the level of performance we should from our personnel or from a given process?" If the answer to this question is "no," the project manager must act immediately to rectify the situation. Poor performance that is allowed to continue will affect every other aspect of the project over time including both budget and schedule.

Some larger engineering and technology firms maintain cost accounting departments to continually monitor projected costs against actual costs including projected work performance against actual performance. These cost accounting departments then use what they learn from the process to help the firm's personnel make more accurate estimates and project managers produce more accurate budgets. In smaller firms that do not have cost accounting departments, project managers and their teams make the necessary comparisons. In either case, project managers should be involved in the cost accounting process for their projects.

If the work for a work package or activity requires more time than experience and the historical record suggest it should, the reason can be poor performance, poor budgeting, or an anomaly that is a special cause. If there is some special cause such as a machine that is critical to the work breaking down, the data used to make future cost estimates and budgets need not be adjusted. If the work falls behind schedule and goes over budget because of poor performance, the project manager must determine if the underlying cause is a lack of talent, lack of motivation, a management issue, or some other cause, and take immediate corrective action. If the work goes over budget because of poor budgeting, the budgeting process must be corrected on future projects.

The Project Management Institute (PMI) lists the following responsibilities of project managers in monitoring and controlling project costs:[1]

- ***Influencing the factors that create changes to the authorized cost baseline.*** It is not enough to just monitor projected versus actual costs. Project managers must use what they learn from monitoring to take the action necessary to influence the various factors that might create changes to the cost baseline. For example, monitoring the work performance of all personnel assigned to the project and taking appropriate corrective action when that performance is not what it should be.
- ***Ensuring that all change requests are acted on in a timely manner.*** When any kind of change is requested that will affect the budget for a project, timely action is essential. Change requests that are either ignored or put off just create larger and more costly problems. Change requests typically result from a problem—not always, but typically. When solving problems, it is always easier to solve them while they are still small because problems tend to just get larger over time.
- ***Managing the actual changes when and as they occur.*** Changes require action on the part of different personnel assigned to the project in question. It is not sufficient to just subject proposed changes to a thorough vetting before approving them. Once they are approved, the actions required by changes must be managed carefully to ensure that they are made effectively and efficiently without creating other problems that will require additional changes.
- ***Ensuring that actual expenditures do not exceed budgeted expenditures in any budget period and for the overall project.*** When budgets are established, specified reporting periods—weekly, monthly, quarterly—are also established. It is important that project managers do the hard work necessary to keep their projects within budget in every reporting period. For example, assume that the project manager reports on the budget monthly and that by monitoring carefully he determines half way through the month that actual expenditures are climbing above the approved budget. The project manager can simply ignore the cost trend and think he will get things under control in the next reporting period or he can take immediate action to bring costs under control in the current reporting period. The appropriate action is the latter.
- ***Monitoring cost performance to isolate and understand variances from the approved cost baseline.*** When performance fails to match projections, it is important to know why. Is there a special one-time cause that is not likely to repeat itself that has affected performance? Are the people and processes assigned to the work in questions performing poorly? Is there a management policy or action that is impeding performance? Project managers must monitor performance on their projects closely enough to

be able to answer these types of questions so that appropriate corrective action can be taken in real time.

- ***Monitoring work performance against funds expended.*** All work required for a project is allotted funds in the project budget to cover the cost of completing the work. If the budget is accurate, the work should be able to be completed within the projected time and budget allocation. If the work is taking more time that projected, or even if it is taking less time, project managers need to know why. Only when project managers know why actual work performance does not match projected performance can they identify the appropriate corrective action. Work expenses that go over or under budget become part of the experience and history that will be factored into future cost estimates and budgets. To make sure the firm's experience and history are accurate, project managers need to know why work performance does not match the funds budgeted for it.

- ***Preventing unapproved changes from affecting the project scope or the resource usage.*** The actual expenditures for a project are used to inform the estimates and budgets for future projects. Consequently, it is important that the information from a project that becomes part of the historical record of the firm be accurate. Once the data from one project becomes part of the firm's historical data it will be used over and over when developing cost estimates and budgets for future projects. Consequently, it is essential that the project managers: (1) Ensure that unapproved changes are not allowed to increase the scope of project (scope creep) and (2) If unapproved changes do make their way into the work of a project, their costs do not become part of the historical data that will be used in developing cost estimates and budgets for future projects.

- ***Informing appropriate stakeholders of all approved changes and their associated costs.*** When a customer requests a change to a project, it is important that the request be placed immediately into the change-order process. This process will determine the cost of the change as well as how it will affect other aspects of the project, including the schedule. Before the change is implemented, it is important to ensure that the customer, personnel assigned to the project, and all other stakeholders who might be affected fully understand what the change will mean to them. Just approving a change and surprising stakeholders with it is bad business. Before any change is implemented, all stakeholders who will be affected by it should understand the change and how it will affect them.

- ***Acting to bring expected cost overruns within acceptable limits.*** Assume that you and your family make a trip to Disneyworld driving the family car. Prior to the trip you develop a budget that includes funds for gas, meals, lodging, tickets to Disneyworld, and miscellaneous purchases. During your week-long vacation, unexpected circumstances cause you to go over budget to the point that it looks like you will run out of money before you complete your vacation. Since the money you budgeted for the trip is all that you can spend on it, you must bring the expected cost overruns under control and get back within your budget. In this case, you would have several options including ending the vacation early, eating at less expensive restaurants, and limiting spending to just the essentials instead of purchasing expensive souvenirs of the trip, to name just a few. Project managers often have to undertake a similar exercise to bring their project expenses within budget. This is why it is so important to monitor budgeted expenses against actual expenses continually and carefully. The earlier that budget problems are identified the more time and flexibility project managers have for correcting them.

Cost Monitoring and Control Methods

There are a number of different methods project managers can use for cost monitoring and control. The most widely used of these, as recommended by the PMI, are summarized as follows:[2]

- Earned value management (EVM)
- Forecasting
- To-complete performance index (TCPI)
- Performance reviews (variance analysis and trend analysis)
- Cost monitoring and control software

EARNED VALUE MANAGEMENT   The PMI's description of the concept of earned value management is explained in this section.[3] Earned value management (EVM) is an approach to project monitoring and control that integrates the three most important factors that must be monitored and controlled: scope, cost, and schedule. EVM allows project managers to monitor actual performance for all activities and work packages along the following three dimensions: (1) Planned Value (PV), (2) Earned Value (EV), and (3) Actual Cost (AC). The integration of these three elements is what makes EVM a powerful approach to monitoring and control.

Planned Value (PV) is the amount of money in the budget planned for a given activity or work package. This budgeted amount becomes a baseline for measuring performance for the activity or work package in question. Because it is used as the baseline for measuring performance throughout a project, the PV is sometimes referred to as the *Performance Measurement Baseline* or PMB. The total PV for a project is referred to as the *Budget at Completion* or BAC. The goal of monitoring and controlling is to ensure that actual performance at the end of the project matches or falls within the PV for each individual activity and work package and within the BAC for the total of all activities and work packages.

Earned value (EV) is the term used to describe the percent of completion of project work at any point during the project. EV can be applied to individual activities and work packages as well as to the overall project. The schedule for a project will project dates at which work should be 25, 50, 75, and 100 percent complete. It is important for project managers to ensure that the actual EV of the work equals or surpasses these checkpoints in the schedule. If the EV of the work on a project is 35 percent complete at the point in the schedule where it was projected to be 25 percent, progress toward completion is ahead of schedule. However, if the EV is 15 percent or any other percentage that is less than 25 when the schedule projection is 25 percent, corrective action is needed.

Actual cost (AC) relating to each activity, work package, or the overall project represents what the work is actually costing. If the PV was based on direct costs, AC measurements should also be limited to direct costs. If the PV was based on direct and indirect costs, AC measurements should also include direct and indirect costs. Measuring the AC against the PV satisfies the need to monitor the cost element of projects. Figure 10.7 shows how AC and EV can be graphed to show how they compare to PV over the course of a project.

Notice in Figure 10.7 that the values being measured—work performance (EV) and cost performance (AC) are plotted on the vertical axis. Time is plotted on the horizontal axis and a specified point in time is chosen for making the measurements in question. At the point in time chosen for comparing the EV and AC to the PV—day 22, week 6, end of third quarter, and so on—the AC line is below the PV baseline. This is where project managers want the AC line to remain throughout their projects because it means that actual costs are

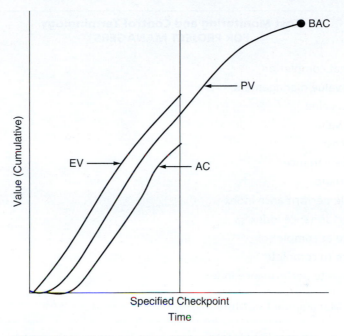

FIGURE 10.7 Company Earned Value (EV) and Actual Costs (AC) to Planned Value (PV) at a specified point in time.

under budgeted costs. At the same point in time, the EV line is above the PV baseline. This is where project managers want the EV to remain for the duration of their projects because it means that work is ahead of schedule.

In addition to the graphical methods of comparing planned versus actual performance, project managers can use the numerical index method. For example, schedule variance can be expressed as a numeric value called the *schedule performance index* or SPI by applying the following formula:

$$SPI = EV/PV$$

An SPI value of less than 1.0 indicates that work is behind schedule and that the project manager needs to take immediate corrective action to get the work on schedule. An index greater than 1.0 indicates work is ahead of schedule. In a similar manner, cost variance can be expressed as a numeric value known as the *cost performance index* or CPI by applying the following formula:

$$CPI = EV/AC$$

A CPI ratio that exceeds 1.0 indicates that costs are under budget. A ratio of less than 1.0 indicates that costs are over budget and the project manager needs to take immediate corrective action.

FORECASTING Forecasting is a monitoring and control method that is used primarily when a project is going to go over budget. In this situation, it is critical for project manager to know just how far over budget their projects will be. The PMI's description of forecasting is

---

Cost Monitoring and Control Terminology
FOR PROJECT MANAGERS

BAC = Budget at completion
EVM = Earned value management
PV   = Planned value
EV   = Earned value
AC   = Actual cost
SV   = Schedule variance
CV   = Cost variance
SPI = Schedule performance index
CPI = Cost performance index
EAC = Estimate at completion
ETC = Estimate to complete
TCPI = To-complete performance index

Source: Project Management Institute.

FIGURE 10.8 Important cost monitoring and control terms for project managers.

explained in this section.[4] Figure 10.8 contains a list of important cost monitoring and control terms project managers should be familiar with. Readers should refer to this list of terms and their abbreviations as they proceed through this section.

The amount of money budgeted for a project is the budget at completion or BAC. If it becomes obvious that actual costs are going to exceed the BAC in spite of the project manager's best efforts to prevent it, the prudent course of action is to estimate how far over budget the project will be. This projection is known as the *estimate at completion* or EAC. The EAC is the product of the most accurate forecast that can be made based on the best information available and the current circumstances surrounding the project.

Another way to view developing the EAC is applying current information that represents actual performance on the project to date to revise the budget and determine a new and more realistic BAC. For example, if a certain type of work is taking more time than was allocated for it and continues to do so in spite of the project manager's best attempts to bring the work into line with the original budget, the work in question will probably continue to exceed its time/cost allocation. Using the time the work is actually taking as the basis the project manager can forecast a new cost and use that new cost to forecast the EAC.

One of the problems faced when developing an EAC is that there is no money in the budget to cover the costs of developing the EAC (unless, of course, a contingency is added to the original budget to cover potential cost overruns). Remember, everything that requires time and labor costs money. Consequently, when developing an EAC it is necessary to add in the developmental costs. In other words, the EAC must include the new cost projections for the work that is behind schedule (ETC) and the cost of developing the projections. Hence, the formula for calculating the EAC is as follows:

$$EAC = AC + ETC$$

When developing an EAC for a project that is over budget, project managers must choose from among several different approaches, the most widely used of which are the following:

- Make ETC calculations based on the original budgeted rate for the work.
- Make ETC calculations based on the cost levels that are causing the budget overrun or cumulative CPI.
- Make ETC calculations based on a combination of the cumulative CPI and cumulative SPI (this approach is best when the schedule cannot be adjusted).

Each of these approaches has its own formula for making the necessary calculations. These formulas are as follows:

- ETC at original budgeted rate: EAC = AC + BAC − EV
- ETC at cumulative CPI: EAC = BAC/cumulative CPI
- ETC at cumulative CPI and SPI: AC + (BAC − EV/cum. CPI X Cum. SPI)

Using any one of these methods, project managers can answer the critical question that arises when a project is going to go over budget: How far over budget will the project be at completion?

TO-COMPLETE PERFORMANCE INDEX   The To-Complete Performance Index or TCPI is an estimate of the cost performance that will be necessary to achieve a specific management goal such as the BAC or EAC. The PMI's description of the TCPI is explained in this section.[5] Assume that revising the BAC is not an option—which is often the case with engineering and technology projects. In essence this means that more work will have to be done in less time. Project management students have to calculate a TCPI all the time.

Assume that an important course project is due on the last day of the term and you have only one hour a day to work on it. You have dedicated one hour a day to the project since the beginning of the term, but it now appears that you are not getting enough work done in the time you have allotted. This means you are going to have to get more work done in the allotted time between now and the end of the semester or miss the deadline for submitting the project. How much more work must be accomplished during each one-hour time allotment is the question. The answer to the how much more work question is the TCPI. The formulas for calculating the TCPI on the basis of the BAC and EAC are as follows:

- BAC: TCPI = (BAC − EV)/(BAC − AC)
- EAC: TCTI = (BAC − EV)/(EAC − AC)

Using these formulas, project managers can determine the TCPI for specific project goals such as the BAC or EAC.

PERFORMANCE REVIEWS   The PMI recommends the following types of performance reviews that can be used for cost monitoring and control:[6]

- ***Variance analysis.*** This type of performance review compares planned performance—cost and/or schedule—to actual performance at any point in the project to identify variances. Variances between planned and actual performance are analyzed and the appropriate corrective action is planned on the basis of the analysis.

• *Trend analysis.* This type of performance analysis is used to determine if performance on a project is improving, remaining static, or declining. By graphing planned performance against actual performance over a specified period of time (see Figure 10.7), project managers can observe trends. A declining trend for cost or schedule should result in immediate corrective action on the part of project managers.

COST MONITORING AND CONTROL SOFTWARE   The cost monitoring and control process has been simplified by the development of project management software. Leading project management software packages typically include cost monitoring and control applications that make monitoring planned performance against actual performance a relatively simple task. Of course, the software cannot take the next and more critical step for project managers: planning and implementing corrective action to get projects back on schedule and within budget.

The list of commercially available project management software packages is long and growing. As it is with any type of software, some of the packages are excellent and some are not. Consequently, it is important for project managers to know how to analyze project management software before recommending the purchase of a specific package. The most effective way to analyze project management software is to observe its capabilities in each of the functional process areas of project management: initiating, planning, executing, monitoring, and closing.

The ideal project management software will have easy-to-use, comprehensive capabilities in each of these process areas. These capabilities should be demonstrable, accessible, convenient, and comprehensive enough to give project managers and their team members the support needed to efficiently and effectively carry out their responsibilities. Much of the software that has been developed for project management applications is dedicated to helping perform a specific process (e.g., cost estimation, budgeting, scheduling). For this reason, some engineering firms are forced to purchase several different software packages to support project managers.

Although purchasing dedicated software packages for specific processes is an acceptable practice, there are problems with this approach. Often the different dedicated software packages are incompatible—they do not interact with each other or share a common database. Incompatibility limits the effectiveness and utility of project management software. The ideal software package is one in which a common database can be created and applied in carrying out the tasks involved in initiating, planning, executing, monitoring/controlling, and closing out projects. Consequently, when considering an investment in project management software, it is wise to look for a package that is comprehensive and detailed in its applications and capabilities.

QUALITY MONITORING AND CONTROL METHODS

Part of the planning process involved developing a quality plan for a specific project (Chapter Six). The plan accommodated both levels of planning: quality management (Big Q) and quality control (Little Q). The overall purpose of the quality plan is to ensure that all deliverables for the project in question meet or exceed the specifications set forth in project documents. The quality plan is composed of several different sections. Among these are the quality objectives for the project, process quality measures, product quality measures, and quality tools to be employed. Both levels of quality—Big Q and Little Q—come into play in this phase of the project.

Big Q Monitoring and Control: Quality Audits

The Big Q aspects of monitoring and controlling quality involve ensuring that the appropriate process and product quality measure are being applied, that the appropriate quality tools are being used, that team members are carrying out their quality-related responsibilities, and that quality performance reports are being regularly prepared, distributed, and acted on. These determinations are made using a monitoring and control method called the *quality audit*.

An effective quality audit will uncover gaps between planned performance and actual performance so that immediate corrective action can be taken. The audit will also identify practices developed for the project that have been so effective in producing the desired results that they should be adopted as best practices for future projects. What is learned as a result of quality audits should become part of the lessons-learned component of project closeout. When quality audits reveal situations requiring corrective action, project managers should update project documents accordingly. The project documents that most commonly require updating as a result of quality audits are the budget, schedule, and quality plans.

Little Q Monitoring and Control: Quality Control

Quality control involves the day-to-day application of the quality tools described in the quality plan for the project in question (Chapter Six). Using such quality control tools as control charts, cause-and-effect diagrams, histograms, Pareto charts, scatter diagrams, statistical sampling, and inspections throughout the course of the project, the project management team can identify quality problems in real time (in process) and take the necessary steps to correct them. In engineering and technology firms that have quality departments, these tools will be applied by quality professionals and their results presented to project managers on a regular basis. However, even though the hands-on quality control work may be done by quality professionals, project managers need to know how to interpret the results produced by these tools so they can take corrective action when it is called for.

USING CONTROL CHARTS   One of the tools frequently used for quality control is the *control chart*. Control charts monitor the performance of processes to identify fluctuations that cause performance to fall outside of acceptable limits. For example, assume that a customer requires that certain tolerances be adhered to in the production of products. A tolerance is a range of acceptable performance typically expressed by a nominal dimension with a plus and minus allowance. For example, assume a machining process is supposed to produce 300 shafts that are 0.75 inch in diameter plus or minus .001 of an inch. The acceptable range for producing the shafts, based on the stated tolerance, is 0.751 at the top end and 0.749 at the low end. Using statistical process control (SPC), production personnel can establish control limits for this machining process that can be observed to ensure that tolerances are being met. Project team members can monitor the performance of the process in question by observing the control chart for the process (see Figure 10.9).

In Figure 10.9, the upper control limit (UCL) would be set at 0.751 and the lower control limit (LCL) would be set at 0.749. As the shafts are being produced, the control chart in Figure 10.9 is monitored constantly to ensure that the actual dimensions of the shafts stay within the set limits. As long as the actual dimensions stay within the prescribed limits, the process is in control and the shafts being produced will meet the specifications. Should the chart show a line that extends above the UCL or below the LCL, the process would have to be immediately adjusted to bring it back into control.

Continuous monitoring using such tools as control charts allows the project management team to engage in prevention of quality problems as opposed to basing compliance on after-the-fact inspections. Inspections as used in the traditional approach to quality control keep noncomplying products out of the hands of customers, but they also result in higher scrap rates. Higher scrap rates, in turn, increase the cost of producing the deliverable. This is why it is so important to monitor the performance of processes in real time and act immediately to get nonconformities out of the process before they can introduce errors that will result in products being produced that do not meet specifications.

USING CAUSE-AND-EFFECT DIAGRAMS   Occasionally a problem will be identified through monitoring for which the observer has no explanation. Since it is necessary to know the root cause of a problem before it can be eliminated, a valuable quality control tool is the *cause-and-effect* diagram (see Figure 10.10). A cause-and-effect diagram can be used by project managers to identify the root cause of a quality problem so that a permanent solution can be implemented.

Assume that an engineering and technology firm is experiencing cracks in the solder applied to printed circuit boards by a wave soldering machine. The cracks in the solder are causing printed circuit board to be rejected as not meeting specifications. To identify the root cause of the problem, the project manager gathers the project ream together and uses a cause-and-effect diagram such as the one on Figure 10.10 to lead team members in a discussion and analysis of the problem.

The team would take each potential cause on the diagram in turn, discussing and analyzing. For example, to determine if the cause of the problem could be faulty material (solder), the solder being used might be tested and the results reported to the project team. Assume the material is determined to be acceptable. The team would next examine the methods being used to operate the wave soldering machine. Are the proper procedures being observed in all cases? Assume that the proper methods are being used. The team would next examine the measurements being taken to make sure they are valid, meaning that there are actually cracks in the solder and that the cracks are sufficient to render the printed circuit boards noncompliant. Assume the measurements are valid.

The project team would next examine the operator. Is the individual who operates the wave soldering machine properly training on the machine in question? Assume the operator does not appear to be the source of the problem. The project team would next examine the wave soldering machine itself to determine if there are defects in it that could be causing the

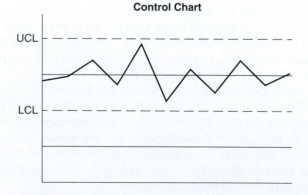

FIGURE 10.9   The process being monitored is in control.

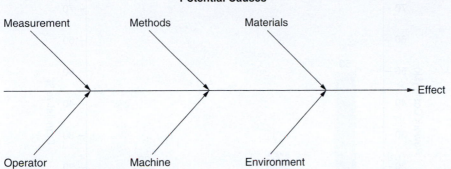

FIGURE 10.10   The Cause-and-Effect Diagram helps project managers identify the root cause of a quality problem.

cracks. Assume the machine has been properly maintained and adjusted. The project team would next discuss the environment in which the wave soldering takes place. Assume that the environment in the wave soldering room is supposed to be maintained at a constant prescribed temperature and humidity. Assume further that spot checks taken at different times during the day reveal that the temperature and humidity exceed prescribed limits during the same two-hour period every afternoon. Using the cause-and-effect diagram as a guide, the project team would be able to narrow the cause of the problem down to an environmental problem relating to temperature and humidity.

USING HISTOGRAMS   A histogram is a bar chart presented in a vertical format that shows how often a given variable occurs. The variables are placed on the X or horizontal axis and the frequency figures are placed on the Y or vertical axis. Histograms are used to illustrate the most commonly occurring problems in projects. Figure 10.11 is an example of a

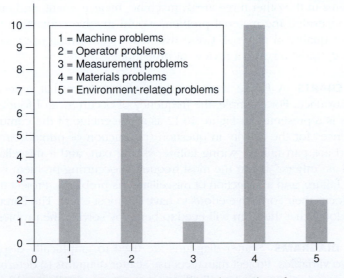

FIGURE 10.11   Histogram summarizing the number of problems that occurred by type during a specific time period.

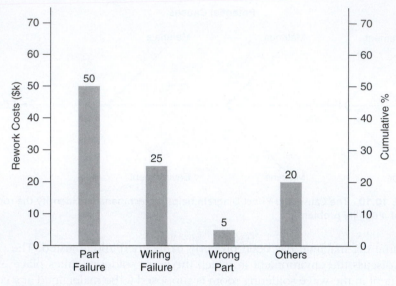

FIGURE 10.12 Pareto chart showing the percent of defects for a project activity.

histogram used to summarize the number of problems that occurred in a project by problem type. Such a chart could be used to summarize problem occurrences over any specified period of time (e.g., week, month, quarter). By displaying graphically what kinds of problems are occurring most frequently, the histogram can show project managers where they need to put the bulk of their corrective actions or which problems need to be attacked first.

In Figure 10.11, the firm in question has experienced a rash of material problems and a high number of operator problems during the time period in question. Consequently, immediate corrective action will be needed in these two areas followed thereafter by actions to correct problems in the other three areas: machine, measurement, and environment. Corrective action concerning the material problem would involve contacting the supplier and obtaining a better quality of material. Corrective action concerning operators might involve providing training, updating, and a review of procedures.

USING PARETO CHARTS A Pareto chart is a type of histogram that orders information by frequency of occurrence. For example, the frequency of occurrence of four different kinds of quality problems is represented in Figure 10.12 as a percentage of the cumulative total of all problem occurrences for the activity in question (production of printed circuit boards). The problems charted are part failure, wiring failure, wrong part, and a miscellaneous collective category labeled as "others." By far the most frequently occurring problem is part failure, followed by wiring failure, and a collection of miscellaneous problems. Project managers need to know where to focus their corrective efforts to have the most effect. The Pareto chart in Figure 10.12 leaves no doubt that the team will need to begin by solving the problem of part failure.

USING SCATTER DIAGRAMS Scatter diagrams are used to graphically display the relationship between two variables. Project managers use scatter diagrams to determine how changing one variable in a project activity will affect another variable. Figure 10.13 is an example

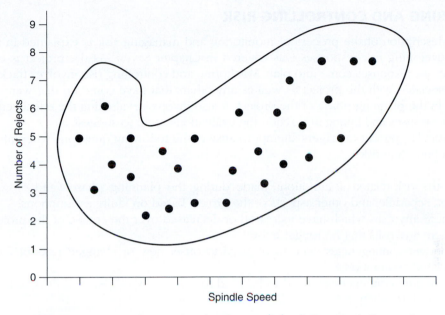

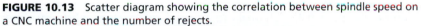

FIGURE 10.13   Scatter diagram showing the correlation between spindle speed on a CNC machine and the number of rejects.

of a scatter diagram. The diagram displays the correlation between spindle speed on a CNC machine and the number of rejects. The correlation in this example is that increasing the spindle speed tends to increase the reject rate. By studying the scatter diagram closely the project manager can also determine the optimum spindle speed (highest speed with lowest number of rejects).

USING STATISTICAL SAMPLING   With large production runs, statistical sampling is a commonly used quality control method. Assume that an engineering and technology firm has a contract to produce 1,000 copies of a given part. Rather than inspect all of the parts—a time consuming and costly process—the firm randomly pulls a statistically valid sample of the parts, say 10 percent or 100 parts, and inspects them. If the parts in the sample pass inspection, the remaining 900 parts also pass inspection. As long as the sample is statistically valid and randomly selected, the laws of statistics suggest that the sample accurately represents the whole.

USING INSPECTIONS   The philosophy of W. Edwards Deming, Joseph Juran, and other quality pioneers was that *you cannot inspect quality into a product*. This, of course, is true. Consequently, inspection, as a quality control method, has acquired a bad name. In spite of this, inspection is still a valid quality control method. It is the after-the-fact inspection that gave this method a bad name. The key to using inspections effectively is to perform them throughout the course of production rather than completing the product and, then, performing after-the-fact inspections. In fact, inspections can be particularly valuable for determining if products that have to be reworked as a result of quality problems finally measure up to specifications.

MONITORING AND CONTROLLING RISK

The PMI's description of the process of monitoring and managing risk is explained in this section.[7] Developing risk responses was covered in Chapter Seven. As the project is executed, those risk responses come into play. Monitoring and controlling risk involves tracking the risks associated with the project as well as any others that have come up that were not considered in the planning phase of the project. It also involves evaluating the effectiveness of the risk responses and taking any action the evaluations seem to suggest.

In particular, project managers attempt to answer the following questions when monitoring and controlling risk:

1. Were the risk-related assumptions made during the planning phase valid or were budget, schedule, and other aspects of the project based on faulty assumptions?
2. Are there any risks which have increased or decreased over the course of the project?
3. Are there any risks that no longer exist?
4. Will budget contingencies need to be used to offset new or changed risks or faulty risk-related assumptions?
5. Will schedule contingencies need to be used to offset new or changed risks or faulty risk-related assumptions?

Faulty assumptions, new risks, changes to original risk factors, and any other risk-related factors that were not anticipated require the project manager to take corrective action. That action might involve developing and implementing new risk strategies, revising the original risk strategies, or using contingencies built into the budget and schedule. Lessons learned concerning faulty risk-related assumptions, changes to original risks, new risks that intruded during the execution process, and any other risk-related factors that may not have been anticipated should be recorded and made part of the lessons-learned component of project closing. In addition, the risk management component of the project management plan should be revised to accurately reflect any and all changes.

The PMI recommends three tools for monitoring and controlling risk: (1) ongoing risk assessment, (2) risk audits, and (3) variance and trend analysis.[8] Project managers and their team members should be familiar with all three of these tools and be able to use them in monitoring and controlling risk. If the engineering and technology firm has a risk management department, these tools will be used by risk management professionals and the results will be shared with the project manager. In smaller firms, project managers may have to apply the tools themselves.

- ***Ongoing risk assessment.*** Risk assessment occurs first during the planning phase. However, risk assessment should be ongoing throughout the course of the project. By regularly assessing risks and the effectiveness of the strategies implemented for minimizing them, project managers can ensure against unpleasant surprises. In addition, ongoing risk assessment will help project managers identify new risks that have been introduced and changes to the original risks. As with all the other aspects of project management, staying out in front of risk is the wise approach for project managers.
- ***Risk audits.*** Risk audits are aimed at measuring the effectiveness of risk responses. Are they working as planned? Are they adequately minimizing or actually eliminating the risks identified? Are the strategies put in place in response to newly introduced

risks having the desired effect? These types of questions are answered by risk audits.

• ***Variance and trend analysis.*** Variance and trend analysis involves establishing risk baselines and graphing actual performance against the baselines. This is similar to the earned value analysis (EVA) method introduced earlier in this chapter. By displaying the risk baseline and actual risk-related results graphically, project managers can observe variations between the planned and the actual and easily determine if the performance trend is toward or away from the baseline.

Anything of significance that comes out of monitoring and controlling risk should be recorded for the purpose of informing the lessons-learned process. In addition, all project documents—including the risk register—should be updated to reflect all changes, revisions, and adjustments made as a result of the risk monitoring and control process.

Project Management Scenario 10.2

Get the work done and worry about the paperwork later

There has been an ongoing disagreement between Marsha Reynolds, Project Manager for the Delta Con Project, and her assistant project manager, Dave Henderson. Reynolds does not care for the paperwork aspects of project management. Her motto is "Get the work done and worry about the paperwork later." She always does an excellent job of bringing her projects in on time, within budget, and according to specifications. Consequently, she is well respected in her firm and has a reputation for getting the work done. Dave Henderson agrees with the first half of Reynolds' motto: "Get the work done…" But he disagrees with the second half: "…worry about the paperwork later." In his experience, paperwork that is put off until later never gets done, which is often the case with Reynolds' projects. Henderson believes that updating project documents continually throughout a project is the better approach.

Discussion Questions

In this scenario, the project manager is good at keeping her projects on time, within budget, and in compliance with specifications. However, she is not good at keeping project documentation updated. Do you see her attitude toward the paperwork aspects of project management as a problem? Why or why not?

SUMMARY

Once the various elements of the project management plan have been developed and execution of the plan is underway, the work of the project must be monitored and controlled. Monitoring and controlling project work involves observing and measuring and, then, making appropriate adjustments and taking appropriate action based on the results of the observations and measurements. Most of the monitoring and controlling work of project managers is aimed at the following project elements: scope, schedule, costs, quality, and risk.

To prevent scope creep, project managers must monitor and control the project scope carefully. Scope creep occurs when work is introduced into the project that is not covered in the scope statements or contract documents. An important tool in preventing scope creep is the change-order process. Monitoring and controlling a project's scope involves ensuring acceptance of deliverables, managing the change-order process, and updating project documents.

Monitoring and controlling the schedule is one of the most important responsibilities of project managers. If a project falls behind schedule, the project manager is forced to respond by choosing one or more of the following options: increase staff, approve overtime, subcontract out some of the work, or miss an activity deadline. There are costs associated with all of these options. Because of this, there is always tension between the schedule and the budget when monitoring the schedule.

Along with the schedule, monitoring and controlling the budget are two of the project manager's highest priorities. The two are inextricably linked. In monitoring and controlling the budget, project managers must do the following: (1) influence the factors that create changes to the authorized cost baseline, (2) ensure that all change requests are acted on in a timely man-

ner, (3) manage the actual changes when and as they occur, (4) ensure that actual expenditures do not exceed budgeted expenditures, (5) monitor cost performance to isolate and understand variances from the approved baseline, (6) monitor work performance against funds expended, (7) prevent unapproved changes from affecting the project's scope or approved resource utilization, (8) keep stakeholders informed of approved changes and their associated costs, and (9) act to bring cost overruns back into acceptable limits. Cost monitoring and control methods include: earned value management (EVM), forecasting, to-complete performance index (TCPI), and performance reviews, including variance and trend analysis.

Monitoring and controlling risk involves validating the original risk-related assumptions, identifying any changes to the original risks, evaluating the effectiveness of risk management strategies, and using budget and schedule contingencies as necessary. The tools that are most widely used in monitoring and controlling risk are ongoing risk assessment, risk audits, and variance and trend analysis.

KEY TERMS AND CONCEPTS

Scope creep
Acceptance criteria
Ensure acceptance of deliverables
Manage the change-order process
Update project documents
Increase staff
Approve overtime
Subcontract out some of the work
Miss an activity deadline

Cost baseline
Earned value management (EVM)
Forecasting
To-complete performance index (TCPI)
Performance reviews
Variance analysis
Trend analysis
Risk audits
Risk assessment

REVIEW QUESTIONS

1. Explain the concept of *scope creep*. Why is it important to prevent scope creep?
2. What three things must project managers and their teams do when monitoring and controlling a project's scope?
3. What is a change order? Why are change orders important?
4. When monitoring and controlling a project's schedule, what are the project manager's options if the project falls behind schedule?
5. What are the project manager's responsibilities in monitoring and controlling costs in a project?
6. Explain the concept of earned value management (EVM).

7. What is forecasting and how is it used to monitor and control project costs?
8. Explain the concept of the to-complete performance index (TCPI).
9. List and explain the types of performance reviews that can be used for monitoring and controlling project costs.

10. What is the best way to evaluate cost monitoring and control software?
11. Explain three tools for monitoring and controlling project risk.

APPLICATION ACTIVITIES

The following activities may be completed by individual students or by students working in groups:

1. Identify an engineering and technology firm that will cooperate in completing this project and ask to see several change orders their personnel have processed on recent projects. Ask your contact in the firm to take you through the change-order process step by step. Develop a report for the class that illustrates and explains the change-order process.
2. Working with your contact from Activity 1, make an annotated list of corrective actions the firm has had to take to keep projects on schedule and within budget. Discuss the budget implications of adjustments that have been made to project

schedules and actions that were taken when a project fell behind schedule but the project completion date could not be altered. Write a report summarizing your findings.
3. Working with your contact from Activity 1, determine which of the following cost monitoring and control methods are used by the firm: earned value management (EVM), forecasting, to-complete performance index (TCPI), performance reviews, and variance analysis. Take careful notes and develop an explanation of what methods are used and how. If the firm uses methods other than those listed, develop an explanation of these methods too. Write a report that summarizes what you learn while completing this activity.

ENDNOTES

1. Project Management Institute, *A Guide to the Project Management Body of Knowledge*, 4th ed. (Newtown Square, Pennsylvania: Project Management Institute, 2008), 179–180.
2. Ibid., 181–187.
3. Ibid., 181–183.
4. Ibid., 184–187.
5. Ibid., 185.
6. Ibid., 186–187.
7. Ibid., 308–312.
8. Ibid., 310.

CHAPTER 11

Project Closeout

The final phase of a project is the closeout phase. Project closeout is a critical phase, but it does not always receive the same amount of attention as the other phases. The tendency of too many engineering and technology firms is to complete the execution phase of a project, take a deep breath, and say "I'm glad that's over with." Sometimes the last thing a project manager wants to do after executing a challenging project is take the time to tie up loose ends and ask hard questions such as: How did we do? What could we have done better? What did we learn over the course of this project? How can this project lead to more business for our firm?

The factors that work against conducting a thorough closeout phase are many, but the most common factors are as follows: (1) pressure to begin the next project, (2) disbursement of project team members to other duties and other projects, and (3) reluctance of project managers and team members to participate in a self-critique. These are understandable reasons, but they do not preclude the need for a comprehensive and thorough closeout process at the end of every engineering and technology project. The cost of conducting such a comprehensive, thorough closeout is less than the cost of failing to conduct one.

To understand both why many project managers like to ignore the closeout phase and why they should never do this, consider the example of having important guests over for a big meal at your house. You need to make a positive impression on your guests so you work for weeks to plan every aspect of the meal down to the smallest details. You go to the grocery store and purchase only the best steaks, vegetables, wine, and other materials. When you get home and unpack your groceries, you find that some of the materials you purchased, although they looked good when you bought them, are not. You have to go back to the store to purchase replacement materials.

When the big day comes, you spend all day preparing the meal, and the preparations are a challenge because things do not go as planned. After working around several minor missteps and overcoming a couple of disasters the meal is finally ready. When your guests arrive and have been seated you serve the meal, all the while holding your breath and hoping they like it. After a long night during which you are constantly on edge, your guests finally leave and you can relax. You are so relieved that your first thought is to just collapse in a chair and pass out from nervous exhaustion. But as tempting as it is to just close the door to the kitchen/dining room and go to bed, doing so would not be wise because you are not done yet. You still have to close out the meal.

Closing out the meal involves washing the dishes and putting them back in the cupboards, wrapping up any leftovers and putting them in the refrigerator, and doing the various other chores necessary to clean up the kitchen/dining room. Then, because you will be required to host other meals for important guests—including the ones who just left—you need to ask yourself several questions while the evening's event is still fresh on your mind. The questions are as follows:

- Was the food up to expectations?
- Were the seating arrangements satisfactory?
- Was the wine up to expectations?
- Were parking accommodations for the guests satisfactory?
- What should I do better or differently next time?

The answers to these questions should be jotted down so that lessons learned tonight can be used to improve future events that you host. Finally, after you get a good night's sleep you need to take some time to celebrate the success of your event. Having done these things, you have closed out the meal and are well prepared to do even better next time.

The costs of overlooking the project closeout process can be high. They include the costs: (1) associated with repeating the same mistakes from project to project, (2) of failing to take advantage of lessons learned, (3) of failing to incorporate new best practices developed into standard operating procedures, (4) of failing to tie up loose ends, (5) failing to gain the customer's support in securing future business for the firm, and (6) failing to ensure that the customer is completely satisfied.

Project Management Scenario 11.1

I am fed up with this project—let's move on

Ly Szu has just finished his first major project as a project manager and the experience has been a challenge to say the least. In Szu's opinion, everything that could possibly have gone wrong did. Now that the execution phase has been completed, Szu wants to move on to his next assignment and not look back. He hopes to never hear another word about his first project—a project he secretly thinks of as the "disaster."

Szu's mentor in the firm, Betina Washington, is encouraging him to go through a comprehensive closeout process. She has explained the various steps in the closeout process and offered to help him with each step. But Szu wants nothing to do with closing out the project. In fact, just this morning he told Washington, "I am fed up with this project—let's move on." Szu wants to get started on his next assignment so that he can begin the process of redeeming himself in the eyes of higher management.

Discussion Questions

In this scenario, Ly Szu has had a bad experience with his first attempt at being a project manager. Consequently, he is resisting engaging in a comprehensive and thorough project closeout. If you were Betina Washington, how would you explain the value of the project closeout process to Szu? How could Szu use the closeout process to begin redeeming himself in the eyes of higher management?

STEPS IN THE PROJECT CLOSEOUT PROCESS

Completing the work on a project and completing the project are two different things. The work required for a project is completed once the execution phase is completed. Returning to the earlier example of hosting a big meal in your home for important guests, the execution phase of the meal was completed once your guests had finished eating and gone home for the night. Once the execution phase has been completed, the final responsibility of the project manager is to close out the project. The steps in the project closeout process are shown in Figure 11.1.

Each of these steps serves a specific purpose and each is important. Taken together, the various steps in the project closeout process ensure that: (1) the project is brought to a satisfactory conclusion for the engineering and technology firm and the customer, (2) all loose ends are tied up properly, (3) lessons learned become part of the firm's standard operating procedures so that the team and the firm benefit from continual improvement, and (4) additional business is generated with the customer in question and through that customer's referrals.

VERIFY THE SCOPE

Verifying the scope of a project at this point means ensuring that all project work has been completed. For example, if the engineering and technology firm contracted to produce 100 low-voltage power supplies for the aircraft industry, does the firm have 100 power supplies completed according to specifications and ready to ship? Verifying the scope in this phase of the project requires that the project manager do the following:

- **_Conduct a functional audit._** The functional audit involves ensuring that all deliverables meet specification before they are shipped. The audit gives engineering and technology firms one last chance to identify and correct any deficiencies before the product is delivered to the customer.
- **_Compare deliverables with their respective product manuals._** It is important to ensure that all applicable product manuals are provided to the customer and that the information in the manuals matches the reality of the completed products that will

---

PROJECT CLOSEOUT CHECKLIST

✓ Verify the scope

✓ Close out the contract

✓ Close out the administrative aspects of the project

✓ Conduct a *lessons-learned* review

✓ Develop the project closeout report

✓ Recognize team members

✓ Complete the final step

FIGURE 11.1   Every step in the project closeout process is important.

be delivered to the customer. Providing inaccurate product manuals to customers is a major customer service error. Differences in the manuals and the finished products sometimes occur because of changes to the product made as a result of change orders. If these changes are not followed closely throughout the project and reflected in the final manuals, the engineering and technology firm will create an ongoing customer service problem for itself—one that is repeated every time the customer opens the manual in question.

- *Conduct a product count.* If the customer ordered 100 items, it is essential to ensure that 100 are delivered, along with the correct number of accessories. Conducting a careful product count before shipping deliverables to the customer will ensure that what was ordered is what is actually delivered. For example, have you ever ordered something from a catalog that had to be assembled only to find that a part was missing or that the manufacturer did not provide enough nuts, bolts, washers, or screws. Such situations are frustrating for the customer. Failing to validate the count on all products and accessories is bad customer service for engineering and technology firms.

Once the project manager is certain that all project work has been properly completed, the project team moves on to the next step in closeout process: closing out the contract.

CLOSE OUT THE CONTRACT

The contract for a project is not closed out until the engineering and technology firm and the customer agree that it is closed out. Securing that agreement is the purpose of the contract closeout part of the process. Contractual loose ends have a way of generating problems up to and including litigation. But the contract closeout component goes beyond just reaching agreement that all aspects of the contract have been satisfied. It also includes ensuring that the relationship between the engineering and technology firm and the customer is one that will lead to additional business.

When an engineering and technology firm completes a contract for a customer the firm wants to ensure that the customer is sufficiently satisfied to do two things: (1) Give the firm another contract for a second time and (2) Recommend the firm to other customers. To this end, the contract closeout part of the process involves the following:

- Completing all required testing required in the contract
- Providing all applicable warranties and guarantees
- Providing all applicable certifications
- Providing all applicable forms, reports, and manuals
- Providing all applicable maintenance agreements
- Providing any required project photographs
- Returning proprietary information to the customer
- Gain written agreement on intellectual property (as applicable)
- Conducting a project review in conjunction with the customer and recording customer feedback for use in lessons-learned sessions

Once the details of the contract have been closed out, the project manager proceeds to close out the internal administrative aspects of the project.

CLOSE OUT THE ADMINISTRATIVE ASPECTS OF THE PROJECT

With every project, there are internal administrative elements that must be closed out. These elements typically include: (1) returning project team members to their functional areas, (2) providing performance feedback for team members, and (3) archiving project information.

- *Return project team members to their functional areas.* As was explained earlier in this text, engineering and technology firms form teams in different ways depending on a variety of factors. Often the teams are cross-functional in that they *borrow* team members from different functional areas in the engineering and technology firm. When this is the case, as it often is, those team members must be released from the project so they can return to their respective functional areas. However, before releasing team members the project manager must ensure that all project work has been completed. It is not wise to release team members back to their functional areas while there is still work to be completed, no matter how minor the work might seem. Once a team member is released from a project, it will be difficult if not impossible to resecure his or her services for tying up loose ends. Once project team members return to their

---

**CONTENTS OF THE
PROJECT ARCHIVES**

- Project charter
- Contract
- Project plan
- Schedule
- Budget
- Change orders
- Status reports
- Time records for team members
- Issues logs
- Acceptance log for deliverables
- Deliverable sign-offs
- Risk management information
- Audit reports
- Correspondence
- Meeting minutes
- Final approval document with signatures
- Lessons-learned documentation
- Miscellaneous reports

FIGURE 11.2 Project archives should be comprehensive and up-to-date.

functional areas they are likely to become involved in other responsibilities, and their supervisors may be reluctant to release them for additional project work.

- **Provide performance feedback for team members.** A project team should never be disbanded before the project manager has completed performance appraisals for team member and reviewed them with the respective members. This is a critical part of the continual-improvement process. Team members need to know what they did well and what they need to improve in terms of their performance. In addition, the functional supervisors of team members need to know how their personnel performed. The performance appraisals of project managers should be used as input when supervisors conduct their annual performance appraisals on personnel who have served on project teams during the year in question.
- **Archive project information.** Throughout a project, there should be a central information collection repository—a central place where all information relating to the project is collected, sorted, updated, and stored. This repository of project information contains the history of the project, an up-to-date audit trail, and a ready reference for future project planning. As part of the closeout process, the project manager should ensure that all project information is present and up-to-date. This information includes all project documents and correspondence relating to the project. Figure 11.2 contains a list of the types of materials that are typically included in the archives of a project.

CONDUCT A LESSONS-LEARNED REVIEW

The *lessons-learned review* is used to promote continual improvement. If performance on the project was poor or mediocre, the firm will want to do better on future projects. If performance on the project was good, the firm will want to do even better on future projects. If performance on the project was excellent, the firm will want to incorporate the best practices from the project into its standard operating procedures. It is important to understand that even if the project was completed on time, within budget, and according to specifications, there remains one more criterion that always applies: customer satisfaction. If the customer is not satisfied, the project is not a success no matter how well the team performed against other criteria. For this reason, determining the extent of customer satisfaction is an important part of the lessons-learned aspect of the closeout process.

There are three parts that make up the lessons-learned review. The three parts are as follows: (1) survey to collect information about lessons learned, (2) face-to-face discussion session of survey results, and (3) incorporation of new best practices into the firm's standard operating procedures. The project manager provides the leadership in all three of these parts of the lessons-learned review.

Survey Team Members

Step one in the lessons-learned review is to develop a survey, distribute it to all team members, collect the completed surveys, and summarize the results. Surveys must be tailored to the project in question—one size does not fit all. However, the sample survey questions in Figure 11.3 are the types of questions that should be asked on the survey for any project. This checklist of sample questions can be used for developing a tailored survey instrument for any project.

QUESTIONS FOR PROJECT REVIEW SURVEYS

- Did all deliverables meet the specifications for the project?
- Was the customer satisfied with all aspects of the project?
- Was the project completed on time?
- Was the project completed within budget?
- Did work mitigation strategies work?
- Was communication effective?
- Was the change-order process properly managed?
- What problems occurred that affected the project?
- What best practices should be incorporated into standard operating procedures or SOP?
- What can be done to improve performance on future projects?
- What can be done to improve customer satisfaction on future projects?

FIGURE 11.3   These types of questions should be included on project review surveys.

Team members should be given a specified period of time—typically one week—to complete their surveys and return them to the project manager. The best approach is to distribute the surveys electronically and have them completed and returned electronically. The project manager then summarizes the team's responses and creates a summary document that will be the basis for the next step: conducting a team discussion of the survey feedback.

Team Discussion of Survey Feedback

The survey is an excellent tool for collecting information from a lot of people in a structured way and in a short period of time. This is the up side of the survey. The down side is that surveys restrict the depth of feedback team members are able to give. Surveys are good at collecting information relating to the "what" aspects of the project but less effective at collecting information relating to the "why" and "how" aspects. Depth and breadth of responses can be improved by conducting a face-to-face session in which each survey question and its summarized responses are discussed.

During the face-to-face discussion session, the project manager leads a brainstorming session in which the summarized responses to survey questions are analyzed, discussed, debated, revised, and adjusted. The team response, as refined by discussion, becomes the response the project manager includes in the summary report of the lessons-learned review. In conducting face-to-face discussions, it is important for project managers to facilitate but not dominate. It is also important that they encourage team members to disagree without being disagreeable. The more frank and open the discussion of survey results, the better.

Incorporation of Best Practices into the Firm's Standard Operating Procedures

With every project there are problems that were not or could not be anticipated. Solving these problems requires innovation and creative responses. These creative responses and innovations often reveal excellent methods that transcend the project in question. When this is the case, it is important to make sure that these best practices are incorporated into the firm's standard operating procedures. Ensuring that this happens is the responsibility of the project manager.

Incorporating new best practices in the firm's standard operating procedures typically requires the following three steps: (1) securing approval of higher management, (2) revising the applicable procedures manual, and (3) organizing training for all personnel who are affected by the change. Securing approval from higher management typically requires demonstrating with hard data that the new practice is better than the old practice it replaces. Updating of procedural manuals should be done in accordance with the engineering and technology firm's approved policy governing procedural updates.

Organizing training for personnel who will use the new procedure is critical and should never be overlooked. People will not automatically accept and adopt a new procedure or practice just because it produces better results than the old practice. They need to be shown not just how to apply the new practice, but also why. Consequently, training for new practices should go beyond just the *how-to* aspects to also include the why aspects.

Challenging Engineering Project

Making Virtual Reality More Realistic

Virtual reality is a technology-based concept that creates the illusion being somewhere else. With virtual reality, the better the design and capabilities of the technology the better the illusion of being somewhere else. In its current state of development, the most common applications of virtual reality are computer games and movies. However, if virtual reality can be made more realistic its applications can be increased exponentially.

Using virtual reality in a training setting is one of the more obvious applications, an application with enormous potential. Its application to training will allow people to learn how to safely perform dangerous tasks without fear of actually being harmed or harming others during the training (e.g., soldiers, surgeons, pilots, electrical workers, chemical workers, etc.). In addition to training applications, virtual reality can also be used to treat psychological and physiological disorders. For example, the concept is already being used to treat people who suffer from selected phobias such as fear of heights, snakes, spiders, and animals of various sorts.

The potential effectiveness of virtual reality in any of its current and potential applications is the *reality* of the virtual reality. The more realistic, more lifelike virtual reality can be made, the more effective a tool it will be. This is the challenge to engineers and technologists—making virtual reality more realistic. In order to make virtual reality more realistic, engineering and technology professionals will have to:

- Develop video displays that have sufficient resolution and fast enough refresh/update rates to create scenes that are actually lifelike and that can change as fast as real scenes change.
- Develop video displays with fields of views that are wide enough and with lighting and shadows that are realistic enough to maintain illusions that appear fully lifelike.
- Develop the ability for a virtual human to interact realistically with a real human through speech recognition, facial expressions, emotion, skin color, and joint movements.
- Determining what level of reality is necessary for humans to accept virtual reality and interact with it in meaningful ways.

• Develop the ability of virtual reality to fully mimic human touch in all of its sensory aspects and capabilities.

Think about all of the various projects that will be generated by the demands to make virtual reality more realistic. Then consider how all of the process and people functions of project management will come into play in those projects.

Source: Based on *National Academy of Engineering.* www.engineeringchallenges.org/cms/8996/9140. aspx?printThis=1

DEVELOP THE PROJECT CLOSEOUT REPORT

The project closeout report is a final report documenting the entire closeout process. Preparing the report is the responsibility of the project manager. The content outline of the report can vary from firm to firm, but should cover the following topics at a minimum (Figure 11.4):

• ***Overall performance on the project.*** This section of the final report contains the project manager's answer to the question: "How did we do?" Although this section will necessarily be subjective, project managers should include hard data wherever possible to document their claims—good or bad. Project managers should avoid the human tendency to accentuate the positive and ignore the negative. Some project managers find it useful to rate overall performance on the project on a scale of 1 to 10 and then explain the rating in this section.

• ***Structure of the project.*** Was the project structured in a way that promoted or inhibited success? Did the project manager have line authority over team members or did team members continue to report to their respective supervisors? Did the team have the right types of positions? Were the right people assigned to the positions? What worked well with the project structure? What did not work well with the project structure? In this section of the final report project managers answer these types of questions and summarize their views concerning how future projects might be structured more effectively.

• ***Most effective/least effective methods used.*** In every project, some things work better than others. This is why the lessons-learned review is an important part of the project closeout process. Some processes turn out to be problematic and have to be

---

Content Outline
FINAL PROJECT REPORT

• Overall performance on the project

• Structure of the project

• Most effective/least effective methods used

• Strengths, weaknesses, opportunities, and threats revealed by the project

• Recommendations for future projects

FIGURE 11.4 The project manager's final report should cover at least these topics.

revised in mid-project. Other processes turn out to be especially effective. This section summarizes what worked well and what did not work well. The information in this section should be used to inform the planning processes for future projects so that project managers are using what works and not using what does not work.

- ***Strengths, weaknesses, opportunities, and threats revealed by the project.*** In this section of the final report, the project manager conducts a Strengths, Weaknesses, Opportunities, and Threats (SWOT) analysis of the firm from the perspective of the project in question. What strengths does the firm and, in turn, the project team have that can be exploited to ensure success on future projects? What weaknesses do the firm and, in turn, the project team have that must be prevented from inhibiting the success of future projects? What opportunities arose during the course of the project the firm should pursue now or that might arise in the course of future projects? What threats to the success of future projects were revealed by the project? An example of a strength that might be exploited on future projects is an especially capable process that performs dependably at a high level. An example of a weakness that will have to be corrected is a production department that is slow and inflexible. An example of an opportunity is a chance to receive additional work from a satisfied customer. An example of a threat is a critical piece of equipment that is aging and will have to be replaced or it will inhibit the success of future projects. The project manager answers these types of questions in this section of the report so that the firm's strengths can be exploited on future projects, its weaknesses corrected, opportunities pursued, and threats mitigated.
- ***Recommendations for future projects.*** The last section in the final report contains the project manager's recommendations for improving performance on future projects. Information relating to improved performance is scattered through all sections of the final report. This section is simply a summary of that information made in the form of recommendations for actions and changes. An effective method is to present the recommendations in the form of an annotated list in priority order.

The project closeout report is written from the perspective of the project manager. Other project managers should be able to use the report as a *how-to* guide in preparing to manage their own projects. Higher management should be able to use the report in developing personnel who have been identified as future project managers. They should also be able to use the report to help improve the firm's overall customer satisfaction effort.

RECOGNIZE TEAM MEMBERS

One of the worst mistakes managers can make is failing to recognize team members who contributed in significant ways to the success of a project. Ultimately, the quality of an engineering and technology firm's performance on a project is determined by the quality of its personnel assigned to the project. Ideally, firms will have a formal process in place for periodically recognizing top performers on projects. If not, projects managers should fill the void themselves. Project managers can use the following strategies for giving credit where credit is due:

- Thank all team members face-to-face for their contributions—large and small—to the team's success.

- Put it in writing. Follow up face-to-face *thank you meetings* with written acknowledgements of the contributions of all team members. E-mail is an acceptable format for conveying appreciation in writing. The key is to give team members a written document they can show others and keep for their records.
- Have an off-site celebration (e.g., cookout, meal at a restaurant) to which team members can bring spouses. Then recognize all team members for their contributions in front of their spouses. This is especially important if completing the project on time has required team members to work extra hours away from their family members.
- Send an e-mail note of appreciation to the supervisors of all team members thanking them for the excellent work of their direct reports.
- Recommend outstanding team members for awards in the firm's formal recognition program.

Giving credit where credit is due and expressing appreciation are critical aspects of the project manager's responsibilities. When recognition is sincerely given to those who deserve it, those individuals will: (1) want to work with the project manager in question again on future projects, (2) strive to do even better on future projects, and (3) set the right example for their peers on future projects.

COMPLETE THE FINAL STEP

The final step in the project closeout process is important and should never be overlooked. This step involves turning the success of the project just completed into future business for the engineering and technology firm. In this step the project manager arranges a closeout conference with his or her counterpart in the customer's firm. During this conference the project manager makes sure the customer knows that he or she would like the project to be just the beginning of a long and mutually beneficial relationship, that future business and business referrals will be appreciated.

Project Management Scenario 11.2

We don't need to recognize people for doing what they are paid to do

Mark Tatum is having a heated discussion with a fellow project manager: Andrea Rodriguez. Rodriguez puts a lot of effort into recognizing her team members after the completion of a project. To show her appreciation, Rodriguez hosts a cookout for her team members and their spouses during which she gives credit where credit is due and says "thank you" to all team members and their spouses. Tatum sees things differently. "Why do go to all of this trouble, Andrea? Your people are just doing what they are paid to do. We don't need to recognize people for doing what they are paid to do. Their paycheck is recognition enough."

Discussion Questions

In this scenario, two project managers have different perspectives on the issue of recognizing project team members after a project has been completed. Who is right in this scenario: Mark Tatum or Andrea Rodriguez? Why?

SUMMARY

The closeout phase of a project is important, but does not always receive the attention it deserves. The factors that work against conducting a thorough closeout phase include pressure to begin the next project, disbursement of team members to other duties, and reluctance of project managers and team members to engage in a self-critique. Unfortunately, the costs of overlooking the project closeout process can be high including the costs of: (1) repeating the same mistakes in future projects, (2) failing to take advantage of lessons learned, (3) failing to incorporate new best practices into standard operating procedures, (4) failing to tie up loose ends, (5) failing to gain the customer's support for securing future contracts, and (6) failing to ensure that the customer is completely satisfied.

Verifying the scope of a project as part of the closeout process involves conducting a functional audit, comparing deliverables with their respective product manuals, and conducting a product count. Closing out the contract for a project involves completing all required testing; providing all applicable warranties, guarantees, certifications, forms, reports, manuals, maintenance agreements, and project photographs; returning all proprietary information to the customer; gaining a written agreement on intellectual property; and conducting a project review with the customer.

Closing out the administrative aspects of a project involves returning project team members to their respective functional areas, providing performance feedback for team members, and archiving project information. There are three parts to the lessons-learned review: survey to collect information about lessons learned, face-to-face discussion of survey results, and incorporation of new best practices into the firm's standard operating procedures. The final closeout report for a project should cover at least the following topics: overall performance on the project, structure of the project, most effective/least effective methods used, strengths, weaknesses, opportunities, and threats relating to the project, and recommendations for future projects. Recognizing the work of team member is important. Recognition strategies include thanking team members face-to-face, recognizing team members in writing, hosting an off-site celebration, sending an e-mail of appreciation to the supervisors of team members, and recommending team members for awards in the firm's formal recognition program.

KEY TERMS AND CONCEPTS

Verify the scope
Functional audit
Product count
Close out the contract
Provide performance feedback
Archive project information
Lessons-learned review

Survey team members
Team discussion of survey feedback
Incorporation of best practices
Develop the final project report
Recognize team members
Complete the final step

REVIEW QUESTIONS

1. What factors work against conducting a thorough closeout phase for a project?
2. List the potential costs of overlooking the project closeout process.
3. What are the project manager's responsibilities in verifying the scope as part of the closeout process?

4. List the things that must be done to closeout a project contract.
5. Explain what is involved in closing out the administrative aspects of a project.
6. List and explain the three parts of the lessons-learned review.
7. What topics should be covered in the final closeout report for a project?
8. Explain how project managers can recognize their team members as part of the closeout process.
9. What are the benefits of recognizing team members for their work?
10. Explain the final step in the project closeout process.

APPLICATION ACTIVITIES

The following activities may be completed by individual students or by students working in groups:

1. Identify an engineering and technology firm in your region that will cooperate in completing this activity. Meet with a project manager and ask that individual to describe how he or she closes out a project. Compare the firm's closeout process to the process recommended in this chapter. Is the firm's process comprehensive and thorough? Write a report summarizing what you learned about the firm's project closeout process.

2. Assume that you accepted a contract to paint a neighbor's house. You have just completed all of the required painting. Summarize in writing all aspects of the closeout process you should now go through.

Project Managers as Team Leaders

Any time people are brought together in a group to accomplish a specific mission such as an engineering and technology project, leadership is required. In fact, the difference between an effective project team and one that is just mediocre is often the quality of the leadership provided by the project manager. People who want to be project managers must understand that processes can be managed, but people must be led. Therefore, leadership is a critical *people skill* for project managers in engineering and technology firms.

Like attitude, leadership is an internal concept that manifests itself in external actions. It is an intangible concept that can produce tangible results. Leadership is both an art and a science. Project teams that are well-led, regardless of their size, are better able to complete their projects on time, within budget, and according to specifications. Consequently, few things will contribute more to a project manager's effectiveness than becoming a good leader. In fact, because project managers often do not have line authority over their team members, the ability to lead rather than order is especially important for project managers.

Even the best process managers—initiating, planning, executing, monitoring/controlling, and closing—will be limited as project managers if they lack good leadership skills because it is the performance of people that will ultimately determine whether a project is completed on time, within budget, and according to specifications. The best project plan ever developed still depends on people to carry it out. The better people are led the better they will carry out the plan.

LEADERSHIP DEFINED

Regardless of the purpose, size, or composition of the project team, leadership is essential to the team's effectiveness. There are a number of different definitions of leadership, primarily because the concept applies to so many different fields of endeavor. The definition presented here applies specifically to leadership for project management in an engineering and technology setting:

> *Leadership is the act of inspiring team members to make a wholehearted commitment to the mission that brought them together.*

The "mission" alluded to in this definition is, of course, to complete the project in question on time, within budget, according to specifications. If this appears to be too simple a definition for such an important concept, look closer. While it is true that some definitions of leadership are longer, do not be misled by the brevity of this one. There is more depth in this brief definition than might be apparent at first. Several aspects of this definition are significant. The first of these is *inspiring team members*.

Other definitions of leadership tend to use the term *motivating* where this one uses the term *inspiring*. While it is certainly true that leaders must be good motivators, in the hierarchy of leadership responsibilities it is a lesser concept than inspiration. Leaders motivate people by providing incentives and rewards that encourage them to perform at their best. This is external motivation. External motivation is an important leadership tool, but it typically provides only temporary results. Consequently, it must be applied regularly. Motivating externally is like filling a car with gasoline. The gasoline will make the car perform, but it burns up quickly requiring frequent refills. Inspiration, on the other hand, can have a long-term or even permanent effect on the performance of people.

Project managers can inspire their team members by being good role models of everything they expect from the team members. When project managers expect their team members to do what is necessary to perform at peak levels and to continually improve, they have to be seen by team members as setting the example. Consistency between words and actions is critical to effective leadership. Project managers who fail to live up to the principles they espouse will not be effective leaders. Team members will be inspired by project managers who exemplify traits they themselves would like to have.

The most effective way for a project manager to inspire team members is to let them know what is expected of them and to be an exemplary role model of those expectations. Project managers inspire team members by being good at their jobs, exemplifying integrity, being able to make difficult decisions, having the courage to do the right thing in difficult situations, being selfless, helping others, caring about the work of the project as well as the people who do the work, being fair and consistent, and being a source of calm in the middle of the inevitable storms that arise during the course of a project.

Another important element in the definition of leadership is "wholehearted commitment." There is a quaint saying relating to commitment that project managers should memorize and put to use: *It is easier to ride a horse in the direction it is going*. This saying has direct application for project managers. It is easier to lead team members when they want to follow, when the project mission is viewed as their mission, and when it becomes just as important to them as it is to the project manager that the project's purpose be accomplished. This is what is meant by "wholehearted commitment." When team members are as committed to completing the project on time, within budget, and according to specifications as the project manager, everybody wins—the team, the project manager, the engineering and technology firm, and the owner.

Project Management Scenario 12.1

The ineffective project manager

John's new title is "Project Manager 1." This means he manages small projects for his company, a large engineering and technology firm. Level 2 project managers oversee medium-sized projects. Level 3 project managers oversee large projects. When he is given a project, John is responsible for leading a team that consists of members from the various applicable departments. He is responsible for securing the commitment of team members; helping develop the cost estimate for the project; developing a schedule; helping minimize all applicable risks associated with the project; helping procure the materials and other resources for the project; and monitoring the work site to ensure that it is completed on time, within budget, and according to specifications. John is a talented engineering and technology professional with excellent credentials, but his former supervisor is concerned that John might not make it as a project manager.

When John was promoted to Project Manager 1, his former supervisor commented to a colleague: "It is as if we have just taken the best hitter on the baseball team and made him the coach. I know John can hit the ball, but I am not sure he can coach the team." This supervisor's concern turned out to be prophetic. As a project manager, John is clearly more comfortable dealing with team members who have the same background as his. When project deadlines create stress, John becomes agitated and panicky. In addition, he struggles with making difficult decisions and often sweeps problems under the carpet rather than dealing with them forthrightly.

Discussion Question

Assume that you are a friend of John's, one he will listen to. Also assume that you are familiar with John's struggles as a project manager. If he asked you for advice on how to become a more effective project manager, what guidance would you give him?

INFLUENCING TEAM MEMBERS

In positions that have line authority over employees, leaders in organizations are able to use the authority of their positions to influence those who report to them. They have the authority of their positions to call on when they need people to perform at their best. Project managers, on the other hand, often do not have line authority over members of their project teams. For example, the typical project team is cross-functional, meaning its members come from different functional departments. Unless the engineering and technology firm is organized in a project structure, the various team members report to their own supervisors rather than the project manager. The project manager has no line authority over these team members.

Because this is often the case with the teams they lead, it is important for project managers to understand how to influence team members in ways that contribute to completing the project on time, within budget, and according to specifications. Influencing team members means convincing them to do what needs to be done without being able to order them to do it. The following strategies will help project managers enhance their ability to influence team members:

- **Understand the sources of the project manager's ability to influence others.** Even in situations where project managers have no line authority, they can still have influence. The ability of project managers to influence others comes from different sources including their ability to: (1) give input to the supervisors of team members concerning salary increases and promotions, (2) make recommendations for assigning individuals to other project teams in the future, (3) provide input to supervisors for the performance appraisals of team members, and (4) call on the support of higher management for the project in question. When leading a team, the project manager's ability to influence is based on his or her ability to marshal the commitment of others to complete the project in question on time, within budget, and according to specifications.

- **Understand how to use the need to complete the project on time, within budget, and according to specifications to influence team members.** One thing that all stakeholders in an engineering and technology project need to understand is the importance of completing a project on time, within budget, and according to specifications. The performance of every team in an engineering and technology company on every project affects the image, profitability, and competitiveness of the entire firm. Consequently, using this fact as the basis for project-related recommendations can give project managers influence. This means that project managers must learn to articulate the positive and negative consequences of completing projects on time, within budget, and according to specifications.

- **Understand what drives individual team members.** To influence individual team members, it is necessary for project managers to understand what drives them. All members of the project team have their individual needs that drive them. Some want recognition; others want a challenge. Some want to earn a raise or promotion, while others want to broaden their job skills. Determining what motivates individual team members can give project managers influence with them.

- **Earn respect, loyalty, and credibility from team members.** Project managers who have the most influence with their team members—regardless of whether or not they have line authority—are those who have earned respect, loyalty, and credibility from them. The next section on the Eight Cs of Leadership explains how project managers can earn respect, loyalty, and credibility from team members.

EIGHT Cs OF LEADERSHIP FOR PROJECT MANAGERS

Many people think that leadership is primarily about image, dressing for success, and charisma. This impression is reinforced by the entertainment industry in movies and television. Even news media outlets promote the image-equals-substance theory. For example, watch network news programs during presidential primaries. Even these supposedly serious programs get caught up in discussions about the looks, image, and gravitas of political candidates.

Because of this emphasis on image, many people who want to be leaders contract with image consultants to advise them on such things as dressing for success, how to look taller, compensating for baldness, developing charisma, how to project confidence, and various other image enhancement strategies. One could easily deduce from this fixation on image that the cover of the book is more important than the contents for people who want to be leaders. This is hardly the case for leaders in general and is most certainly not the case for project managers.

This is not to say that image does not matter. It does. Image is important because it often takes an attractive cover to convince people to open a book or, in other words, to give leaders a chance to prove themselves. The point to understand here is that the amount of attention devoted to image can be of proportion to its importance. After all, once a book has been pulled off the shelf, the reader's attention quickly moves from the cover to the contents. No matter how attractive the cover, people will not read a book that lacks substance or is poorly written. This same principle applies to leadership in a project management setting. No matter how attractive project managers might appear on the outside, their inner substance will soon reveal itself—for better or worse. An empty box, no matter how nicely decorated, is still empty.

Rather than putting too much of their effort into image enhancement, project managers who want to be good leaders should focus on developing the basic characteristics that are essential to effective leadership. The author calls these essential characteristics the *Eight Cs of Leadership*. The Eight Cs of Leadership are leadership characteristics that make people want to follow an individual. These characteristics (see Figure 12.1) are as follows:

- Caring
- Competence
- Character
- Communication

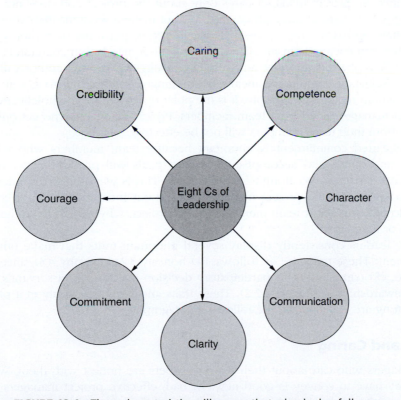

FIGURE 12.1   These characteristics will ensure that a leader has followers.

- Clarity
- Commitment
- Courage
- Credibility

Each of these characteristics is important to project managers who want to be effective leaders.

CARING AND LEADERSHIP

Project managers who are good leaders inspire team members to make a wholehearted commitment to completing the project on time, within budget, and according to specifications. This is important because team members are people and people have agendas, egos, and personality quirks. The only way to keep agendas, egos, and personalities from getting in the way of progress is to gain a wholehearted commitment to the project from all team members.

Few things will inspire team members to follow the lead of project managers more than knowing that they care—about them and the work to be done. Caring is an important ingredient in the formula for inspiring members of project teams to commit fully to the project. When team members know that the project manager cares about both the project and them, they will be more open to following his or her lead. On the other hand, if team members think the project manager cares only about the project and does not care about them or, worse yet, cares about neither, they will not make a wholehearted commitment.

Of course, in cases where project managers are in positions of authority over team members, they can coerce them into carrying out their demands, but coercion is not leadership. In fact, it is just the opposite and typically results in what the author calls *reluctant compliance*. Reluctant compliance means begrudgingly going along to get along—doing only the minimum required to comply. It is the polar opposite of the wholehearted commitment project managers need from team members. Project managers who get only reluctant compliance from their team members will not be effective.

Wholehearted commitment is important because team members who willingly put their hearts and minds into accomplishing project goals will produce better results than those who just reluctantly go along to avoid trouble. This is why it is so important for team members to know that the project manager cares about them and the project. If project managers do not care about team members or the project, why should team members care about them or the project?

Caring leaders consistently display several important traits that make others willing to follow them. These traits are as follows: (1) honesty, (2) empathy, (3) sincere interest, (4) patience, (5) commitment to participatory decision making, (6) a servant's heart, and (7) good stewardship (see Figure 12.2). These traits and how they apply in a project management setting are explained in the following paragraphs.

Honesty and Caring

Project managers who care about their team members are honest with them. Whether the message they have to convey is good news or bad, effective project managers show they care by telling the truth. Honesty is one of the most fundamental leadership traits. When

---

**Checklist of Caring Traits
for
PROJECT MANAGERS**

✓ Honesty

✓ Empathy

✓ Sincere interest

✓ Patience

✓ Commitment to participatory decision making

✓ A servant's heart

✓ Good stewardship

FIGURE 12.2 Caring leaders engender loyal followers.

communicating with an honest project manager, team members can take comfort in knowing that the message, whether good news or bad, is the truth. What makes this fact so important is that members of project teams will not wholeheartedly support a project manager they do not trust, and nothing dampens trust faster than lies, prevarication, obfuscation, deceit, or failure to follow through on promises.

Team members who think they are being lied to or are having information withheld from them will not be inspired to make a wholehearted commitment to completing the project on time, within budget, and according to specifications. Team members want to be informed, and they want to know that they can trust the message. Whether the news is good or bad, members of project teams want and need the truth. They want to be able to trust that what they are told is reliable, complete, accurate, and up-to-date.

Consequently, effective project managers are honest with their team members at all times. However, a caveat is in order here. Being honest, especially when the news is bad, does not mean being tactless. Project managers who care about their projects and their team members make a point of being tactful when conveying unwelcome news. Using tact can be viewed as driving in the nail without breaking the board. It should not be construed to mean withholding or even minimizing information that might be unwelcome or hurtful. Effective project managers are empathetic and tactful, but they fully, honestly, and accurately convey all information, including unwelcome messages.

This ability to tactfully deliver unwelcome news is important when trying to lead low-performing team members to perform at peak levels. For example, project managers often find it necessary to give constructive criticism and corrective feedback to the people who do the hands-on work on their projects. These are situations where driving in the nail without breaking the board is crucial. Assume that it is necessary to give corrective feedback to a team member who has fallen behind schedule. One way to deliver the message would be to say: "The schedule we established is just that—a schedule. It's not a suggestion. Your job description doesn't say stay on schedule if you feel like." Another way to deliver the message would be to say: "You did some outstanding work on our last project. Let's talk about how you can do an equally good job of staying on schedule with this project."

Both versions of the message make the point that falling behind is unacceptable, but the first does so in a tactless manner, a manner likely to break the board. The second is better. It is firm but tactful. Rather than use hurtful sarcasm, the second approach acknowledged that the team member had done well in the past and then offered the constructive criticism about staying on schedule. As a result, the second approach is more likely to be received in a positive manner that will lead to improved performance.

Empathy and Caring

Empathy means identifying with and understanding another person's needs, concerns, fears, and circumstances. Caring project managers are empathetic. They try to put themselves in the shoes of stakeholders when making suggestions, recommendations, and decisions that will affect them. Empathy is about putting oneself in the shoes of the other person and trying to see things from his or her point of view. It should not be confused with sympathy. They are not the same thing. Sympathy is about sharing the sorrow of another person. Empathy is about trying to see things from others' point of view.

Returning to the examples of giving constructive criticism to the poorly performing team member in the previous section, which approach was the more empathetic of the two? The second approach—the tactful approach—was more empathetic because it took into account how the constructive criticism would affect the individual. Being tactful is by definition an empathetic act. Project managers can ensure that they are empathetic by asking themselves the following question before making suggestions, recommendations, and decisions or making statements that affect others: *How would I want this to be said if I were in the other person's place?*

Sincere Interest and Caring

Because they care about team members, the most effective project managers take a sincere interest in them. This means they take the time to find out about their needs concerning the project. In short, they get to know them. Taking an interest in team members is critical for project managers because it helps them match the goals of team members with those of the project. Further, by getting to know team members project managers can improve their ability to assist them in ways that will enhance their commitment to the project.

Patience and Caring

The most effective project managers are patient with team members. This does not mean that they accept inappropriate behavior or condone a lack of commitment. Never forget that the *caring* aspect of leaders means caring about the project as well as the individuals assigned to the project. The patience called for in project management means forbearance in dealing with changes and other inconveniences without developing a negative attitude. This kind of patience can manifest itself in numerous ways including staying positive when changes are made during the course of the project or being willing to listen when a team member makes an unwelcome recommendation.

It can also mean being willing to mentor a member of the team who, in spite of trying, is falling short of expectations rather than just firing him. Patience requires self-discipline, empathy, and a willingness to maintain a positive attitude when things are not going exactly as planned with the project—which often happens. Patience does not mean tolerating poor performance or disruptive behavior. Rather, it means staying positive and focusing on solutions when problems arise rather than losing control and becoming angry and uncooperative. For many people, patience comes hard. But patience is a leadership skill that can be learned by project managers and should be learned by them.

Team members who are treated with patience are more likely to maintain their wholehearted commitment to the project. Patience is especially crucial when providing constructive criticism to team members. Project managers who lose their patience with team members on a regular basis will not win the support they need to be effective leaders of project teams. Impatience on the part of project managers often results in reluctant compliance or even behavior that is purposefully counterproductive on the part of team members.

Commitment to Participatory Decision Making and Caring

Project managers must make many decisions before work on a project begins and many more during the course of the project. Engaging team members when making decisions is called *participatory decision making*. It does not mean that project managers allow others to make the decisions they should make and it does not mean that they take a vote. Rather, it means that before making a decision project managers solicit input from the team members who will have to carry out the decision. Participatory decision making encourages *ownership*—sometimes referred to as *buy-in*—on the part of team members while at the same time contributing to better decisions.

Different team members will have different perspectives about problems that require decisions. By asking for the input of team members, project managers gain the benefit of their perspectives, experience, and knowledge. Asking team members for their opinions concerning decisions is an excellent way to show that they are part of a team. This simple act shows team members that their opinions matter and, by extension, that they matter. Team members who know they matter are more likely to maintain their commitment to the project.

Participatory decision making is an approach in which the stakeholders who are closest to the problem in question and who will have to implement the decision once it is made are included in the decision-making process. For example, when a problem arises that could affect the work of a given team member, that individual would be included in the decision-making process. This does not mean that the project manager lets the team member make the decision—far from it. Rather, it means involving him in the decision-making process for the purpose of making a better decision and for securing his or her support.

Team members who are given a voice in the decision-making process, even if the eventual decision is not the one they recommended, are more likely to commit to its successful implementation than those who are left out. As mentioned earlier, this phenomenon is known as *buy-in*. Buy-in is a major benefit of participatory decision making. Another benefit is that it gets the team members who are closer to the problem than the project manager focused on finding the best solution. Project managers are often one or two levels removed from the hands-on aspects of the problems they have to deal with. Consequently, involving team members who are closer to the problem in determining how to solve it can lead to a better solution.

Servant Leadership, Stewardship, and Caring

Effective project managers are servant leaders and good stewards. They know that team members are more likely to commit to the project if the project manager is a good steward and a servant leader. Project managers who are servant leaders put the best interests of the project ahead of their personal agendas. When team members see the project manager working late to solve a problem that has popped up at the last minute or sacrificing his personal time on a weekend to keep the project on schedule, they will know that he is a servant leader who cares about the project. When other team members hear that the project manager did something thoughtful like buying supper for a team member who had to work late, they will know that he is a servant leader who cares about them.

Project managers who are good stewards take care of the resources—human, physical, and financial—entrusted to them. When team members see the project manager setting an example of fiscal responsibility and wise use of project resources, they will know that he is a good steward who cares about the project. When team members see the project manager making sure that they have the resources needed to do their jobs, they will know he is a good steward. Nothing shows that a project manager cares about them and the project more than servant leadership and good stewardship.

COMPETENCE AND LEADERSHIP

Project managers must have three distinct sets of skills in order to be competent. The first skill set consists of those needed to be effective in their professional field (i.e., engineering or technology). People who want to be project managers should be well-versed in their engineering or technology field through formal education, work experience, or both. A good rule of thumb for prospective project managers to remember is this: *Step one toward becoming a project manager is to become proficient in your professional field.*

The second skill set needed by project managers consists of the process skills of project management. These include all of the processes that go into initiating, planning, executing, monitoring/controlling, and closing out projects. The third skill set consists of the people skills of project management. These skills include leadership, teambuilding, motivating, time management, change management, diversity management, conflict management, and perseverance.

This book is designed to help prospective project managers in the field of engineering and technology develop the second and third skill sets—those that relate directly to project management. However, it is important to note that professional knowledge and experience are not just important—they are vital. A project manager who does not appear to be well-versed in his or her professional field will not inspire confidence in team members.

CHARACTER AND LEADERSHIP

One of the absolute prerequisites of effective leadership is trust. Customers, team members, higher management, and all other stakeholders in an engineering and technology project must know they can trust the project managers they work with. Project managers must be trustworthy. Stakeholders will not follow nor will they be influenced by project managers they do not trust. Further, when dealing with the typical problems that arise during a project, stakeholders will not give the benefit of the doubt to project manages they do not trust.

Team members are more likely to trust project managers who consistently exemplify strength of character. Character is what allows project managers to recognize the right course of action in any situation and the courage to take it in spite of pressure and temptations to the contrary. Project managers who consistently tell the truth, follow through on promises, treat team members with dignity and respect, are fair and equitable in human relations, and do the right thing even when it does not serve their personal preferences are more likely to be able to influence team members in a positive way. On the other hand, when team members observe or even just sense a lack of character in a project manager, they will not make a wholehearted commitment to the project. Nor will they be influenced by the project manager. Unimpeachable character is the foundation of a project manager's credibility with team members and all other stakeholders.

COMMUNICATION AND LEADERSHIP

Project managers who are effective leaders are good communicators. They have to be. Communication is the oil that lubricates the gears of human interaction. At any given time in an engineering and technology project, there are a lot of things happening and a lot of people involved. Effective communication is necessary to keep all of the wheels turning in the right direction and everyone involved fully informed so they can do their parts to keep the project moving on schedule.

The most important communication skill for a project manager is listening. Listening to people empowers them. This, in turn, makes them more willing to do their part to complete projects on time, within budget, and according to specifications. Listening, verbal communication, and nonverbal communication are such important skills for project managers that Chapter Fourteen is devoted to these concepts.

CLARITY AND LEADERSHIP

Team member assigned to a project team are just like people on a long trip. They want to know where they are going—where the project manager proposes to take them. When people are unclear about their direction they feel as if they are peddling a stationary bicycle rather than making progress toward a definite destination. This is why project managers must be clear in establishing the mission of the project and making sure that team members understand it. Part of the mission is always to complete the project on time, within budget, and according to specifications. Team members must understand this as well as the more specific aspects of the project's mission (e.g., design and develop a prototype for the new landing gear for the F-35 jet).

In addition, team members in a project need to feel that there is meaning in their work. They need to know that the project they have been assigned to is important. If the people who do the work in a project wonder if their work matters they might begin to wonder if they matter. If this happens morale will plummet and so will performance. Consequently, project managers should develop a clear mission statement for their projects, explain it to team members, and make sure they understand it.

COMMITMENT AND LEADERSHIP

To be effective leaders, project managers must be committed to their projects and to helping team members do their part to complete projects on time, within budget, and according to specifications. Commitment means more than just trying or even trying hard. It means being willing to sacrifice to get the project completed on time within budget, and according to specifications. It means being willing to go the extra mile to make the project a success.

There is an amusing illustration of the concept of commitment that has circulated over the years in leadership circles. This illustration demonstrates the difference between just being involved and being committed: *With a breakfast of bacon and eggs, the chicken is involved, but the pig is committed.* This humorous illustration makes the point that commitment means more than just trying hard. It means being willing to sacrifice to make the project a success. This is an important distinction for project managers because mediocre performance results more often from a lack of commitment than a lack of talent.

When one commits to the success of a project, there will be sacrifices. It might be necessary to sacrifice time that could have been spent doing something else, resources that could have been used for another project, or the comfort of not having to confront factors that undermine the progress of a project. When team members see that the project manager is committed, they are more likely to make a commitment themselves. Of course, the obverse is also true. Commitment is one of those characteristics that must be exhibited by example.

It is not uncommon to have people involved in a project who are afraid to make a commitment because they think doing so will require too much of them. Such people tend to sit back and let others do the hard work involved in completing projects. People assigned to a project who are lukewarm in their commitment must be either motivated or removed from the team. Motivating team members is the subject of Chapter Thirteen.

COURAGE AND LEADERSHIP

Courage is essential to project managers who want to be effective leaders. In the context of project management, the type of courage required is not physical courage. Rather, it is moral courage—the courage to do the right thing even when it hurts. This is because project managers are often required to deal with difficult situations involving conflict, high expectations, limited resources, pressing deadlines, demanding customers, nonperforming subcontractors, and delinquent material suppliers. Further, project managers sometimes have to tell important people who are above them in the organizational hierarchy things they do not want to hear and influence people to do things they do not want to do. In doing so, project managers may be putting their credibility or even their job security at risk. Putting oneself at risk to do the right thing requires moral courage.

Being a project manager can be a frightening prospect when one considers the level of responsibility and the consequences of failure. It takes courage to face the sometimes daunting responsibilities of the project manager. When things do not go as planned with a project, when problems arise, the project manager is the person higher management will hold accountable. Because the onus of responsibility and accountability can be frightening burdens to bear, project managers will occasionally experience fear. Consequently, students who wish to be project managers need to understand that courage is not a lack of fear. Rather, it is a willingness to do the right thing in spite of fear. It is a willingness to go ahead and do something that is difficult because it is the right thing to do.

For example, a project manager might understandably fear the consequences of failing to complete an important project on time, within budget, and according to specifications. He or she might understandably fear the prospect of telling a demanding customer that his proposed change cannot be accomplished without delaying completion of the project. Project managers must have the courage to overcome the fear of confrontations and to sacrifice their sense of security to do what is best for their projects, even when there is pressure to do otherwise. The most effective project managers are those who learn to put their fears aside and focus on getting the job done rather than being paralyzed by contemplating the consequences of failure.

CREDIBILITY AND LEADERSHIP

Credibility is what project managers have when stakeholders believe in them and see them as leaders who are worthy of confidence. Credibility is earned by exemplifying the first seven of the Eight Cs of Leadership. In addition to exemplifying these characteristics, another important strategy for earning credibility is to continually improve in all aspects of project management—the process skills and the people skills. Project management is like any other endeavor in that those who become good at it will earn credibility from the stakeholders they hope to lead. Credibility, in turn, will make it easier to lead and influence stakeholders.

Project Management Scenario 12.2

This team needs some leadership

Jane Evans faces the biggest challenge of her career. The CEO of her company just informed Evans that the project manager for the company's most important project just resigned. She has been selected to replace him. That's the good news. The bad news is that the project is behind schedule and overbudget, the morale of the project team is at rock bottom, and the customer is beginning to ask some hard questions. The CEO was frank in telling Evans that "…this team needs some leadership."

Discussion Question:

In this scenario, Jane Evans has inherited either a mess or a great opportunity. She wants to turn it into an opportunity. How can Jane Evans use the Eight Cs of Leadership to turn the project around and get it back on track? What should she do first? What should she do after that? Are there things she should avoid?

SUMMARY

Leadership in project management is the act of inspiring stakeholders to make a wholehearted commitment to the success of a project. The Eight Cs of Leadership are caring, competence, character, communication, clarity, commitment, courage, and credibility. Caring leaders consistently display several important traits that make others willing to follow them. These traits are as follows: honesty, empathy, sincere interest, patience, commitment to participatory decision making, servant's heart, and good stewardship.

Competence in one's professional field as well as in project management is essential. Consequently, before becoming project managers, engineering and technology professionals need to develop professional knowledge or gain on-the-job experience or both. Commitment means more than just trying hard. It means being willing to sacrifice to complete projects on time, within budget, and according to specifications. It also means striving to improve performance all the time and going the extra mile to ensure that projects are a success. Credibility is what project managers have when stakeholders believe in them and when they are seen as being worthy of confidence. Credibility is earned by exemplifying the other seven Cs of the Eight Cs of Leadership.

KEY TERMS AND CONCEPTS

Leadership

Caring

Competence

Character

Communication

Clarity

Commitment

Courage

Credibility

Honesty and caring

Empathy and caring

Sincere interest and caring

Patience and caring

Participatory decision making and caring

Good stewardship and caring

Buy-in

REVIEW QUESTIONS

1. Define the term "leadership."
2. What is the significance of the term "inspiring" in the definition of leadership?
3. How does the adage "It is easier to ride a horse in the direction it is going" apply to leading project teams?
4. How does the concept of caring affect a project manager's ability to lead?
5. List and briefly explain the traits displayed by caring leaders.
6. Why is it important that project managers be competent in their engineering or technology field?
7. Why is character so important to those who hope to lead others?
8. Describe the importance of communication skills to project managers.
9. What is meant by the term "clarity" as it relates to leading others?
10. What is meant by the term "commitment"?
11. How does courage apply to project managers?
12. Explain how a project manager can earn credibility with stakeholders.

APPLICATION ACTIVITY

This activity may be completed by individual students or by students working in groups. Think of a person you are familiar with who is or was in a leadership position (executive, manager, supervisor, coach, military leader, elected official, historical figure, etc.). This person can be someone you know or someone you have read about. Analyze this person on the basis of the definition of leadership and the Eight Cs of Leadership. Is or was this individual an effective leader? Why or why not? How does this person exemplify or fail to exemplify the Eight Cs of Leadership? Would this person make a good project manager? Why or why not?

Project Managers as Motivators

Project managers need to be good motivators. They must be able to motivate members of project teams to perform at peak levels and continually improve. The motivational techniques presented in this chapter can help project managers in engineering and technology firms motivate their team members to do their parts to complete projects on time, within budget, and according to specifications. These motivational techniques supplement the leadership techniques presented in the previous chapter. It is important that prospective project managers understand that good leadership provides the foundation for motivating team members. Without good leadership, the motivational techniques presented in this chapter will have a much less positive effect.

Motivating others is an important skill for project managers because highly motivated team members can be surprisingly persistent and innovative at completing assignments, meeting challenges, and overcoming obstacles. Motivating team members is about creating an environment and providing incentives that, together, encourage team members to give their best on behalf of the project. A highly motivated team member will pursue the team's mission with determination and persistence, even in the face of roadblocks, setbacks, and adversity. Highly motivated team members will typically find a way to get the job done, including working with each other in mutually supportive ways, even when they do not especially like each other. Consequently, motivating team members to adopt the team's mission as their own is a worthwhile endeavor for project managers.

For project managers, the short-term goal of motivation is to encourage team members to do their best to help complete projects on time, within budget, and according to specifications. The long-term goal of motivation is to help team members become self-motivated. Encouraging self-motivation is important because self-motivated team members are more valuable to project managers than those who must be continually motivated and remotivated. Project managers can create a motivational environment and provide incentives, but in the long run team members must learn to motivate themselves. Fortunately, under the right circumstances and with proper leadership, most will motivate themselves. When team members become self-motivated, striving for successful completion of projects becomes normal behavior. When team members are self-motivated, doing their best to help accomplish the team's mission becomes a personal goal. Self-motivated team members require little prodding, convincing, or supervision and can be depended on to do the right thing for the project when problems arise, and thus, decisions must be made.

MOTIVATION DEFINED

When it is said that people are motivated, it means that they are driven to do something. This drive can be internal, external, or a combination of both. The ideal state is for motivation to be internal. When project managers speak of motivation, they mean the drive in team members to strive for the successful completion of a project. A highly motivated member of a project team will consistently strive to do what is best for the project in all situations. Team members who do this are self-motivated. Hence, developing self-motivated team members should be an ever-present goal of project managers.

The goal of developing self-motivated team members can be accomplished by project managers who are willing to do the following: (1) provide a positive example consistently, (2) create an environment that encourages peak performance and continual improvement, and (3) provide incentives that encourage peak performance and continual improvement.

When the team members to be motivated report directly to the project manager, the performance appraisal form can be an excellent motivational tool for the project manager. Even if team members do not report normally directly to them, project managers are often tasked with completing performance appraisal for the work they do when assigned to projects.

Often, the performance appraisal form used by project managers to evaluate team members will contain a criterion that asks how well a team member works without supervision. This is an important criterion because project teams need members who will do their best, not just when project managers are watching, but also when they are not. Team members who need close and constant supervision can take up so much of a project manager's time and energy that there is little left for other responsibilities. When this happens, project managers are no longer leading, managing, guiding, or assisting. They are babysitting. Having to hover over team members to ensure that they are working is counterproductive and can undermine a project manager's effectiveness.

MOTIVATIONAL CONTEXT

When learning specific motivational techniques, it helps to view them in a context that allows for systematic study. Abraham Maslow established a workable context for motivational techniques with his well-known *Hierarchy of Human Needs*.[1] Maslow posited that people are motivated by basic human needs that can be categorized according to level—from the lowest level of need to the highest level. An important point to understand about motivating team members is that there are two sides to the equation. One side involves proactively doing things that motivate them. These things include establishing a supportive work environment, providing incentives, and recognizing/rewarding high-performance behavior. The other involves eliminating factors that undermine their motivation—factors that are demotivators. Demotivators are factors in the workplace that lead team members to believe that performing well is of no use because it will render no benefit to them. For example, when promotions, bonuses, and salary increases are awarded on the basis of longevity, seniority, friendship, cronyism, or any factor other than performance, the fact will be a demotivator.

When studying the motivational techniques explained in this chapter, remember that there will be times when the best way to motivate team members will be to remove demotivators. Demotivators are factors that rob team members of their drive to do what is necessary to complete projects on time, within budget, and according to specifications. All motivational techniques presented in this chapter are explained within the context of Maslow's hierarchy of needs. The levels of human need set forth by Maslow from lowest to highest are as follows:[2]

- Basic survival needs
- Safety and security needs
- Social needs
- Esteem needs
- Self-actualization needs

Maslow posited that in any situation people will focus on their lowest unmet needs. Although his work concerning human needs applies to life in the broader sense, it can be applied specifically to project teams. Maslow's five levels of human need provide a context for systematically learning techniques that can be used for externally motivating members of project teams. Team members who are externally motivated to give their best efforts and who receive the recognition and rewards they are due for performing well are more likely to become self-motivated.

BASIC SURVIVAL NEEDS AND MOTIVATION

The basic survival needs of a human being include air, water, food, clothing, and shelter (see Figure 13.1). In Maslow's hierarchy, these are the lowest human needs. Unless these basic needs are satisfied, people will focus on little else but these basic unmet needs.

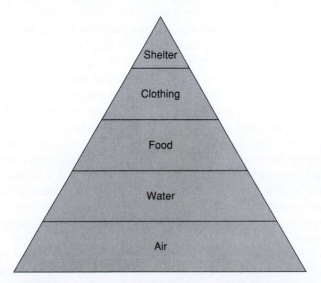

FIGURE 13.1 The most basic survival needs of humans.

For example, in the days before a natural disaster such as a hurricane, tornado, or earthquake, team members assigned to engineering and technology projects are focused on such concerns as meeting project milestones, staying within budget, and meeting specifications. However, when their basic survival needs are threatened by a natural disaster or any other factor, these work-related concerns suddenly go by the wayside. The focus of people whose lives are turned upside down by a natural disaster becomes basic survival needs, and what motivates them is anything that will contribute to meeting those needs.

At first glance, it might seem that basic survival needs have nothing to do with an individual's job performance, but a closer look will reveal direct ties between an individual's performance at work and the quality of the air, water, food, clothing, and shelter available to that individual and his family. This is because how well people perform their jobs can have much to do with how secure their jobs are. There are two sides to the job security issue: (1) people who perform their jobs well are more likely to keep their jobs and (2) by performing their jobs well people help the engineering and technology firm to remain competitive which, in turn, means the firm will continue to have jobs to offer them.

Job security is important to people because without jobs they have no way to provide the basic needs for themselves and their families. This tie between job performance and basic survival needs can be used to help motivate members of project teams. People who understand basic survival needs are more likely to appreciate the significance of performing well on their jobs.

Motivational Technique Based on Basic Survival Needs

Project managers need to understand that the key to using basic survival needs to motivate members of teams is found in helping them make the connection between their jobs and providing for these needs. More specifically, it involves helping members of teams understand the connection between performing well on a given project and their job security.

Some people take life's basic needs for granted. As long as they have a good job, they give little or no consideration to basic survival needs such as air, water, food, clothing, and shelter. But people who have experienced the debilitating effects of long-term unemployment have a different perspective. They know firsthand how being without a job can change their circumstances and threaten their basic survival needs. Consequently, using the basic survival needs to motivate works better during challenging economic times than during times of economic prosperity. However, it can be used at anytime regardless of the state of the economy or the engineering and technology firm's competitiveness.

During the great recession that began in 2007, many people who had been corporate executives and highly paid professionals one day found themselves in the unemployment line the next. People who had never had to do without anything suddenly found themselves worried about securing the most basic of necessities. This connection between employment and life's basic needs can be used to motivate team members if it is used appropriately and tactfully. Using basic survival needs to motivate is simply a matter of helping members of teams see, in a nonthreatening but factual manner, the connection between basic needs and their jobs. Team members who understand the connection are less likely to take their basic needs, their job security, and their performance on projects for granted.

Talking about the connection between basic human needs and performing well on projects during team meetings and in one-on-one conversations can enhance the value of having a job and performing well on projects in the eyes of team members. Further, it can eliminate the tendency of people to take life's basic needs and their jobs for granted. An effective and nonthreatening way to initiate conversations about the connection between performing well on a project and basic human needs is to explain the connection between performing well and getting future contracts. Customers are much more likely to award future contracts to firms who perform well on current contracts.

How a firm is viewed by customers who need something designed, prototyped, tested, and/or manufactured depends to a large extent on the performance of the firm's personnel and project teams. The key to helping team members make the connection without feeling threatened is being tactful. Think of tact as driving in the nail without breaking the board. Of course, connecting basic survival needs to how well team members perform on a given project works better during difficult economic times than during times of prosperity. It also works better in engineering and technology firms whose personnel understand that the firm depends on repeat business from satisfied owners for its survival. Helping team members understand this important connection is the project manager's responsibility.

Project Management Scenario 13.1

My team members just don't seem to get it

Max Renfroe was delighted when he was selected as the project manager for the huge Delta Project for his company. The Delta Project represented the largest contract his company had ever won. If his team did a good job, there was no question there would be future contracts from the customers, not to mention the boost it would give to his career. This was the good news. The bad news was that things weren't going very well with his project team. The best word Renfroe could come up with to describe the problem was "complacency." It was as if his team members took their jobs and the future of the company for granted. In discussing his concerns with another project manager in the firm, Renfroe said: "My team members just don't seem to get it. There is no urgency on their parts to show the customer they can do a good job on the Delta Project." Renfroe feels like he is the only person on the team who cares about getting the project done on time, within budget, and according to specifications. He is not sure how to go about helping his team members understand how important their performance is on the Delta Project.

Discussion Questions:

In this scenario, the project manager understands the need for the team to perform well but his team members do not seem to share his sense of urgency. Have you ever worked in a situation where the people involved seemed to take things for granted as Renfroe's team members apparently do? What can Renfroe do to motivate his team members to take their responsibilities more seriously?

SAFETY AND SECURITY NEEDS AND MOTIVATION

The next level of human needs in Maslow's hierarchy is safety and security. Safety and security needs include safety from physical harm and security from crime, health problems, and financial adversity. For most people, these needs relate directly to their employment. By working, people are able to earn the income necessary to provide a measure of safety and security for themselves and their families. The better the job and the higher the income, the more the individual can invest in safety and security. A home in a safe neighborhood, a security system, financial investments, quality health care, and various types of insurance are all things people use to gain a measure of safety and security for themselves and their families. Of course, all of these things cost money.

Most people must work to earn the income necessary to provide a measure of safety and security for themselves and their families. Consequently, for working people job security is a major concern. The correlation is simple: no job—no security. This fact allows project managers to use job security coupled with safety and security needs as a tool for motivating their team members. The key to using safety and security needs to motivate team members is helping them understand the role their individual performance can play in securing their own job security.

In order to provide jobs for people, engineering and technology firms must be able to not just compete in their markets, but win the competition on a consistent basis. For some, the competition is local. For others it might be regional, national, or global. Regardless of the nature of the competition, all firms face the daily challenge of having to outperform the competition in order to survive. Employees in engineering and technology firms can contribute much to their own job security, a fact that can be used to motivate them.

People place a high value on job security because their job is the principal vehicle for satisfying their other security needs. This is why project managers must make sure that their team members understand how their individual performance affects the overall performance of the firm. The connection that team members need to understand is as follows: (1) the better they perform, the better the firm performs; (2) the better the firm performs, the more competitive it is; and (3) the more competitive the firm is, the more job security it can offer its employees (see Figure 13.2).

Project managers should never make the mistake of assuming that the connection between employee performance and employee job security is intuitively understood by their team members. This is a bad assumption. Not only do engineering and technology firms have employees who do not understand the connection, but such employees are surprisingly common. However, once employees understand the connection between their individual performance and their personal job security, the connection will be motivational for most. Consequently, explaining the connection between employee performance and job security can be an excellent motivational technique.

SOCIAL NEEDS AND MOTIVATION

Project managers should always remember the simple fact that humans are social beings. It is part of the nature of most people to want and need relationships with others. The social needs of people include having positive relationships with family members, friends, and

---

JOB SECURITY CHECKLIST

✓   When individuals in a team perform well, the firm is able to perform well.

✓   When a firm performs well, it is more competitive.

✓   When a firm is competitive, it is better able to provide job security for its personnel.

FIGURE 13.2   Individuals in engineering and technology firms contribute to their job security.

colleagues as well as satisfying the natural human desire to belong, to be part of a group. The human need to be part of a group can be seen in the way people join clubs, fraternities, sororities, and become rabid fans of college and professionals sports teams. Another way to meet the individual's need to be part of a group is through teams at work. An individual's job can go a long way toward meeting the social needs project team members. On the other hand, if a job fails to help meet these needs, it can become 40 hours a week of high stress and drudgery. Many of the relationships in people's lives are tied to their jobs.

Positive relationships with other team members and the project manager are important to people in engineering and technology firms because these relationships help satisfy their social needs. People who have negative relationships at work often experience low morale and high stress. Low morale and high stress, in turn, rob people of their motivation to strive for excellence. Positive relationships, on the other hand, can help motivate employees to perform at their best. Project managers who understand the importance of positive relationships with their team members can use relationship building as a motivating strategy.

Using Relationships to Motivate Employees

For project managers, relationship building has the following three components: (1) establishing positive working relationships with team members, (2) facilitating relationship building among team members, and (3) repairing damaged or broken relationships among team members (see Figure 13.3). Project managers can establish positive working relationships with their team members by doing the following:

- Communicating with them often and well
- Being good listeners
- Being honest and trustworthy
- Encouraging input and feedback from team members
- Being fair and equitable
- Treating team members with respect
- Being a positive and consistent example of the behavior expected of team members
- Sharing the same challenges, burdens, and circumstances team members face
- Recognizing team members for doing a good job
- Caring about team members and the work to be done
- Being an advocate for team members with higher management
- Forthrightly apologizing when wrong

In addition to building positive relationships with their team members, project managers must be attentive to facilitating the establishment, maintenance, and repair of relationships

---

Relationship Building Checklist
FOR PROJECT MANAGERS

✓ Establish positive working relationships with all members of the team.

✓ Help team members develop positive working relationships among themselves.

✓ Help repair damaged or broken relationships within the team.

FIGURE 13.3 Relationship building can be an effective motivational strategy for project managers.

among team members. This can require the application of both teambuilding and conflict management skills. Both of these topics are covered in detail elsewhere in this book. At this point, what is important to understand is that positive peer relationships at work motivate while negative peer relationships demotivate. Creating opportunities for team members to get to know each other as people and interceding when necessary to resolve conflicts will help motivate team members while, at the same time, minimizing the demotivating effects of conflict.

Other relationship-oriented techniques that can be used to motivate team members grow out of the concept of *affiliation*. Some people are achievement-oriented, while others are affiliation-oriented. Affiliation-oriented people are motivated by feeling as if they belong to some type of defined group such as a team. This is one of the reasons college students join the various student organizations on campus. They are motivated by the positive association with others and, in the case of teams, by the esteem they receive from helping the team perform well. Affiliation-oriented people are natural team players. Every team member can and should learn to be a good team player, but affiliation-oriented people are natural team players—they do not have to learn the concept.

The needs of affiliation-oriented team members can be used to motivate them by applying the following techniques:

- Provide opportunities for social interaction with their teammates (e.g., after-hours social events).
- Ask affiliation-oriented employees to help other team members improve in specific areas.
- Ask affiliation-oriented employees to keep their fingers on the pulse of the team's morale and share any problems they think might be developing.

A word of caution is in order here. Affiliation-oriented team members will sometimes choose team harmony over team performance. Consequently, project managers should make a point of talking with affiliation-oriented team members one-on-one and reminding them that going along to get along will rarely improve team performance and that a happy team is not necessarily a productive team.

ESTEEM NEEDS AND MOTIVATION

People have an inherent need for self-esteem as well as for the esteem of others. Esteem relates to the self-respect, worth, and dignity people feel. People with self-esteem feel good about themselves. They have self-respect and self-worth. Typically, people who lack self-esteem also lack self-respect and self-worth. A lack of self-esteem can result in feelings of inadequacy, and such feelings can be powerful demotivators. Consequently, as with most motivational techniques, when using esteem needs to motivate team members it is important to be attentive to identify and remove demotivators.

In helping team members build self-esteem and win the esteem of their peers, a variety of techniques are available in the following areas: (1) achievement, (2) competition, (3) the potential for promotions, (4) incentives, and (5) legacy (see Figure 13.4). All of the techniques explained in the following paragraphs can help team members build self-esteem and win the esteem of peers.

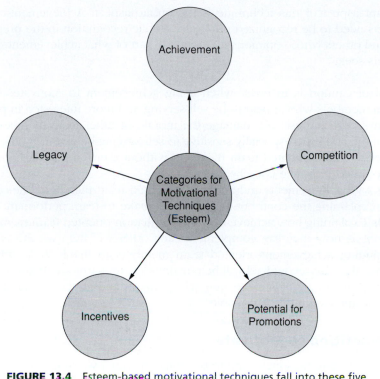

FIGURE 13.4 Esteem-based motivational techniques fall into these five categories.

Using Achievement Activities to Motivate

Achieving something important can give team members a sense of accomplishment. Accomplishing tasks that are both important and difficult will build self-esteem. Many people have an inherent need to achieve, although some do not. Because there are people who are not motivated by achievement, the first step in applying the techniques explained in this section is to identify those team members who are achievement-oriented. Such people are usually easy to recognize. They tend to be task- and goal-oriented. They typically need continual positive reinforcement and focus intently on evaluations of their performance. Achievement-oriented people like to collect physical evidence of their achievements, such as certificates, trophies, plaques, and other recognition memorabilia.

The message conveyed to project managers is that recognition is important to achievement-oriented people. Project managers who understand this can use the following techniques to motivate achievement-oriented employees:

- Tell them specifically what they can do to help accomplish the team's mission.
- Put them in charge of specific high-priority tasks. Let them know what needs to be done and give them a deadline. Then step back and give them room to work. Do not micromanage achievement-oriented personnel.
- Recognize them whenever they accomplish a goal or a specific assigned task. The recognition need be nothing more than a public pat on the back from time to time. The

important aspect of this technique is the public aspect. Achievement-oriented team members need to be recognized publically. Public recognition in the presence of their peers and others whose opinions they value is part of what achievement-oriented team members seek.

A word of caution is in order when using achievement to motivate. Achievement-oriented team members who appear to be self-serving and more interested in personal glory than the team's best interests can damage the morale of other team members. Affiliation-oriented team members are especially sensitive to self-serving behavior in others. To get the most from achievement-oriented team members without damaging team harmony, project managers should confront the issue head on.

Talking with achievement- and affiliation-oriented team members one-on-one and as a group and explaining the contributions each can make to team performance is an effective approach. Explaining how achievement- and affiliation-oriented team members are both needed as well as how they can complement each other will help get the best from both groups. Reminding achievement-oriented team members to thank those who supported them whenever they are recognized will help maintain team harmony. It also helps if project managers can maintain a balance between achievement- and affiliation-oriented people on project teams to the extent this is possible.

Using Competition to Motivate

Most people like to win personally and to be affiliated with a winning team. Winning builds self-esteem. Watch children as they play games. They like to outperform other children who challenge them. A child's competitive spirit is nurtured through play and sometimes through participation in organized sports. Project managers can use the natural competitive spirit many people have to motivate team members, but caution is the order of the day when doing so. In order to have the desired effect, competition must be carefully planned, closely monitored, and scrupulously controlled.

Competition that is not properly planned, monitored, and controlled can go awry and result in a win-at-any-cost mentality that actually undermines performance. When this happens, team members can lose sight of the real goal of doing what is necessary to complete projects on time, within budget, and according to specifications. If winning at any cost becomes the goal when competing against other teams, undermining the work of the other teams becomes an effective strategy. Out-of-control competition can lead to cheating and hard feelings among team members. If this is allowed to happen, the competition will do more harm than good.

The following tips will help project managers ensure that competition is properly controlled and that it contributes to improved performance:

- Involve the team in planning the competition, and explain to them that making sure the project is completed on time, within budget, and according to specifications is the purpose of the competition. Be frank in letting team members know that behavior that undermines another member's performance or another team's performance is unacceptable and will only hurt them, their team members, and everyone else in the firm.
- To the extent possible, plan the competition so that it is between a team and a set of performance benchmarks rather than another team or individuals within the team.

Competition between project teams or individuals within teams can quickly become personal and counterproductive. Ideally, the competition would be against another team's record that is already established. For example, say that team XYZ holds the firm's record for completing a project two weeks ahead of schedule, 10 percent below budget, and in accordance with all applicable criteria in the specifications. This record would be an excellent basis for the competition.

- Make sure the competition is as fair as it can possibly be. For example, do not ask a team of rookies to compete against a team of experienced members or the record of such a team.
- Be specific in selecting the basis of the competition—decide what will actually be measured. With teams this will usually mean competing in such areas as beating deadlines and budget projections. Of course meeting or exceeding specifications is always the controlling criterion. A team that beats a scheduled deadline but does not meet or exceed customer specifications has achieved nothing.

This section is not intended to argue against using competition to motivate. Rather, it is provided to remind project managers to take human nature into account when using competition. Once the competition has been planned, but before starting the competition, conduct a *roadblock analysis*. This is done by brainstorming with other project managers and team members about what roadblocks or unintended consequences might arise that could derail the competition and undermine performance. Once all potential roadblocks and unintended consequences have been identified, find ways to eliminate, mitigate, or control them before beginning the competition.

Using the Potential for Promotions to Motivate

People in engineering and technology firms are just like most others in that they want to advance up the career ladder in their professions. Ambitious, hard-working people hope that over the course of their careers they will be promoted to increasingly higher levels positions that have correspondingly higher pay. When this happens, it builds self-esteem and helps earn the esteem of colleagues. Consequently, opportunities for promotions can be used to motivate team members who are career-minded.

However, like most motivation strategies, promotions can have either a positive or negative effect depending on how they are handled. The two basic approaches to filling vacant positions in engineering and technology firms are promoting from within and hiring from outside the firm. Of the two, promoting from within is the approach most likely to be a motivator, provided the promotion process is handled properly. This can be important for project managers because they are sometimes asked to make promotion recommendations for personnel who have served on one or more of their project teams.

The following rules of thumb will help ensure that when project managers make recommendations for filling vacant positions by promoting from within, the result is positive (Figure 13.5):

- ***Never promote solely on the basis of seniority.*** Experience is important, but it does not necessarily make an individual the best qualified person for an open position. Seniority is a legitimate factor to consider when making promotions, but it should not be the only factor. It is better to use seniority as a tie breaker when the other factors considered

---

Checklist for
USING POTENTIAL PROMOTIONS TO MOTIVATE

✓   Do not promote solely on the basis of seniority.

✓   Do not promote on the basis of popularity.

✓   Do not promote on the basis of friendship.

✓   Do make performance the key factor in promotion decisions.

FIGURE 13.5 Project managers should base promotion recommendations primarily on performance.

are equal. If an individual with seniority is promoted over a less senior but more qualified person, morale will suffer and the promotion process will be a demotivator.

• ***Do not promote on the basis of popularity.*** Personal popularity is no guarantee of effectiveness in a new and higher level position. It is not uncommon for a team member to be popular for reasons that have nothing to do with performance. Even if an individual is well-liked, he or she will still have to be able to do the new job and do it well. If the popular individual is not able to perform the new job effectively, he or she will not be popular long, and the credibility of the promotion process will suffer. If this happens, the promotion process will be a demotivator.

• ***Do not promote on the basis of friendship.*** Friendship should never be the reason behind a promotion. Promotions that are viewed by other team members as being influenced by friendship are doomed from the outset. Further, allowing friendship to influence promotions will not only serve to demotivate employees, but it will also undermine the promotion process.

Challenging Engineering and Technology Project

Securing Cyberspace

Most of the challenges engineering and technology professionals deal with require not just technical skills, but also project management skills. A particularly challenging problem facing engineers and technologists is securing cyberspace. Identity theft is perhaps the best known security problem in cyberspace, but securing cyberspace is a much bigger enterprise than just protecting personal information. It also involves such things as protecting bank records, national security, college records and transcripts, and the nation's physical infrastructure to name just a few.

Serious breaches of cyber security are a common problem. Viruses and various other types of cyber attacks happen all the time. Computer hackers are never far behind in finding ways to get around the latest security procedures, most of which are really just after-the-fact cyber-patches applied once a problem surfaces. Firewalls that are supposed to prevent breaches in cyber security are really nothing more than temporary expedients that will soon be overcome by smart and determined hackers.

Engineering and technology professionals have major roles to play in securing cyberspace. These roles include: 1) developing better ways to authenticate hardware, software, and data in computer systems, 2) establishing better ways to verify user identities, 3) developing more secure software that has protections built into the code, 4) developing better means for protecting the data that flow over the Internet, 5) establishing better methods for monitoring and detecting security compromises, and 6) developing methods that take the whole system into account rather than just its individual components.

The various roles engineering and technology professionals must play in securing cyberspace will require them to work in teams. These teams will undertake hundreds of on-going individual projects—projects that will have to be managed. In other words, securing cyberspace will require smart engineering coupled with effective project management. Securing cyberspace will require problem solvers to have not just advanced technical skills, but project management skills too. Before proceeding with the remainder of this chapter, stop here and consider how all of the process and people skills of project management might be used in meeting the challenge of securing cyberspace.

Source: Based on *National Academy of Engineering*, Grand Challenges for Engineering, "Secure Cyberspace." http://www.engineeringchallenges.org/cms/8996/9042.aspx.

To ensure that the possibility of being promoted is a motivator, project managers must tie promotions to performance. Team members must know that if they consistently meet or exceed performance expectations, they have a realistic chance of being promoted. If they believe the process is tainted by seniority, popularity, or friendship, the promotion process will be a demotivator.

Using Incentives to Motivate

Like most other organizations, one of the most common topics of informal conversation among people in engineering and technology firms is money or, more specifically, salaries and wages. There is more than just economics to an individual's paycheck. There is also the issue of esteem. Generally speaking, the more people earn the more they are esteemed by their peers and the greater their self-esteem. One can certainly debate the advisability of tying self-esteem to income, but there is no question that people do it.

One way for people in engineering and technology firms to increase their income is by earning incentive pay. For those motivated by money, incentive pay can be doubly motivating. First, there is the obvious motivation of receiving the extra money. But, this is not the only motivational benefit of properly managed incentive programs. Incentives also give people opportunities for achievement. Just the fact that people receive incentives over and above their normal pay—irrespective of the actual amount—can be a motivator because doing so represents an achievement. This is one of the main reasons that achievement-oriented people typically respond well to incentives.

In order to gain the motivational benefits of incentives, engineering and technology firms must plan and manage incentive programs carefully. Poorly planned and managed incentives can quickly become demotivators. Consequently, project managers must be prepared to provide input that will help bolster the effectiveness of incentive programs. The following strategies can help make incentives programs effective (Figure 13.6):

- **Define the objectives.** The overall purpose of an incentive program is to motivate people to higher levels of performance. This should be understood by project managers

Checklist for
MOTIVATING WITH INCENTIVES

✓ Define the objectives of the incentives.

✓ Set a positive example of the incentivized behaviors.

✓ Award incentives to teams rather than individuals.

✓ Involve team members to ensure that incentives are meaningful to them.

✓ Establish specific performance criteria.

✓ Communicate the objectives of the incentives to team members as well as the performance criteria.

FIGURE 13.6 Project managers can use these strategies to make incentives effective.

as well as by all personnel who might have opportunities to earn the incentives. But just stating this overall purpose is not enough. Project managers must ensure that the firm takes the next step and defines specifically what is to be accomplished. Does the firm want to improve productivity? Quality? Safety? All of these? In addition to making people aware of the overall purpose of the incentive program, it is important to make them aware of the specific objectives of it.

• *Set a positive example.* By offering incentives, engineering and technology firms establish high performance expectations for members of project teams. It follows then that project managers must make a point of exemplifying the behaviors that are expected of their team members. Project managers should always remember that one of the most fundamental principles of leadership is to set a positive example of exemplifying the behaviors they expect of team members.

• *Award incentives to teams whenever possible.* Awarding incentives to teams can be more effective than awarding them to individuals. Project managers should always be prepared to make this case with higher management. Most work in engineering and technology firms is done in teams. In project teams, people depend on each other to complete projects on time, within budget, and according to specifications. Because of this interdependence, team members who have contributed to the team's success might, understandably, resent just one member receiving incentives. If this happens, the incentive program can backfire and do more harm than good.

• *Make incentives meaningful.* For an incentive program to be effective, the incentives provided must be meaningful to potential recipients. Giving team members rewards they do not value will not produce positive results. This is a critical point that project managers must be prepared to make when involved in designing incentive programs for engineering and technology firms. An effective way for ensuring that incentives are meaningful to people is to involve them in developing the list of incentives that will provided. The people who are to be motivated by incentives know better than anyone else what types of incentives will motivate them and what types will not. In other words, if you want to know what types of incentives will motivate team members, ask them. Then use this feedback to develop the list of actual incentives that will be provided.

- ***Establish specific criteria.*** If the purpose of incentives is to encourage peak performance—beating scheduled deadlines and budget projections or exceeding productivity, quality, and safety benchmarks—there must be specific criteria for measuring performance. On what basis will incentives be awarded? Specific criteria define the levels of performance that are to be rewarded. For example, assume that one of the areas to be measured is the project's budget. The incentive goal relating to budget is to complete the project in question at least 10 percent below budget. Project managers might ask that incentive dollars be earned for savings that exceed 10 percent on the basis of a certain amount per percentage point. It is important for team members to know specifically how performance improvements will be measured. This means establishing benchmarks against which performance will be measured and establishing intermediate as well as overall performance targets.
- ***Communicate, communicate, communicate.*** Team members must be completely informed about the incentive program if the incentives are going to motivate. Team members should know the purpose of the incentive program, its specific objectives, the performance criteria, when incentives will be awarded and how often, and anything else that will help maintain the program's effectiveness and credibility. With incentive programs, there should be no surprises and no confusion concerning the details.

Using the Legacy Question to Motivate

People in engineering and technology firms go through stages in their careers. In each stage, they tend to focus on different concerns. For example, when young people first begin their careers, money is one of the most important of their work-related concerns. This is the *income phase* of their career. People in this phase are concerned primarily with income because they are at the bottom of the pay scale in their respective positions and are struggling to get established financially. People who have just begun their careers are often shocked to learn how much it costs to buy or rent a home and pay for utilities, groceries, and upkeep. This is why they tend to focus so intently on income and why they are typically good candidates for being motivated by monetary incentives.

Once people reach the point in their careers where they are past the initial shock of how much it costs to live from day to day and they have moved up in their firm's salary schedule, they begin to focus on whether or not they like their jobs. This is the *personal satisfaction* phase of their career. People in this phase want their job to provide them with a sense of personal satisfaction and enjoyment. Of course, all jobs have their good days and bad days, but people in this stage want to like their jobs, at least generally speaking, after balancing the good days with the bad ones. In this phase, they are concerned less with money because they are earning a sufficient income and more with whether or not they enjoy their jobs. For example, there are many people who hate their jobs even though they make plenty of money. On the other hand, there are people who love their jobs in spite of earning less than they would like in those jobs.

In the final phase of their careers, people who are relatively satisfied with their income and generally like their jobs begin to focus on the *legacy question*. The legacy question grows out of a need people have to know that their work matters—that their work is important, that it has meaning, and that it allows them to make a difference. For most people, knowing that their work is important builds self-esteem. Knowing that others think their work is important adds to the esteem. When they believe that their work matters, people are more likely to believe that they matter. Philosophers will argue that people have value irrespective of their work—and they certainly do. However, there is no getting around the fact that people tend to tie their self-worth to the relative worth of their jobs. People with this perspective consider their life's work a major part of their legacy.

This human desire to matter and to leave a worthwhile legacy can be used to motivate team members who are in the *legacy phase* of their careers. Legacy-minded people will work harder and smarter when they believe their work matters. Consequently, project managers can motivate legacy-minded team members by helping them see the importance of their work. This is not a difficult task for project managers in engineering and technology firms. Engineering and technology professionals produce things that improve the lives of individuals as well as the quality of life in communities. Whether they work in design, testing, or manufacturing, people in engineering and technology fields make a significant difference. Project managers can use this legacy-related fact to help motivate team members.

SELF-ACTUALIZATION NEEDS AND MOTIVATION

Self-actualization refers to the human need to achieve one's full potential. In order to achieve self-actualization, people must first satisfactorily meet all of their other needs: basic survival, safety/security, social, and esteem. In reality, few people ever reach the level of self-actualization. However, with self-actualization the pursuit may be more important than the accomplishment, at least from the perspective of motivation. The U.S. Army was appealing to the human need for self-actualization when its recruiting slogan was: "Be all that you can be."

The key to using self-actualization to motivate is the concept of potential. Project managers can use the concept of potential to motivate team members who want to climb the career ladder as well as those who want to broaden their career horizons. Broadening one's career horizons in the current context refers to learning new career skills or even a new job through cross-training. The key to using potential to motivate is to tie both concepts—advancement and expansion—to performance. Those who consistently perform at peak levels must be the ones who advance the fastest in the firm and who are given opportunities to expand their horizons by learning new job skills. Project managers use potential to motivate by letting their team members know that the ones they will recommend for advancement and broadening opportunities are those that exceed performance expectations.

The fact that advancement and expansion should be tied to performance would seem to go without saying, and it should. However, there are engineering and technology firms that tie advancement and expansion to nonperformance factors such as seniority. Further, there are cases in which advancement and expansion decisions are influenced by such factors as friendship, cronyism, who knows whom, and favoritism. Self-actualization needs can be used to motivate only if the potential for advancement or career broadening is tied directly to performance.

DEVELOPING PERSONAL MOTIVATION PLANS

There is no one-size-fits-all strategy project managers can use to motivate their team members. Because motivation is based on appealing to individual needs—needs that vary from person to person—project managers must be prepared to personalize their motivational strategies. What will motivate a given team member depends on where that person is in Maslow's Hierarchy of Needs as well as other factors specific to the individual. This is why project managers must get to know their team members well enough to understand where they fit into Maslow's hierarchy and to be aware of other motivation-related factors affecting them at any given time.

An effective approach for personalizing motivation strategies is to develop Personal Motivation Plans (PMPs) for individual team members. A PMP is a brief plan containing strategies for motivating a project team member that takes into account that individual's specific human needs. If the individual in question is achievement-oriented, his or her PMP should be based on meeting those kinds of needs. If the individual is concerned about his legacy, the strategies in the PMP should be based on meeting legacy needs. The key is to personalize the strategies in the PMP for the individual in question rather than applying the same one-size-fits-all strategies for everyone on the team.

ADDITIONAL MOTIVATION STRATEGIES

In addition to the motivational strategies explained earlier in this chapter, there are several others that have proven to be effective in helping motivate members of project teams. Project managers may wish to employ the following strategies for motivating their team members:

- ***Explain the benefits of the project in their terms.*** With this strategy, project managers explain why it will be good for the team and its members to complete the project on time, within budget, and according to specifications. If doing so might result in additional contracts, higher profits, or any other benefit, explain how these things will also benefit the team members (e.g., better job security, potential for incentive bonuses, possible raises and promotions in the future).
- ***Eliminate the fear factor.*** When first beginning a project, team members may feel overwhelmed. They might fear that the project will involve more work than they can possibly get done or skills they do not have. When this is the case, the fear factor must be eliminated. This can be done by making the project appear more feasible by: (1) breaking it into smaller subprojects; (2) sharing the schedule with team members and discussing it; (3) allowing team members to voice their concerns, ask questions, and make suggestions; and (4) calmly expressing confidence in the team's ability to get the project done on time, within budget, and according to specifications.
- ***Give continual feedback.*** Team members tend to focus on their individual assignments when working on a project. They do not always see the big picture or how the overall project is progressing. Giving them continual feedback on how the project is progressing can reassure team members and help them see that the team can actually complete the project on time, within budget, and according to specifications.
- ***Recognize the work of team members continually in real time.*** Project team members are just like anyone else in that they need to know how they are doing and how they need to be encouraged. Both of these needs can be satisfied by providing continual informal recognition to team members and the team as a whole. For example, sending e-mail notes of appreciation to individual team members and the team as a whole when project milestones are met, giving public pats on the back to team members when they do a good job on some aspect of the project, and recommending team members for various types of rewards and bonuses are all effective techniques for encouraging team members. The key is to do these things in real time throughout the project's duration rather than waiting until the project is completed.

A prerequisite to motivating people in teams is to develop an understanding of human needs. People are motivated by actions that meet their specific individual needs. Although the specific needs of individuals differ, there are generic needs common to most people. Maslow summarized and categorized these generic human needs in his hierarchy of needs. Project managers who learn to tie the needs Maslow identified to motivational strategies can become effective at motivating their team members to perform at their best.

Project Management Scenario 13.2

I don't know how to motivate these people

Marie played organized sports from the time she was a child right through college. In fact, she was an All-American volleyball player and team captain at Northwest Florida Institute of Technology where she earned her engineering degree. Consequently, Marie knew plenty about motivating team members. At least she thought she did, but lately she has begun to have doubts. The motivational techniques that always worked so well with her volleyball teammates do not seem to be working with the members of her project team.

The XYZ Project is her first as a project manager, and Marie is determined to make a good impression on her company's vice president. But the motivational techniques she is using do not seem to be having the desired effect. The various members of her project team just do not respond the way her volleyball teammates from college did. Marie is at a loss as to how to get her team motivated.

Discussion Questions

In this scenario, Marie is using the one-size-fits-all approach in trying to motivate her team members and it is not working. Have you ever been involved with a team in which the coach or team leader used the same motivational techniques for everyone? If so, describe how that worked out. If Marie asked for your advice concerning how to motivate her team members, what would you tell her?

SUMMARY

When it is said that people are motivated it means they are driven to do something. Motivation can be internal, external, or a combination of both. When project managers speak of motivation, they mean the drive that team members have to do their best to ensure that projects are completed on time, within budget, and according to specifications. Ideally, members of teams will become self-motivated. A good context for motivational techniques is Abraham Maslow's Hierarchy of Human Needs. The levels of human need set forth by Maslow from lowest to highest are basic survival, safety and security, social, esteem, and self-actualization needs.

Basic survival needs include air, water, food, clothing, and shelter. These needs can be used to motivate team members by tying them to their jobs and how well they perform them. Safety and security needs include safety from physical harm and security from crime, health problems, and financial adversity. Again, these needs can be used to motivate team members by tying them to their job security and job performance. Social needs include having positive relationships with family members, friends, and colleagues. By facilitating positive relationships among team members and by repairing damaged relationships in the team, project managers can use social needs to motivate.

People have an inherent need for self-esteem and for the esteem of others. Esteem relates to the self-respect, worth, and dignity people feel. Project managers can use achievement, competition, the potential for promotions, incentives, and the legacy question to build self-worth and self-esteem in team members. This, in turn, will motivate them. Self-actualization refers to the human need to achieve one's full potential. "Potential" is the key to using the need for self-actualization to motivate team members. When team members believe they have the potential to move up in terms of salary, ben-

efits, positions, and perquisites based on performance, most will be motivated by the fact. However, if they think that moving up in the organization is based on nonperformance factors such as seniority, friendship, cronyism, or favoritism the result will be demotivation and a loss of morale.

There is no one-size-fits-all strategy project managers can use to motivate team members. Because motivation is based on appealing to individual needs—needs that vary from person to person—project managers must be prepared to personalize their motivation strategies. An effective approach for personalizing motivation strategies is to develop Personal Motivation Plans (PMPs) for team members. A PMP is a brief plan containing strategies for motivating an individual team member and it takes into account that individual's specific human needs.

KEY TERMS AND CONCEPTS

Motivation
Motivational context
Hierarchy of human needs
Basic survival needs
Safety and security needs
Social needs
Esteem needs
Self-actualization needs
Affiliation

Achievement
Competition
Potential for promotions
Incentives
Legacy
Seniority
Popularity
Friendship
Personal Motivation Plans (PMP)

REVIEW QUESTIONS

1. Define the term "motivation."
2. What is meant by the term "motivational context."
3. Explain Maslow's Hierarchy of Human Needs.
4. Explain one motivational strategy from each category of need in Maslow's hierarchy.
5. For project managers, what three components comprise relationship building in a project team?
6. List three strategies project managers can use for establishing positive working relationships with their team members.
7. Explain the concept of "affiliation" and how it can be used to motivate certain members of teams.
8. Explain how achievement can be used as a motivational technique.
9. Explain how competition can be used as a motivational technique.
10. How can project managers keep competition from getting out of hand and undermining team performance?

11. Explain how the potential for promotions can be used as a motivational technique.
12. Explain how offering incentives can be used as a motivational technique.
13. Explain how the legacy question can be used as a motivational technique.
14. Explain three rules of thumb that will help ensure that when project managers make recommendations for filling vacant positions by promoting from within, the result is positive.

15. Explain how organizations can ensure that performance incentives will actually be effective at motivating project team members.
16. Explain how the concept of self-actualization can be used to motivate project team members.
17. Explain what Personal Motivation Plans are, why they are necessary, and how to develop one.

APPLICATION ACTIVITIES

The following activities may be completed by individual students or by students working in groups:

1. Assume that you are a new project manager and you want to motivate your team members to perform at their best. Develop one strategy for each of the categories of human need in Maslow's hierarchy. Tell specifically how you will use each specific category of need to motivate team members.

2. Think of things that would motivate you personally to perform at your best as a member of a team. Use this information to develop a PMP for yourself. If this activity is done in groups, pair off and have one student serve as the project manager and one as the team member for whom the PMP is being developed. Discuss the things that motivate you and why.

ENDNOTES

1. Saul McLeod, "Maslow's Hierarchy of Needs." Retrieved from www.simplypsychology.org/ Maslow.html on July 26, 2013.

2. Ibid.

CHAPTER 14

Project Managers as Communicators and Negotiators

Effective communication is an essential skill for project managers. Engineering and technology projects involve a lot of different people, all of whom have their own concerns, goals, agendas, responsibilities, egos, and attitudes. There are a lot of moving parts in an engineering and technology project. To keep all of the parts moving properly and working together, effective communication is essential. Communication skills help project managers gain commitments from team members; understand the views, perspectives, and messages of others; clearly, accurately, and succinctly convey work instructions to team members; offer constructive criticism in a positive, helpful manner; convince others of the veracity of their ideas; promote teamwork, cooperation, and collaboration; enhance interpersonal relationships; ensure that team members understand the big picture and where they fit into it; and prevent and resolve conflict.

Good communication is critical to the success of engineering and technology projects. In fact, if a team can be viewed as a machine, communication is the oil that keeps it running smoothly. Of all the skills needed by project managers, effective communication may be the most important. It is certainly one of the most important. Good communication is an essential people skill for project managers. This chapter explains how to develop the communication skills needed to help effectively lead project teams.

BEGIN WITH A COMMUNICATION PLAN

Project managers are responsible for keeping a number of different stakeholders informed about their projects. Some of the stakeholders are part of the project manager's firm while others are from outside the firm. Internal stakeholders include the engineering and technology firm's higher management team and members of the project team. External stakeholders include the customer, subcontractors, and—at times—government officials. All of these stakeholders will want to be fully informed about the specific aspects of the project that apply to them, how the project is progressing, what problems have been encountered, what is being done to overcome problems, and so on.

It is incumbent on project managers to keep all members of their project teams fully informed of progress, problems, budget issues, and anything else they need to know to be fully informed, contributing team members to do their parts to ensure that the project is completed on time, within budget, and according to specifications. An ill-informed team member is not equipped with the information needed to do the best possible job in this regard.

Since there are so many different stakeholders who will want to stay informed about projects, it is important for project managers to develop a communication plan so that communication is well-organized and systematic. A communication plan for a given project contains the following information:

- A list of stakeholders—internal and external
- The types of communication to be provided for each stakeholder (written report, team meeting with verbal communication, stand-and-deliver presentations to selected groups, e-mail updates, etc.)
- Frequency of communications with each stakeholder
- Standard content to be contained in each type of communication
- Who is responsible for collecting and providing the information for each type of communication and when it should be provided (e.g., members of the project team are often tasked with preparing draft reports of various kinds for the project manager to distribute)

Higher management and other stakeholder groups within the organization will probably receive periodic written reports. The same is true of the customer, although project managers can expect to occasionally be asked to make stand-and-deliver presentations to higher management and the customer. Team members will typically receive verbal updates presented during team meetings along with a written summary of the updates. The types of communication that will be expected of project managers are typically specified for them based on negotiations that occur during the initiation phase of the project.

COMMUNICATION SKILLS CAN BE LEARNED

Communication is a human process. Hence, it is an imperfect process. Doing it well takes work. This is because the quality of communication is affected by so many different factors. These factors include speaking ability; hearing ability; listening ability; language barriers; differing perceptions and meanings based on age, gender, race, nationality, and culture; attitudes; nonverbal cues; writing ability; and the level of trust between senders and receivers to name just a few. Because of these and other factors, communicating effectively can be difficult. Regardless of the difficulty, good communication skills are essential for project managers. Fortunately, most people can learn to be effective communicators.

With sufficient training and persistent practice, most people—regardless of their innate capabilities—can learn to communicate well. Project managers must be good communicators.

Consequently, those who want to be project managers should strive to develop their communications skills.

Project Management Scenario 14.1

I'm just not a good communicator

Nancy Powers is becoming increasingly frustrated. She worked long and hard to become a project manager, and was pleased and proud when she was promoted. But now less than six months later Powers is beginning to think she made a mistake. Her problem is communication. Powers does not seem to be able to communicate clearly with her team members. Too many mistakes are being made in her team as the result of poor communication. Even when she is sure her team members heard what she told them, they do not seem to get things right. There always seems to be confusion over who is supposed to do what and when. Yesterday, her supervisor asked why there seemed to be so many problems with her current project. A frustrated Powers responded: "I am just not a good communicator, and I never will be."

Discussion Questions

In this scenario, Nancy Powers is having problems moving the work of her project team forward because of poor communication. Have you ever known someone whose poor communication skills inhibited progress by creating confusion or other problems? If so, relate your experience. If you were Nancy Powers' supervisor, how would you respond to her belief that she will never be a good communicator?

COMMUNICATION DEFINED

Project managers who want to be good communicators should never confuse *telling* with *communicating*. Unfortunately, some do. When a problem develops these confused project managers are likely to protest, "I don't understand why he didn't get this done. I told him what to do." In addition, some project managers confuse *hearing* with *listening*. When there is a problem, these confused project managers are likely to say, "That isn't what I told her to do. I know she heard me. She was standing right next to me!"

The project managers in these examples did not understand the concept of communication. Project managers should always remember that what they say is not necessarily what the other person hears, and what the other person hears is not necessarily what they intended to say. Often, the missing ingredient when communication goes awry is comprehension. Communication may involve telling, but it is not *just* telling. It may involve hearing, but it is not *just* hearing. The following definition clarifies the concept of communication:

> Communication is the transfer of information that is received and fully understood from one source to another.

A message can be sent by one person and received by another, but until the message is fully understood communication has not occurred. This qualifier applies to spoken, written, and nonverbal communication.

COMMUNICATION IS A PROCESS

Communication is a process. As such it has several components: *sender, receiver, method, medium,* and the *message* itself. The sender is the originator or source of the message. When a project manager conveys work instructions to a team member, he or she is a sender. The receiver is the person or group for whom the message is intended. When team members receive instructions from a project manager they are receivers. The message is the information that is to be conveyed, understood, and acted on. When a project manager conveys work instructions, they are the message. The medium is the vehicle by which the message is carried (e.g., telephone, e-mail, social networking software). The method is the type of communication chosen for conveying the message.

There are four basic categories of communication methods: *verbal, nonverbal, written,* and *graphic* (see Figure 14.1). Verbal communication takes place in face-to-face conversation and telephone calls, but it also takes place during speeches, public announcements, press conferences, and other venues for conveying the spoken word. Nonverbal communication includes facial expressions, voice tone, body poses, gestures, and proximity. Written communication includes letters, memorandums, billboards, bulletin boards, manuals, books, e-mail, and all of the other electronic means of conveying the written word. Graphic communication involves using pictures and nonalphanumeric symbols to convey a message.

Technological developments have significantly enhanced the ability of people to send and receive information, but not necessarily their ability to communicate. Technological development in the broad field of communication include the Internet, e-mail, social networking, word processing, satellites, telephones, cellular phones and an ever-growing variety of other handheld devices, answering machines, facsimile machines, and pocket-sized dictation devices. Project managers should make a point of becoming skilled at using

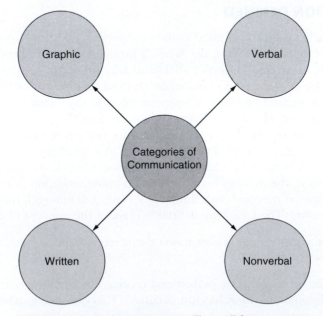

FIGURE 14.1 Project managers will use all four categories of communication.

the various technological aids that are readily available to enhance communication in their project teams. However, a word of caution is in order here. Just sending more messages using more technological aids does not necessarily enhance communication.

NOT ALL COMMUNICATION IS EFFECTIVE

When the information conveyed from one source to another is received and understood, communication has occurred. However, understanding alone does not guarantee effective communication. *Effective communication* occurs when the information received and understood is accepted and acted on in the desired manner. For example, a project manager might ask a team member to retrieve some old files from the company's warehouse. The team member in question verifies that he received and understood the message. However, rather than make the short walk to the company's warehouse, he decides to put the task off. He is right in the middle of another assignment and does not want to put the current task aside until it is complete. By the time he is ready to make the trip to the warehouse, it is time to go to lunch. He decides to retrieve the file needed by the project manager after lunch.

Unfortunately, by the time he returns from lunch he has forgotten all about the file in the warehouse. When the project manager drops by his office to pick up the file—a file he needs for a meeting that begins in just five minutes—the embarrassed team member has to admit he does not have it. There had been communication when the project manager gave the team member his assignment, but it was not effective communication because it did not result in the desired action.

Effective communication is a higher level of communication because it requires not just understanding but acceptance and action by the receiver. The acceptance aspect of effective communication may require project managers to apply influence, persuasion, and monitoring. Since acceptance of the message is essential to effective communication, project managers need to know how to gain acceptance of the messages they communicate.

The first step toward ensuring acceptance of messages is to gain credibility with those who receive the messages. Project managers who have credibility with team members will find it easier to influence them in a positive way. The more influential project managers are, the more likely it is that their messages will be accepted and acted on in the desired manner. When credibility and influence are lacking, receivers tend to question or even doubt the veracity of the message. They may not voice their doubts, but they will make them known by their hesitance to accept the message and their corresponding reluctance to respond to it.

Persuasion can be an important factor in gaining acceptance of messages. Project managers can be more persuasive by explaining: (1) the *why* behind their messages, (2) the benefits of accepting their messages and acting on them in the desired manner, and (3) the consequences to the team member, the team, and the firm of failing to accept their messages and act on them in the desired manner. This is why a dictatorial approach to communication that says "do what I say and don't ask any questions" does not work very well. Of course, in crisis situations there may not be time to explain. When this is the case, acceptance of the message can still be gained provided the sender has gained credibility with the receiver(s). This is another reason why credibility is essential to effective communication—project managers will not always have time to explain. However, when there is time to do so, explaining the reasons for compliance as well as the consequences of noncompliance will help gain acceptance of the message.

Finally, monitoring is an effective way to ensure acceptance of the message and that the desired action is taken. For example, in the earlier example in which the individual failed to retrieve the file for his project manager, the embarrassed team member would probably have acted differently if his supervisor had monitored him. Instead, the project manager simply told the individual to retrieve the file he needed. He did not call or e-mail in the interim to ensure that the task had been completed. This is an example of both poor communication and poor leadership.

FACTORS THAT CAN INHIBIT COMMUNICATION

There are several factors that can inhibit communication. Project managers should be familiar with these inhibitors and understand how to overcome them. Factors that can inhibit communication include the following (Figure 14.2):

- **Differences in meaning.** Differences in meaning are inevitable in communication because people have different backgrounds, experience, and levels of education. In a country as diverse as the United States, project teams are likely to mirror that diversity (e.g., different races, cultures, and nationalities). Because of this diversity, the words, gestures, and facial expressions used by people can have altogether different meanings. To overcome this inhibitor, project managers must invest the time necessary to get to know their team members and learn what they mean by what they say.
- **Insufficient trust.** Few factors can inhibit communication more than insufficient trust. If team members do not trust their project manager, they are not likely to believe what she tells them. If they do not believe what she tells them, they are not likely to accept it and act on it in the desired manner. Team members who do not trust their project manager often question the motives behind her messages. They will tend to concentrate on reading between the lines and looking for a "hidden agenda." In fact, they might focus so intently on reading between the lines that they miss the real message.

Checklist of Factors That Can INHIBIT COMMUNICATION

✓ Differences in meaning

✓ Insufficient trust

✓ Information overload

✓ Interference

✓ Condescending tones

✓ Listening problems

✓ Premature judgments

✓ Inaccurate assumptions

✓ Technological glitches

FIGURE 14.2 Project managers must learn how to overcome these inhibitors.

Project managers should understand this and, as a result, strive to build trust with their team members.

- *Information overload.* Because advances in communication technology have enabled and encouraged the rapid and continual proliferation of information, members of project teams can find themselves dealing with more information than they can process effectively. This is known as *information overload*, and it can easily cause a breakdown in communication. Project managers can protect their team members from information overload by screening, organizing, summarizing, and simplifying the information conveyed to them. For example, project managers should never take the reports they receive from higher management and just hand them over to their team members. This concept is known as an *information dump*. Instead, project managers should take the time to extract or at least highlight the information that is pertinent so that their team members do not get bogged down wading through superfluous information.

- *Interference.* Interference is any external distraction that inhibits effective communication. It might be something as simple as background noise or as complex as atmospheric interference. Regardless of its nature, interference can distort or even completely block out communication. Consequently, project managers should be attentive to the setting and the environment when trying to communicate with team members. The author once had to move an entire audience of 100 people when giving a speech in a resort on the Gulf of Mexico. The beautiful emerald waters of the Gulf were not the problem. Rather, a contractor was doing renovations to the resort and one of its workers was using a jack hammer. Sometimes, to eliminate interference it is necessary to change the setting in which communication will be attempted.

- *Condescending tones.* Communication problems created by condescension result from the tone rather than the content of the message. People do not like to be talked down to. If team members sense that the project manager is talking down to them, they might respond by tuning out. Worse yet, they might express their resentment by intentionally ignoring the message.

- *Listening problems.* Listening problems are one of the most common inhibitors of effective communication. They can result from the sender not listening to the receiver and vice versa. To be good communicators, project managers must be good listeners. This topic is important enough to warrant a section of its own later in this chapter.

- *Premature judgments.* Premature judgments by the sender or the receiver can inhibit effective communication. This inhibitor exacerbates listening problems because as soon as people make a premature judgment they stop listening. One cannot make premature judgments and maintain an open mind, and an open mind is essential to effective communication. Therefore, it is important for project managers to listen nonjudgmentally and avoid making premature judgments when receiving messages.

- *Inaccurate assumptions.* Perceptions are influenced by assumptions. Consequently, inaccurate assumptions can lead to inaccurate perceptions. Here is an example. Janice will go to great lengths to avoid participating in presentations her project team has to make to the customer or her firm's higher management. Having never worked with Janice, her project manager assumes that she is lazy. As a result, whenever Janice makes suggestions in team meetings, the project manager either ignores her input or simply tunes out. The project manager is making an inaccurate assumption about Janice. She is actually a highly motivated team member who works hard to help the team. Her

reluctance to help make presentations is the result of fear not laziness. Janice is morti-
fied at the thought of getting up in front of an audience, but she is embarrassed to
admit it. Because of an inaccurate assumption, Janice's project manager is missing out
on the suggestions of a talented team member. In addition, his misperception points
to a need for trust building. Perhaps if Janice trusted her project manager more, she
would be less embarrassed to discuss her fear of public speaking.

- ***Technological glitches.*** Software bugs, computer viruses, dead batteries, power out-
ages, holes in coverage, and software conversion problems are just a few of the technological glitches that can interfere with communication. The more dependent project
managers and their team members become on technology for conveying messages, the
more often these glitches will interfere with and inhibit effective communication.

LISTENING WELL IMPROVES COMMUNICATION

Hearing is a physiological process, but listening is not. A person with highly sensitive hearing can be a poor listener. Conversely, a person with impaired hearing can be an excellent
listener. Hearing is the physiological process of receiving sound waves, but listening is
about perception. Project managers should never confuse hearing with listening.

Understanding the following definition of listening can help project managers become
better listeners:

> Listening is receiving a message, correctly decoding it, and accurately perceiving what is
> meant by it.

Notice that this definition contains three elements—all of which are critical and must
be present for effective listening to occur. These elements are as follows: (1) receiving the
message, (2) correctly decoding the message, and (3) accurately perceiving what is meant
by the message. If even one of these elements is missing, effective listening will not occur.

Inhibitors of Effective Listening

Communication will not occur if a receiver hears but does not accurately perceive a message. Several inhibitors can cause this to happen. The most common inhibitors of effective
listening include (Figure 14.3):

- Lack of concentration
- Preconceived notions
- Thinking ahead
- Interruptions
- Tuning out

To perceive a message accurately, it is necessary to concentrate on what is being said,
and how it is being said—verbally and nonverbally. Nonverbal communication is explained
in the next section. This section focuses on listening to verbal messages. *Concentration*
requires that distractions be either eliminated or mentally shut out. When project managers
concentrate, they clear their minds of everything but the message being conveyed and focus
on the team member who is sending the message.

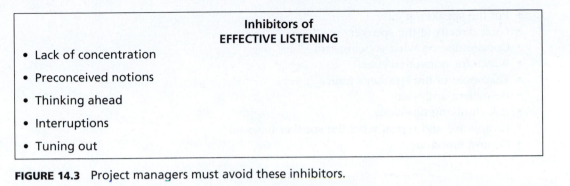

Inhibitors of
EFFECTIVE LISTENING

- Lack of concentration
- Preconceived notions
- Thinking ahead
- Interruptions
- Tuning out

FIGURE 14.3 Project managers must avoid these inhibitors.

Preconceived notions can also inhibit listening by causing people to make premature judgments. Making premature judgments shuts down listening. Project managers should practice being patient and listening attentively. People who prematurely jump ahead to where they think the conversation is going may get there only to find that the speaker was going somewhere else.

Thinking ahead is typically a response to being impatient or in a hurry. Project managers are often in a hurry. Project deadlines always loom large, and they are unforgiving. Consequently, it is understandable that project managers would get in a rush and think ahead when team members try to convey a message. However, project managers need to understand that it takes less time to hear someone out than it does to start over after jumping ahead to a preconceived conclusion that is wrong. The time-saving approach is to listen attentively and get the message right the first time.

Interrupting can be especially harmful in that it can inhibit effective listening and frustrate the speaker. Consequently, it is doubly bad to interrupt someone who is speaking to you. If clarification is needed during a conversation, project managers should make a mental note of it and wait for the speaker to reach a stopping point. Mental notes are preferable to written notes. The act of writing can, itself, distract the speaker or cause the listener to miss the point. If it is necessary to make written notes, project managers should keep them short. They should also avoid the temptation to interrupt and should not allow cellular phones or other people to interrupt.

Tuning out inhibits effective listening. A person who has tuned out will typically appear distracted or have a far-away look on her face. However, some people become skilled at using facial expressions and body language to make it appear they are listening when in fact they are not. Project managers should avoid the temptation to tune out during conversations with team members and to engage in nonverbal ploys to make it appear they are listening when they are not. An astute team member might ask the project manager to repeat what was said. At any point during a conversation, the project manager should be able to paraphrase and repeat back to a team member what she has said.

Strategies for Promoting Effective Listening

Project managers can improve their listening skills by applying the following strategies for promoting effective listening:

- Use the five-minute rule
- Remove all distractions

- Put the speaker at ease
- Look directly at the speaker
- Concentrate on what is being said
- Watch for nonverbal cues
- Take note of the speaker's tone
- Be patient and wait
- Ask clarifying questions
- Paraphrase and repeat what the speaker has said
- Control emotions

THE FIVE-MINUTE RULE   The *five-minute rule* is really a self-defense mechanism for project managers. Consider the case of Luke. As a newly minted project manager, he wanted to maintain an open-door policy for team members and an open ear for their problems, concerns, complaints, and recommendations. Having come up through the ranks, Luke knew firsthand how it was to work for supervisors who were not accessible. Consequently, Luke was determined to be just the opposite.

On the other hand, listening to the problems, concerns, complaints, and recommendations of team members can be time consuming. Like all project managers, Luke had other duties that needed his attention. Before long his open-door policy had him spending most of his time in the workday doing nothing but listening to the input of his team members. Luke was making an "A" in listening, but an "F" in attending to his other duties.

Luke's open-door policy was popular with his team members and did produce some positive results beyond just the morale boost it gave them. On the other hand, he often found himself in the office late at night trying to finish his other duties that were interrupted by drop-in visits from his team members. Clearly, Luke needed to find a way to retain his open-door policy without allowing team members to monopolize all of his time. The answer that eventually solved his problem was what the author calls the *five-minute rule*.

The five-minute rule works like this. Project managers let their team members know that—within reason—they can have five minutes on a drop-in basis any time they have a complaint, recommendation, problem, or any other type of input to offer. However, the time for these drop-in visits will be limited to five minutes. Lest the reader think this policy is too restrictive, five minutes is actually plenty of time to explain a problem or make a recommendation provided the team member has thoroughly considered what she wants to say. Preparation is the key. Spending time listening to a team member who rambles on because of poor preparation is wasting time. In addition to saving time, the five-minute rule helps team members learn how to organize their arguments and prepare brief, succinct, but comprehensive explanations that get right to the point without wasting time. This is a skill that will serve them and the team well.

The allotted time is not to be used for thinking out load or brainstorming. There is a time and place for these things, but it is not during five-minute rule sessions. During the allotted five minutes, the team member is expected to explain his problem and offer a recommended solution that is realistic. Recommending poorly conceived solutions is a major faux pas on the team member's part. Proposing a $100 solution to a $10 problem is not acceptable. Team members who ask for five minutes are expected to have already conducted a cost-benefit analysis for the solution they plan to propose or recommendation they plan to make.

The cost-benefit analysis might amount to just carefully thinking through the recommendation to be made, but even this will help team members realize that some solutions are better than others. Nothing is free. There is a cost associated with everything. Consequently, team members who make recommendations should: (1) be aware of the costs associated with their recommendations and (2) make sure the potential benefits of their recommendations outweigh the costs. The cost-benefit analysis requirement of the five-minute rule can prevent time from being wasted considering unrealistic solutions. It can also help make team members better problem solvers.

Not all issues in teams can be properly dealt with in five minutes. Issues that are too complex to fit into the five-minute format should be handled in the normal manner (i.e., the team member makes an appointment and asks for as much time as is needed). The five-minute rule is a strategy for facilitating effective listening while allowing project managers to maintain an open-door policy. It is not intended as a replacement for traditional problem-solving methods such as brainstorming, focus groups, or team meetings.

OTHER LISTENING-IMPROVEMENT STRATEGIES AND THE FIVE-MINUTE RULE   To gain the most from five-minute sessions with team members, project managers should apply the other listening-improvement strategies explained earlier. The strategy of removing distractions and giving full attention to the speaker is important. Anyone who has ever tried to talk with someone who was distracted by other concerns will understand why. Removing distractions typically involves such things as turning off cellular phones, putting a temporary hold on landline calls, allowing no other visitors to drop in, and getting away from distracting paperwork on the desk top.

An easy way to get away from distracting clutter without having to clean off the desk is to have two chairs in front of or away from the desk. Project managers should not try to sit at their desks and listen to team members. Trying to concentrate on what someone is saying while pressing paperwork on the desk beckons takes more self-discipline than most people have.

Before asking the speaker to begin, put her at ease—particularly if you sense nervousness or discomfort. Asking about something unrelated to the job such as children, grandchildren, sports, or hobbies will usually suffice. Then, once the speaker begins, look directly at him and concentrate on what is being said. Project managers should never waste a moment of their team members' time by being inattentive. Rather, they should concentrate on what is being said and learn to listen not just with their ears, but also with their eyes. In other words, project managers should learn to watch for nonverbal cues. Nonverbal communication is explained in the next section.

Project managers should learn to avoid interrupting or pushing the team member along. One of the keys to effective listening is to be patient and wait. When there are hesitant pauses in a team member's explanation it can mean that he or she is trying to decide: (1) how to say what is really on his or her mind or (2) if he or she is even going to say what is on his or her mind. If project managers interrupt or try to prompt a hesitant speaker, they risk preempting the speaker and, as a result, missing out on the real reason he or she asked for five minutes in the first place—especially if what the team member has to say is sensitive or embarrassing. Better for project managers to give hesitant team members a positive, affirming facial expression and then be patient and wait.

Once a team member has stated his or her case, the project manager should ask clarifying questions to gain a more complete and accurate understanding. Once the project manager has a complete and accurate understanding, an effective strategy is to paraphrase what the team member has said and repeat it back to him or her. Paraphrasing can be beneficial in two ways. First, it shows the team member that the project manager listened. Second, if the message has been misperceived the team member can clarify further. Paraphrasing can prevent a situation in which the project manager wastes time trying to solve the wrong problem or dealing with something that is not a problem.

The final strategy—control emotions—is critical. A good rule of thumb for project managers to remember about communication is this: When dealing with team members, losing one's temper will undermine communication and trust. One of the differences between being a project manager and a member of a project team is that there are higher expectations for the project manager. When engineering and technology professionals step up to leadership positions such as project managers, it is not just their pay that increases, but it is also their level of responsibility and behavioral expectations that increase.

Project managers who lose their tempers when team members bring them unwelcome information soon find themselves without messengers or messages. This is one of the worst things that can happen to project managers because the more unwelcome the message the more likely it is that they need to hear it. They may not want to hear it, but they need to hear it and the sooner the better. Bad news that goes unattended has a way of turning into even worse news.

NONVERBAL COMMUNICATION

Nonverbal messages represent one of the least understood but most powerful modes of communication. Nonverbal messages can reveal more than verbal messages for those who are attentive enough to observe them. Nonverbal communication is sometimes called *body language*, an only partially accurate characterization. Nonverbal communication does include body language, but body language is only part of the concept. There are actually three components of nonverbal communication: body factors, voice factors, and proximity factors (Figure 14.4):

Body Factors

An individual's posture, facial expressions, gestures, and dress—in other words his or her body language—can convey a variety of messages. Even such factors as makeup or the lack of it, well-groomed or messy hair, and clean or scruffy shoes can convey a message. Project managers who learn to be attentive to these body factors can improve their communication skills markedly. The key is to understand that nonverbal messages should agree with, support, and enhance verbal messages. For example, when someone says "yes" with a smile and a nod of the head, the verbal message—yes—and the nonverbal—a smile and a nod—all agree. However, if a person says "yes" but frowns and shakes his head from side to side in the universal signal for "no" there is disagreement between the verbal and nonverbal messages. In this example, the disparity between the verbal and the nonverbal is obvious. However, this is not always the case. Differences between the verbal and nonverbal can be subtle and require careful attention to perceive.

The key to understanding nonverbal messages lies in the concept of consistency. In a conversation with another individual, are the spoken messages and the corresponding nonverbal messages consistent with each other? They should be. In a conversation, if nonverbal

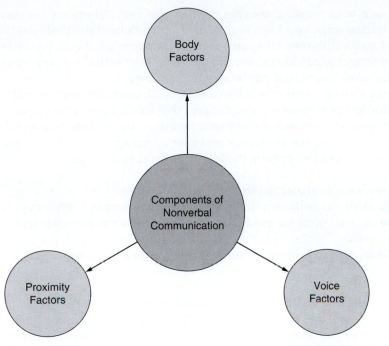

FIGURE 14.4 Nonverbal communication is more than just body language.

messages do not seem to match the verbal message, something is probably wrong and it is a good idea to find out what it is. An effective way to deal with inconsistency between verbal and nonverbal messages is to tactfully but frankly confront it. A simple statement such as, "Andrew your words say that you put the XYZ File back in the drawer yesterday, but your body language says you didn't." Such a statement can help project managers get to the truth in conversations with team members.

Voice Factors

Voice factors are also important elements of nonverbal communication. In addition to listening to the words team members speak, it is important to listen for voice factors such as volume, tone, pitch, and rate of speech. These factors can reveal feelings of anger, fear, impatience, uncertainty, interest, acceptance, confidence, and so on. As with body factors it is important to look for consistency when comparing words and voice factors. It is also advisable to look for groups of nonverbal cues. A single cue taken out of context has little meaning. But as one of a group of cues, each appearing to validate the others, a given nonverbal cue can take on significance.

Proximity Factors

Proximity factors range from the relative positions of people in conversations to how an individual's office is arranged, the color of the walls, and the types of decorations displayed. A project manager who sits next to a team member during a conversation conveys a different message than one separated from him by a desk. Coming out from behind a desk and sitting next to a team member conveys the message that, "There are no barriers between us—I want to hear what you have to say." Remaining behind the desk sends a message of distance and standoffishness. Of course there are times when this is precisely the message the project manager wants to convey, but the point here is that it is important to be aware of the nonverbal messages that can be sent by proximity.

A project manager who makes his office a comfortable place to sit and talk is sending a message that invites communication. On the other hand, a project manager who maintains a cold, impersonal office sends the opposite message. To send the nonverbal message that team members are welcome to stop by and take advantage of the five-minute rule, project managers should consider applying the following strategies:

- Have comfortable chairs available for team members.
- Arrange chairs so as to be able to sit beside team members rather than behind a desk.
- Choose neutral colors for the walls of the office rather than harsh, stark, overly bright, or busy colors.
- If possible, have refreshments such as water, coffee, tea, and soda available for team members.

Some people like to turn their offices into to personal shrines displaying their achievements. In offices that are shrines, visitors will find trophies, plaques, photographs taken with important people, award certificates, and various other career mementoes displayed prominently. There is nothing wrong with a project manager having a "look-at-me" wall in his office, but the concept can be overdone. To make a positive impression on team members, evidence of career and personal achievements can serve a valuable purpose. However, when trying to encourage team members to open up and reveal their concerns, issues, and problems, it is helpful to have a more inviting place to meet—one that is comfortable and inviting. A good rule of thumb for decorating your office is this: two walls for visitors and two for the occupant.

VERBAL COMMUNICATION

Effective verbal communication ranks close in importance to effective listening. Even in the age of technology, talking is still by far the most frequently used method of communication. This is why project managers should strive to continually improve their verbal communication skills. Being attentive to the following factors will help project managers improve the quality of their verbal communication (Figure 14.5):

- *Interest.* When speaking with team members, project managers should show interest in their topic, that they are sincerely interested in communicating the message in question. Project managers should demonstrate interest in the team members—the receivers of the message—as well. It is a good idea to look listeners in the eye, or if in a group, spread eye contact evenly among all receivers. Project managers who sound bored, ambivalent, or indifferent concerning their own message cannot expect receivers of the message to be enthusiastic about it.
- *Attitude.* Maintaining a positive, friendly attitude enhances verbal communication. This is because people are more open to listening to someone who is friendly and positive. A caustic, superior, condescending, disinterested, or argumentative attitude

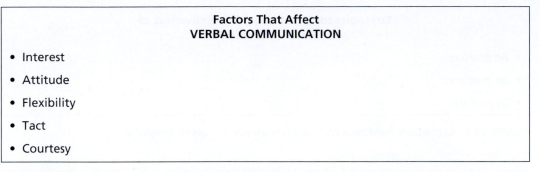

Factors That Affect
VERBAL COMMUNICATION

- Interest
- Attitude
- Flexibility
- Tact
- Courtesy

FIGURE 14.5   Project managers can use these factors to improve verbal communication.

will shut down or, at least, inhibit communication. To increase the likelihood that their messages will be received in a welcome or at least open-mined manner, project managers should make an effort to be positive and friendly.

- *Flexibility.* Project managers who are dogmatic and dictatorial in their verbal communication increase the likelihood that their messages will be rejected by receivers. Flexibility and a willingness to hear other points of view will usually improve the chances of having a message received in a positive manner. For example, if during a team meeting the project manager presents a case for solving a problem, she should let team members know that their views and opinions will be welcomed, heard, and appreciated. Even if no one has an alternative idea to propose the fact that the project manager is open enough to ask for other opinions and flexible enough to listen to them will improve communication. In fact, an effective communication tactic is to ask team members for their ideas first. When the project manager presents his or her ideas first, some team members may be reluctant to appear to disagree.

- *Tact.* Tact is an important factor in verbal communication, particularly when delivering a sensitive, potentially controversial, or unwelcome message. Using tact can be thought of as hammering in the nail without breaking the board. The key to tactful verbal communication is to think before speaking. Tact does not mean being less than forthright. Rather, it means finding a way to candidly say what has to be said without adding insult to injury.

- *Courtesy.* Being courteous means showing appropriate concern for the needs and feelings of the receiver. Calling a meeting as team members are leaving for home on a Friday evening is inconsiderate and will inhibit communication. Courtesy also dictates that project managers avoid monopolizing conversations. When communicating verbally, they should give receivers ample opportunities to ask questions for clarification and to state their own points of view. Project managers are wise to remember that one-sided conversations are not conversations at all—they are broadcasts.

COMMUNICATING CORRECTIVE FEEDBACK

Project managers occasionally need to give corrective feedback to team members. In fact, this is an important responsibility for project managers. If project teams are going to their projects on time, within budget, and according to specifications their members must be willing to continually improve their performance. Corrective feedback is given to help team

---

**Strategies for Enhancing the Effectiveness of
CORRECTIVE FEEDBACK**

- Be positive

- Be prepared

- Be realistic

FIGURE 14.6 Corrective feedback can backfire unless it is given properly.

members learn to improve and perform better. Effectively given corrective feedback will do this. But in order to be effective, corrective feedback must be received in a positive manner by those at whom it is directed. This means it must be communicated properly and effectively. The following guidelines will help project managers enhance the effectiveness of their corrective feedback (Figure 14.6):

- ***Be positive.*** To actually improve performance, corrective feedback must be accepted and acted on by the team member. This is more likely to happen if it is delivered in a positive and tactful manner. Corrective feedback that is delivered in a less than tactful manner may cause the receiver to become defensive. If this happens, project managers are more likely to get excuses than improved performance. Project managers should give the team member being corrected the necessary feedback, but avoid focusing exclusively on the negative. Rather, they should try to find something positive to say. For example, assume a team member has arrived late for work twice in one week. One approach would be to confront the tardy team member and say, "I'm glad you could bother to show up today. I certainly hope this job isn't interfering with your social life." The tardy team member would certainly get the message about coming to work on time, but he might also be offended by the sarcasm. Also, if there is some legitimate reason for the tardiness, he might resent the assumption that his social life is the cause. A better approach would be to say, "Is everything OK? I noticed that you have been late twice this week. Is there anything I can help you with?" This approach lets the team member know that his tardiness has been noticed, but without boxing him into a corner where his only option is to become defensive. Another positive approach would be to say, "Mark I am really proud of how you helped solve that problem we had last week. Now let's talk about how you can do an equally good job of getting to work on time." The latter two examples let the individual know that his tardiness is unacceptable but without creating resentment or defensiveness. Project managers in these situations should remember that the goal is to correct and improve, not to punish.

- ***Be prepared.*** Before giving corrective feedback, project managers should do their homework. They should avoid kneejerk reactions and gather the facts. Then they should give corrective feedback based on the facts. It is always best to give specific examples of the poor performance or unacceptable behavior that needs to be corrected or improved. A common response to corrective criticism is for team members to view it as just plain criticism and become defensive. Return to the earlier example of the team member who was late for work two days in one week. If this team member becomes defensive he might try to deny being late. However, if he knows that the project manager has the facts, this is less likely to happen. A project manager who is poorly prepared might say, "Mark, you have been late a couple of times this week." The vagueness of this statement might encourage Mark to challenge it. However, if the project manager says, "Mark, you were 30 minutes late on Monday and 45 minutes late on Tuesday," he will know better than to challenge the statement.

- **Be realistic.** Project managers who are trying to improve the performance of team members can be forgiven for wanting to see a change for the better right now if not sooner. This is understandable since they are always are under pressure to meet deadlines, stay within budget, and comply with specifications. Unfortunately, developmental activities seldom produce immediate results. Improving performance can take time, and often does. Consequently, when giving corrective feedback for improving performance it is important to be realistic. First, project managers should use their experience to determine how much improvement is realistic to expect over a given period of time. They should then be guided by this knowledge in setting improvement goals. Second, project managers should make sure the performance they are trying to improve is within the control of the team member. Project managers should never make the mistake of expecting team members to correct something over which they have no control. An effective approach is to explain the situation to team members, ask for their feedback, and listen carefully. If there is an organizational inhibitor—policy, procedure, or practice—standing in the way of the desired improvement, removing that inhibitor is the project manager's job.

WRITTEN COMMUNICATION

Project managers can expect to be required to prepare written reports of the team's progress for different stakeholders. The following techniques will help project managers prepare written reports that are concise but comprehensive, readable, and understandable:

- **Begin by identifying the audience(s) for the report.** A report is written for a specific audience. That audience might be the engineering and technology firm's higher management team, the customer, the project team, or local government officials. The audience determines the approach that should be used for preparing the report, the language that can be used, the format, and the point of view.
- **Choose a format that will allow the report to be concise but comprehensive.** The key to making a report comprehensive enough to cover all necessary information, but concise enough to be read, understood, and remembered is to choose the right format. A good format for project reports is to present the information to be reported in categories such as the following: (1) results to date (how much of the project work has been competed at present), (2) comparison of scheduled progress versus actual progress, (3) comparison of budget projections versus actual expenditures, and (4) explanation of problems and proposed solutions.
- **Use graphics where appropriate.** It is true that a picture can be worth a thousand words, especially if it is a well-developed graphic. If information can be displayed on a simple graph or chart rather than using several written paragraphs of text, project managers should do so.
- **Use language that is appropriate for the audience.** Determine how much technical terminology can be used in a report by considering the makeup of the audience. Project reports are provided to different stakeholders. Consequently, they should be

written in language that can be understood by the stakeholder in question. The same report modified for the audience in question can be an effective way to solve the language problem.

- **Highlight actions that need to be taken by readers of the report.** If a project report contains actions that must be taken by the stakeholder reading it, these actions should be highlighted to call the stakeholder's attention to them. Two effective ways to accomplish this include: (1) by having the actions in question typed in boldface in the report or (2) by adding an "actions needed" component at the end of the report that contains the list of required actions and who is responsible for each.

- **Keep reports as short as possible.** As a rule, people are inhibited by lengthy reports. Add to this that they are busy and the need to keep reports short becomes apparent. People are more likely to read short reports than long. Historians claim that during World War II, in spite of the complexity and scope of the issues he had to deal with President Franklin D. Roosevelt would not read a report of more than one page. Project reports will not always fit into a one-page format, but they should be kept as short as possible.

Communication is an imperfect, but essential process. Without effective communication, members of project teams cannot do their part to complete projects on time, within budget, and according to specifications. Consequently, investing the time and effort necessary to become an effective communicator is a worthwhile endeavor for project managers. Few things will serve project managers better than learning to: (1) be effective listeners, (2) communicate well verbally, (3) understand and make use of nonverbal communication, (4) provide constructive criticism that is helpful and tactful, and (5) develop written project reports that are concise, comprehensive, readable, and understandable.

INFLUENCING AND NEGOTIATING IN PROJECT MANAGEMENT

Two important ways that project managers use their communications skills are in influencing others and in negotiating on behalf of their project. The best project managers are influential and they are good negotiators. They have to be. Every project has a number of different stakeholders, all of whom see the project from their own self-interested perspectives. The project manager needs the cooperation and support of all these stakeholders. Getting cooperation and support from stakeholders can require influence and the application of negotiating skills. In fact, being an effective negotiator will enhance a project manager's influence.

To influence stakeholders, project managers apply some of the various people skills explained in Part Two of this book (i.e., leading by example, communicating effectively, managing time well, managing adversity, and managing diversity). Stakeholders can also be influenced through effective negotiations. Negotiating is a skill that must be learned and can be learned. Like any skill, the more it is practiced the better one becomes at it. The word "become" is important here because being good negotiators is not something that people *are,* it's something they *become* through hard work and persistent practice.

NEGOTIATION DEFINED

It is important for students and professionals who want to be project managers to understand what *negotiation* means before beginning to develop negotiating skills. This is because people tend to misconstrue the concept to mean Party A cleverly out maneuvering

Party B in ways that give him all the value while leaving Party B empty-handed and wondering what happened. This view of negotiating is both inaccurate and shortsighted. A better way to define negotiation is as follows:

> Two or more parties working together to reach a mutually beneficial agreement that serves them well in the present and leaves the door open for future agreements between them.

This is a simple definition, but it is loaded with meaning. The words "working together" are used to make it clear that the two sides in a business negotiation should not be enemies, adversaries, or even opponents. Rather, they should be partners. The reason for this is found in the "mutually beneficial" element of the definition. For example, a project manager for a large engineering and technology firm might need to negotiate with one of his colleagues to *borrow* a certain highly skilled employee for one of his projects. To make the agreement mutually beneficial, he offers to help his colleague solve a problem he is currently facing. In this way, the eventual agreement is mutually beneficial and it leaves the door open to future agreements. Project managers who trick or coerce others into one-sided agreements put future agreements at risk. Nobody wants to deal with a dishonest or one-sided negotiator.

An agreement negotiated today between an engineering and technology firm and a customer should lay the groundwork for future agreements with the same parties. Mutually beneficial relationships and repeat business are critical to engineering firms that operate in a competitive environment. Negotiating lopsided agreements is a sure way to undermine future agreements with the *victim* of the one-sided negotiation. Consequently, negotiating for a lopsided result can mean that an engineering firm "wins" in the short run, but loses in the long run. When a negotiation is concluded, there should be no victims. This is true whether the negotiation is between an engineering and technology firm and a customer or project managers and personnel within their organizations.

CHARACTERISTICS OF EFFECTIVE NEGOTIATORS

Good negotiators are not slick dealing, domineering types who out maneuver their witless opponents; nor are they indecisive, submissive, unimaginative types afraid to advocate for their point of view. Rather, good negotiators are typically patient, fair, well informed, cooperative, innovative, imaginative, and intuitive. These are the characteristics that are most likely to lead to agreements that are fair, balanced, and mutually beneficial.

Project managers who develop these characteristics will typically display the following characteristics or behaviors during negotiations:

- Quick to see through the fog of debate to the heart of a matter.
- Solve problems in real time before they can derail the process.
- Think clearly and quickly under pressure.
- Depersonalize comments that are made and see through emotionally charged language to the issues in question.
- Listen carefully and with patience.
- Approach the process with an open-mind knowing there is always more than one way to achieve a desired result.

- Develop alternative options and solutions quickly and on the spot.
- Watch for verbal and nonverbal cues and use them to assess people.
- Think critically (i.e., recognize assumptions, rationalizations, justifications, and biased information that are presented as facts).
- Take the long-term/repeat business perspective.
- Give themselves and the party they are negotiating with room to maneuver (i.e., they avoid boxing themselves and others in).
- Maintain a sense of humor and a positive attitude throughout the negotiating process.
- Consider issues from the other side's point of view as well as their own.
- Understand that timing is important to success in negotiating.
- Prepare, prepare, prepare.

PREPARATION AND SUCCESSFUL NEGOTIATIONS

The behaviors explained in the previous section are all important to successful negotiations. But none is more important than the last one—preparation. Going into a negotiation unprepared is a sure way to guarantee a bad result. The following questions will help ensure that project managers are well prepared for negotiations with customers, subcontractors, colleagues, higher management, government officials, and other stakeholders:

- What do we want out of this negotiation? What does the other party want?
- What are we willing to give up in the negotiation in order to get what we want? What is the other party willing to give up?
- What is at risk here for us and for the other party? In other words, if we cannot reach an agreement, what do we lose and what does the other party lose?
- How much do we know about the other side and their needs? How much do they know about us?
- Do we have any "hot-button" issues? Does the other party have "hot-button" issues?
- What don't we know about the other party, and who can help us learn what we don't know?
- Are there factors that might affect the outcome that we or the other party have no control over? What are those factors?
- What is our bottom-line—at what point do we just walk away? What is the other party's bottom-line?
- What are the easy issues we can use to generate early agreement? What are the other party's easy issues?

Good preparation is essential to a successful negotiation. Before beginning a negotiation—whether with customers, subcontractors, or internal personnel—project managers must know what they want and what they do not want, what they can accept and what they cannot accept, what they are willing to give up in order to get what they want and what they are not willing to give up. It is also important for project managers to understand that there is more than one way to get what they want.

The story of Janice illustrates this last point. Janice was a project manager for a large engineering and technology firm. As the result of a divorce, she was a single mother. Child care costs were making it difficult for Janice to make ends meet. Consequently, she arranged a meeting with her supervisor to ask for a raise. The supervisor respected Janice, admired her work, and was sympathetic to her plight. Janice was an excellent project manager, and the supervisor wanted to help her. Unfortunately, his hands were tied when it came to giving Janice a raise.

The company they worked for had a salary schedule that paid project managers based on their level of education, years of experience, and performance. Janice's performance was excellent. Consequently, she would certainly receive a raise, but not for about a year. At the time in question, Janice had just been promoted to a new level, meaning she had recently gotten a raise. The company's policy was that an employee had to perform well at a new level for at least one full year before receiving the next raise.

The supervisor told Janice that if she could hang on for another 11 months, she was virtually guaranteed the raise she wanted. However, Janice was adamant that she needed another raise immediately. In an attempt to solve the problem without violating company policy on raises and promotions, the supervisor offered Janice a counter-proposal. The company had a new program that would help pay the child care costs of employees who qualified for it. If Janice would accept this program instead of a raise, her child care costs would be reduced by an amount equal to almost 75 percent of the raise she was demanding. But Janice was so intensely focused on the raise that she failed to see compensation for child care costs as an equally viable solution. Out of anger and frustration, Janice quit.

Her resignation was a problem for the company. Janice was managing an important project at the time she quit. But the resignation turned out to be an even bigger problem for Janice. Having given up her only source of income and needing a job quickly, Janice was compelled to accept a lower-paying job. She had received better job offers, but all of them required her to relocate to another state—something she could not do. Janice needed a local job because her divorce decree required that she live within 75 miles of her ex-husband to accommodate his visitation rights.

This is an example of why it is so important for project managers to understand what they really need out of a negotiation. Janice thought she needed a raise when, in fact, what she really needed was financial relieve. As it turned out, accepting the offer of a company-provided child care subsidy would have benefited Janice almost as much as the raise she demanded after tax considerations were factored in.

CONDUCTING NEGOTIATIONS

Once project managers have prepared to negotiate, there are strategies they can use to help the process work better. These strategies fall into the following broad categories: (1) negotiate in stages and consider timing, location, and image, (2) create favorable momentum, and (3) observe certain behaviors during negotiations. Strategies in each of these categories are presented in this section.

Negotiate in Stages

People who are unskilled in negotiating typically want to jump right in, make an agreement, and quickly close the deal. Closing the deal is certainly an important part of the process, but it is the third stage in a process that has three distinct stages. A good rule of thumb is to go through each stage, even if a given stage takes only a short time. This is because the first stage sets up the second and the second sets up the third. Project managers who skip a stage are likely to find themselves going backwards in the process just when they thought they were done.

The author created the *bridge analogy* to describe the stages in a negotiation. At the beginning of the process a river runs between the two parties to a negotiation that keeps them apart. The project manager is on one bank and the other party is on the opposite bank. In order to get together, they must build a bridge. Stage 1 in the process involves building the foundation. Stage 2 involves building the skeletal structure for the bridge. Stage 3 involves adding the finishing touches. Once Stage 3 has been completed, the project manager and the other party can come together at the center of the bridge.

Based on this analogy, the three stages in the negotiating process can be labeled as follows: Stage 1—Building the foundation; Stage 2—Erecting the structure; and Stage 3—Completing the bridge. What occurs in each of these stages is explained in the following paragraphs:

STAGE 1—BUILDING THE FOUNDATION.   Building the foundation involves laying the groundwork for the negotiating process. This is the stage in which project managers convince the other party to listen to their propositions by explaining how a successful negotiation will have mutual benefits. If this stage cannot be successfully completed, there is no reason to go to the next. Both parties must agree that there is value in negotiating before proceeding with the remainder of the process. For this reason, it is important for project managers to practice explaining the need for a negotiation from the perspective of how the other party will benefit.

STAGE 2—ERECTING THE STRUCTURE.   Once it is apparent that the other party accepts the need to negotiate, the project manager can get into the specifics. The specifics include coming to an understanding of the expectations of both parties, an explanation of how both parties will benefit from what is being proposed, and specifics concerning the needs and concerns of both sides.

For example, when buying a car, this is the point in the process where the buyer explains the details of what she wants (e.g., color, features, size) and the seller explains what is available. This is not the step in which price is discussed. When both parties have agreed on expectations, needs, concerns, and potential benefits, then they can move to the final stage.

STAGE 3—COMPLETING THE BRIDGE.   At this point both parties have agreed that they would like to come to an agreement and both have explained their needs and expectations. The final stage—the one most often associated with negotiating—involves bargaining over the specifics of the agreement such as price, deadlines, specifications, and other negotiable factors. This step will be much easier to conduct if a thorough job has been done in stages 1 and 2.

Create Favorable Momentum

Baseball teams try to score at least one run in their first time at bat. Football teams try to score a touchdown on their first possession. Tennis players try to win the first game or break the other player's serve during the first game. In all of these examples, the goal is to create favorable momentum. Momentum is the impetus or tendency of something to go in a certain direction. If project managers can get negotiations moving in the right direction from the outset, they will tend to keep moving in the right direction the rest of the way. For project managers, it is wise to create favorable momentum from the outset.

Additional Strategies for Effective Negotiating

Once a negotiation has commenced and is proceeding in the right direction, it is important to keep it going in the right direction. Momentum gained can be quickly lost if participants fail to do what is necessary to keep the ball rolling. The following strategies can be used during negotiations to keep things moving in the right direction:

- ***Think critically.*** Do not confuse facts with opinions or issues with positions. Tactfully insist on facts to back opinions and be quick to point out that issues are not positions. Issues can be resolved. Positions are negotiated.

- ***Listen to what is said and what is not said.*** Unfortunately, the other party in a negotiation might not begin the process as a partner. Consequently, project managers must be prepared to bring the other party along and turn that individual or team into a partner. To do this, project managers should listen attentively not just to what is said, but also to what is not said during negotiations. If the other party appears to be holding back, withholding information, or putting a certain "spin" on proposals and counter proposals, this could be evidence of a hidden agenda. Do not be afraid to tactfully say, "Something seems to be missing in the discussion. Can you clarify?" It might take a while to earn sufficient trust to convince the other party to drop his guard and be a partner. But if he is negotiating in good faith, he will eventually come around. If the other party is not negotiating in good faith, continuing the negotiation will be counterproductive.

- ***Keep the other party's needs and hopes in mind.*** When preparing for a negotiation, determinations are made concerning what the other party wants to achieve. Project managers should keep the other party's needs and hopes in mind during the process. Before making a proposal or a counter proposal, ask how it might affect the other party's needs and hopes. Can the proposal be made in a way that will achieve the goals or a sufficient enough portion of them for both parties? Making a proposal that stomps on the needs and hopes of either party is not a good negotiating strategy.

- ***Be patient.*** Do not rush negotiations. Rushing will create suspicion in the mind of the other party. Be patient. Give things time to develop. Remember, this principle can be applied only if the negotiation has been scheduled early enough that there is no need to rush. Consequently, keep scheduling in mind when arranging the timing of negotiations.

- ***Ignore personal comments.*** Negotiations can become heated. Occasionally the other party might make a comment that is offensive. When this happens, project managers should simply ignore the negative comments. Project managers should practice being objective and refusing to take things personally. It is likely in these cases that the other party is just a poor negotiator and does not know how to make proposals or counter proposals without getting personal. However, it could be that the other party is purposefully trying to agitate as a way to gain an advantage. Project managers should stay calm, depersonalize, and stay focused. Negotiators who use personal remarks to gain advantage are trying to break the other team's focus. Once they see that their tactic is not working, they will drop it.

- ***Leave room to maneuver.*** Avoid stating bottom line positions. Stating bottom line positions can paint a negotiating team into a corner, and make its members look foolish if, after stating a bottom line position, they are forced to change positions. There are usually a number of different ways to solve the same problem or meet the same need. Although project managers attempt to anticipate these various ways to meet needs, it is not possible to anticipate all problems and corresponding solutions. Consequently, it is wise to remain open to a better idea the other party might propose.

AFTER AGREEING—FOLLOW THROUGH

The negotiation process does not end once an agreement is made. As soon as one negotiation is concluded, it is time to begin paving the way for future agreements with the same party. Remember, negotiations have more than one purpose. The first, of course, is to arrive at an agreement concerning the issue in question at the moment. The second is to pave the way for future agreements with the other party. For example, assume that a project manager found it necessary to negotiate with a colleague to have a certain employee serve on his project team.

This is a situation that will arise from time to time in most engineering and technology firms. Consequently, it is important to leave the door open for future negotiations with this colleague. The best way to leave the door open for future negotiations, assuming a win-win agreement has been concluded, is to follow through and do what was agreed to. The best negotiators refuse to make promises they cannot keep and always follow through on the promises they make.

Whatever a project manager promises to do during a negotiation must be done both properly and on time. If unanticipated problems arise, the project manager should make the other party aware of them immediately. The trust and credibility developed during the negotiation will now be either reinforced or lost based on how well the project manager performs and keeps his promises. Hence, project managers should stay in touch with the other party, keep him informed, solicit feedback, and always be available and responsive to him. The better the project manager performs on the current agreement, the easier the negotiations will be for future agreements. Following through on agreements that were negotiated is important regardless of whether the negotiations are with customers, team members, colleagues, subcontractors, or higher management.

Project Management Scenario 14.2

Your problem is that you have no tact

John has a strong grasp of the process aspects of project management. He is an excellent scheduler and does a good job of monitoring and controlling. But when it comes to the people aspects of project management, he has some glaring weaknesses. One of them is in providing constructive criticism to team members. John can certainly criticize, but his criticism is seldom constructive. In fact, it is usually downright offensive—something John seems oblivious to. As a result, John's team members often resent his attempts to improve their performance or correct their behavior. The resentment manifests itself in a variety of different ways, but all of them are negative.

In an attempt to learn why his team members always seem to react so negatively to his "constructive criticism," John approached a fellow project manager, one who had been a team member on several of John's teams. His colleague was blunt: "John, I would rather take a beating than get constructive criticism from you. Your problem is that you have no tact." After learning how it felt to receive a dose of his own medicine, John asked his colleague how he could improve.

Discussion Questions

In this scenario, John is oblivious to the damage done by his tactless attempts at constructive criticism. Have you ever worked with a person who was tactless when dealing with others? If so, how was this person received by others? If you were John's colleague in this scenario, what advice would you give him concerning how to improve?

SUMMARY

Because project managers are responsible for keeping different individuals and constituent groups up-to-date concerning progress and problems, it is important for them to develop a communication plan for each project they manage. A communication plan identifies the individuals and constituent groups that need to be kept informed, the types of communication that will be used to keep them informed, the frequency of communication, the content of each type of communication, and the individual responsible for collecting and providing the information for each type of communication.

Communication is the transfer of information that is received and fully understood from one source to another. Communication is a process that has the following components: sender, receiver, method, medium, and the message. There are four basic types of communication: verbal, nonverbal, written, and graphic. Effective communication occurs when the information that is received and understood is accepted and acted on in the desired manner. Communication can be inhibited by a number of factors, including differences in meaning, insufficient trust, information overload, interference, condescending tones, listening problems, premature judgments, inaccurate assumptions, and technological glitches.

Listening means receiving a message, correctly decoding it, and accurately perceiving what is meant by it. Inhibitors of effective listening include a lack of concentration, preconceived notions, thinking ahead, interruptions, and tuning out. The five-minute rule allows project managers to maintain an open-door policy for listening to their team member's complaints, suggestions, recommendations, and problems. It means that, within reason, team members can have five minutes of the project manager's time at any time to discuss a problem. However, during the five minutes the team member must convey the information about the problem and recommend a viable solution.

Nonverbal communication consists of body factors, voice factors, and proximity factors. The key to understanding nonverbal communication is to look for agreement or disagreement between what is said verbally and what is said nonverbally. To improve the quality of their verbal communication, project managers can show interest in their topic, maintain a positive attitude, be flexible in making their points, use tact in delivering the message, and be courteous. When communicating corrective feedback, project managers should be positive, prepared, and realistic.

Written communication can be improved by: (1) identifying the audience before writing reports, (2) choosing a format that will allow the report to be concise but comprehensive, (3) using graphics wherever it is appropriate, (4) using language that is appropriate for the audience, and (5) highlighting actions that need to be taken by readers of the report.

Project managers must be good negotiators. Some of the factors necessary for completing a project in on time, within budget, and according to specifications must be negotiated.

Important considerations when negotiating include the following: (1) negotiate in stages and consider timing, location, and image, (2) create favorable momentum, and (3) observe certain behaviors during negotiations.

During the negotiating process, project managers should think critically, keep the other party's hopes and needs in mind, be patient, ignore personal comments, and leave room to maneuver. After an agreement has been reached, project managers should follow through and do what they agreed to do.

KEY TERMS AND CONCEPTS

Communication plan
Communication
Sender
Receiver
Method
Medium
Message
Effective communication
Differences in meaning
Insufficient trust
Information overload
Interference
Condescending tones
Listening problems
Premature judgments
Inaccurate assumptions
Technological glitches
Listening
Lack of concentration
Preconceived notions
Thinking ahead
Interruptions
Tuning out

Five-minute rule
Nonverbal communication
Body factors
Voice factors
Proximity factors
Verbal communication
Interest
Attitude
Flexibility
Tact
Courtesy
Corrective feedback
Written communication
Negotiate in stages
Bridge analogy
Building the foundation
Building the structure
Completing the bridge
Think critically
Be patient
Ignore personal comments
Follow through

REVIEW QUESTIONS

1. What is a communication plan, why is one needed, and what does one contain?
2. Define the term "communication."
3. What are the components of the communication process?
4. List the four basic categories of communication methods.
5. Distinguish between communication and effective communication.
6. List and briefly explain the factors that can inhibit communication.
7. Define the term "listening."
8. List and briefly explain the inhibitors of effective listening.
9. Explain the five-minute rule. Why is it needed and how does it work?
10. Explain briefly the three components of nonverbal communication.

11. List and briefly explain the factors that will help project managers improve their verbal communication.

12. How can project managers ensure that their corrective feedback does not make team members defensive?

13. Briefly explain how project managers can ensure that their written reports are concise, comprehensive, readable, and understandable.

14. Explain the stages in the negotiating process.

15. Summarize the bridge analogy and explain how it applies to negotiating.

16. Explain how the following factors can affect a negotiation: timing, location, and image.

17. What is momentum and how can a project manager create it during the negotiating process?

18. List and explain five strategies that can be helpful during the negotiating process.

19. How should a project manager handle the following situation during a negotiation: The other party begins to make negative personal comments?

20. Why is it important to follow through after negotiating an agreement?

APPLICATION ACTIVITIES

The following activities may be completed by individual students or by students working in groups:

1. Assume that you are a new project manager. You will be required to keep higher management, your project team members, and the customer fully informed about progress and problems. Develop a communication plan for your project.

2. For a week, observe people who are having conversations. Write down the various forms of nonverbal communication exhibited by people in these conversations. Make note of how people use nonverbal cues to emphasize points, convey agreement, convey disagreement, and so on.

3. Assume that you are a project manager who has to give corrective feedback to a team member who is coming to work late on a regular basis.

Write down what you will say to this team member and how you will say it. If working in a group, pair off and complete this activity verbally.

4. Assume that you are a project manager for a major project. After the contract has been signed and the project is underway, the customer calls and says he needs to negotiate some changes to the contract including a new deadline for completion. You have been tasked with putting together your firm's negotiating team. The members of your team all have the necessary discipline-specific knowledge to participate, but they have never been involved in a negotiation. Therefore, your first task is to develop a brief seminar to teach them how to play a positive role during the negotiation. Develop that seminar.

CHAPTER 15

Project Managers and Personal Time Management

Time is always of the essence in engineering and technology projects. Consequently, time management is an important project management concept. Usually, when the term *time management* is used in the context of project management, it refers to managing the time scheduled for the project. Effectively managing a project's schedule is essential to success in project management, but there is another kind of time management that is also important for project managers: managing their own time.

Chapter Four explained how to develop a schedule for a project so that the time devoted to the project can be managed. Chapter Ten explained how to monitor and control the work of the project to keep it on schedule and complete it on time. This chapter explains how project managers can manage their own time and help team members better manage theirs. This is important because project managers who cannot manage their own time cannot manage the project's time, or anything else for that matter. In fact, project managers who fail to effectively manage their own time are not likely to effectively manage project time. This, in turn, means that their teams are not likely to complete projects on time, within budget, and according to specifications.

Time management on the part of project managers and their team members affects all three of the basic success criteria for projects: time, cost, and quality. Poor time management can cause a number of problems, including wasted time, added stress, lost credibility, missed deadlines, poor follow through on commitments, inattention to detail, ineffective execution, and poor stewardship. Further, project managers who are poor time managers will not be able to help their team members be good time managers. Clearly, effective time management is an important skill for project managers.

When a project team becomes rushed because of poorly managed time, its members often respond by cutting corners on their work. When the time of project teams is not managed well, its members cannot meet their deadlines. Finally, when project teams become rushed because of poorly managed time, additional resources in the form of time, personnel, and technology have to be added to the project in order to get it completed on schedule. Resources cost money. Hence, adding them to a project as the result of poorly managed time just adds to the cost of a project.

POOR TIME MANAGEMENT AND TEAM PERFORMANCE

The following scenario shows how poor time management can undermine the performance of a project team. David is a project manager for his engineering and technology firm. Although he has excellent credentials, David's projects never seem to meet their deadlines or stay within budget. The main reason for the mediocre performance on his projects is poor time management. As is often the case, when the leader of a team is a poor time manager, David's poor time management snowballs making it difficult for other team members to manage their time well and causing various other problems.

Because he is always running late, David never gets to spend much one-on-one time with his team members. He claims to have an open-door policy, but in reality David is always so busy trying to catch up that he has no time to listen when his team members have problems, recommendations, or concerns. As a result, he typically just ignores their requests for face-to-face meetings. His most frequent response when team members need a few minutes of his time is: "I'll get with you later. I don't have time right now."

Unfortunately for David, complaints that are ignored often become problems that cannot be ignored. The longer they are ignored, the bigger the problems become. David and his team suffer from the snowball effect all the time, which means that the engineering and technology firm suffers from it too. Poor time management is a major inhibitor when it comes to completing projects on time, within budget, and according to specifications.

COMMON TIME MANAGEMENT PROBLEMS AND THEIR SOLUTIONS

Project managers often face a common situation. They prepare their "to-do" lists for the day, but never get to even the first item on the list. Instead, they spend their day dealing with one crisis after another. The concept is known as *putting out fires*. By the time they have put out all the fires that have popped up, the work day is over and their to-do list remains untouched. For many project managers, this type of day is not just common, it's typical. The reasons some project managers spend too much time putting out fires and too little time working on planned activities vary, but most of them are predictable.

The most common causes of time management problems for project managers are unexpected crises (fires that must be put out), telephone calls, taking on too much (failing to say "no" when appropriate), unscheduled visitors, poor delegation, disorganization, and inefficient meetings (see Figure 15.1). Project managers must learn how to deal effectively with these common time wasters.

Reducing the Number of Crisis Situations

Crisis situations are a part of the job for project managers. Critical materials are not delivered on time, an important employee leaves the firm in the middle of the project, machines break down, software develops bugs, or a unionized workforce goes on strike. These types of crisis situations happen all the time on engineering and technology projects. Even the best project managers will never completely eliminate crisis situations—there are just too many

---

**Checklist of Common Causes of
TIME MANAGEMENT PROBLEMS**

✓   Unexpected crises

✓   Telephone calls

✓   Taking on too much—failing to say "no"

✓   Unscheduled visitors

✓   Poor delegation

✓   Disorganization

✓   Inefficient meetings

FIGURE 15.1   Project managers must learn how to overcome these time wasters.

causal factors that cannot be controlled. However, there is a correlation between planning and the number of crisis situations that must be dealt with. This correlation can be stated as follows: *the better the planning the fewer the crises*. This is the good news. The bad news is that even with good planning, crisis situations will still occasionally occur.

A crisis, by definition, is a situation that must be dealt with right away. Consequently, when crises occur they take precedence over other obligations. This is why it is important to limit the number of crises and, in turn, the amount of time devoted to dealing with them. The following strategies can help project managers minimize the amount of time they must devote to putting out fires (Figure 15.2):

- **End each day by planning the next.** One of the best ways to prevent crisis situations is through good planning. By planning, project managers can take control of their days rather than letting their days control them. Part of being an effective project manager is developing a plan and a schedule for completing projects on time. This is the big picture plan. Wise project managers also develop small picture plans. The process of developing small picture plans can be as simple as developing a "to-do" list for tomorrow before leaving the office today or a schedule for the upcoming week. For daily to-do lists, an effective approach is to list everything that is planned for tomorrow down one side of a sheet of paper in time order (e.g., 8:00 A.M.—Task 1, 9:30 A.M.—Task 2). On the other side of the sheet of paper list the same tasks but

---

**Strategies for Minimizing Time Devoted to
PUTTING OUT FIRES**

- End each day by planning the next

- Schedule loosely

- Avoid being an amateur psychologist

- Remember that most tasks take longer to complete than planned

---

FIGURE 15.2   Putting out fires is a common time waster for project managers.

in priority order. The latter step is done to accommodate the fact that even with the best planning, emergencies can still occur that throw the project manager's planned schedule into disarray. When this happens, it is important to know which tasks on the daily "to-do" list are the most important since the emergency might leave only limited time in the day to do other things. Spending the last 15 minutes of each day planning for the next will not eliminate emergencies, but it will benefit project managers in two important ways by: (1) allowing them to adjust more efficiently and effectively to emergencies and (2) preventing the creation of emergencies that occur because the project manager forgot something he was supposed to do.

- **Schedule loosely.** Anyone who has been to a doctor's office is familiar with the concept of overbooking. Physicians tend to overbook. They schedule appointments so tightly that more often than not by mid-morning their work is hopelessly backed up. Too often, the doctor's appointment log is more of a dream sheet than a realistic schedule. Physicians who schedule too tightly are guilty of ignoring the management adage that *most things take longer than you think they will.* All it takes is an unexpected turn of events with just one patient, and the rest of the day is thrown off schedule. Project managers are often like physicians in that they schedule more appointments and activities in a day than they can realistically handle—a practice that creates crises. The solution to this problem is to schedule loosely. This means scheduling more time than you think appointments will require and, then, trying to complete the appointments on schedule. If an appointment should take 15 minutes, schedule 20 or even 30 minutes for it. Also, if someone says, "I just need a few minutes of your time," do not take this literally. People rarely take just a few minutes. Schedule 10 or 15 minutes. Scheduling loosely also means leaving catch-up time in between appointments. Often, there is follow-up work that needs to be done at the conclusion of an appointment. Leave sufficient time between appointments to get this work done immediately. Follow-up work that is left undone will add up throughout the day and become tomorrow's crisis.

- **Avoid being an amateur psychologist.** Project managers must be open to listening when employees have ideas, concerns, recommendations, or complaints relating to their jobs. Allowing project managers to maintain an open-door policy is the purpose of the five-minute rule that was explained in Chapter Fourteen. However, it is not uncommon for members of project teams to ask for time to discuss personal problems—problems that are not work-related. Few things can rob project managers of productive time faster than the personal problems of their team members. Project managers who allow themselves to get bogged down in the personal problems of team members typically end up devoting too much of their time to personal counseling and too little to project management. The predictable result of this imbalance is that the project falls behind schedule and crises begin to occur. Consequently, when team members bring their personal problems to project managers, the appropriate response is for the project manager to make a helpful referral. Making a referral is appropriate, but it is as far as the project manager should go in such cases, especially during work hours. It is also appropriate to tell team members that you will hear their personal problems after work. Project managers should avoid the temptation to play amateur psychologist when team members want to discuss personal problems. There are several good reasons for this and some are listed as follows: (1) project managers are not qualified to give advice on personal problems, (2) providing personal counseling is not in the project manager's job description, (3) if the project manager's advice exacerbates the problem in question, the team member will blame him, and (4) playing amateur psychologist is a time-consuming exercise, and time is a precious asset to project managers. Problems that relate to the job should receive the full and immediate attention of project managers, but the personal problems of team members should be referred to human resource professionals who are better equipped to deal with them.

- *Remember that most tasks take longer to complete than planned.* This is a good rule of thumb to follow. No matter what has been planned, experience shows it will probably take longer than expected. Consequently, it is wise to build a little extra time into your schedule. For example, if a task should take 30 minutes, allow 45 minutes, and then try to finish in 30 minutes. In this way there will be extra time, if it is needed, without having to rush. If the task is completed in less time than is allotted, the extra time gained can be put to good use returning telephone calls or getting a head start on other obligations.

Project Management Scenario 15.1

All I ever do is put out fires

Sherry Jackson is ready to pull her hair out. She has an excellent plan for completing the Tele-Tec Project on time, within budget, and according to specifications. She also has an excellent project team to work with. The problem is that she never seems to be able to get around to executing the plan or to get her team members engaged. Instead she spends all of her time putting out fires. Jackson has all the qualifications to become a good project manager except one: She is a poor time manager. She does not control the events in her days—they control her. In frustration, Jackson told a colleague: "I can't get this project going because all I ever do is put out fires."

Discussion Questions

In this scenario, Sherry Jackson is frustrated because she never seems to be able to get things moving in the right direction. When it comes to project management, Jackson may be her own worst enemy because she is a poor time manager. Have you ever worked with someone in any setting who was a poor time manager? How did poor time management affect this individual's work? What advice would you give Jackson about how to reduce the number of crises she has to deal with everyday?

Making the Telephone a More Efficient Tool

The telephone can be a time saving device for project managers, but it can also be a time waster. Spending time on hold, listening to irritating messages on answering machines, playing telephone tag, and engaging in dialogue unrelated to the purpose of the call are all common time wasters for project managers.

Cellular phones and other handheld communication devices with their almost infinite list of applications have tended to compound the problem. On the other hand, if used wisely handheld communication devices can be time savers rather than time wasters. The following strategies will help project managers minimize the amount of time they waste on the telephone (Figure 15.3):

**Strategies for More Efficient Use of
THE TELEPHONE**

- Use e-mail instead of the telephone whenever appropriate.

- Categorize calls as important, routine, and unimportant.

- Use a cellular phone to return calls between meetings and during breaks.

- Block out time on the calendar for returning telephone calls.

- Limit unrelated dialogue during telephone calls.

FIGURE 15.3   The telephone can be a time saver or time waster.

- *Use e-mail instead of the telephone whenever appropriate.* One of the best ways to avoid wasting time on hold, playing telephone tag, or listening to recorded messages is to use e-mail instead of the telephone whenever appropriate. Of course, e-mail is not always an appropriate option. However, when it is project managers can simply type a brief message, click "Send," and move on to their next task. With e-mail there is no pressing one for this option or two for that option, no talking to answering machines, and no being put on hold. In addition, people are often more prompt about returning e-mail messages than they are about returning telephone calls or responding to telephone messages.

- *Categorize calls as important, routine, and unimportant.* Project managers will find that time invested in helping administrative assistants learn to distinguish between important and unimportant telephone calls and between important and routine calls will be time well spent. One of the ways to do this is for project managers to provide administrative assistants with a list of people they always want to talk to. Within reason, these calls should always go through. On others, the administrative assistant can take a message, let the caller leave a recorded message, or even suggest sending an e-mail message. If administrative assistants are going to take written telephone messages, train them to be comprehensive and detailed. In addition to who called and when, a good telephone message will contain the caller's telephone number, reason for calling, and a good time to return the call. This final item—a good time to return the call—is important because it will help project managers avoid the time wasted playing telephone tag.

- *Use a cellular phone to return calls between meetings and during breaks.* On the one hand, cellular phones can be obnoxiously intrusive. Being interrupted by the untimely ringing of a cellular phone during a conversation, in a restaurant, during a meeting, in a movie, or even during a funeral has become a common annoyance. On the other hand, cellular phones can help project managers turn time that might otherwise be wasted into productive time. Project managers can save valuable time by taking telephone messages to meetings and using a cellular phone to return them during breaks and between meetings. Project managers can also use cellular phones to return calls from their cars, provided they are off the road and parked. Causing an accident by trying to send text messages while driving with one's knees will ultimately cost project managers more time than the texting would have saved them.

- *Block out time on the calendar for returning telephone calls.* Telephone tag is one the project manager's most persistent and frustrating time wasters. Assume that a project manager really needs to talk with someone and e-mail is not an appropriate communication medium. She places the call, but the person in question is not available. She leaves a message. The individual in question calls back, but now the project manager is tied up in a meeting. The caller leaves a message. The project manager calls back, but just misses him. This frustrating situation is known as "telephone tag," and it repeats itself many times every day in engineering and technology firms. To minimize the amount of time wasted playing telephone tag, project managers can block out times on their calendars for returning calls and let the times be known to callers who leave messages. An effective approach is to schedule three blocks of time each day: one in mid-morning, one in mid-afternoon, and one near the end of the day. Project managers should block these times out on their calendars as if they are appointments. Then they should make sure that the administrative assistants who take their messages let callers know that these are the times during which they typically return calls. The times should also be included in recorded telephone answering messages if possible. In this way, if callers really need to talk to the project manager, they will make a point of being available during one of these times. Another version of this time-saving strategy is to send an e-mail or a text message that asks either: (1) What is a good time for me to call you today? or (2) Will you call me at (specify the time) today?
- *Limit unrelated dialogue during telephone calls.* One of the reasons telephone calls are such time wasters is the human propensity to engage in dialogue about matters unrelated to the purpose of the call. Project managers can save a surprising amount of time on the telephone by simply getting to the point, and by tactfully nudging callers to do the same. There is certainly nothing wrong with a few appropriate comments on the latest ball game, movie, or news item, but the amount of time devoted to unrelated issues should be kept to a minimum. Project managers should practice being focused, staying on task, and tactfully nudging callers do the same.

Cutting Back on Nonessential Responsibilities

Taking the initiative and seeking responsibility are traits of effective project managers. There is both good news and bad news in this fact. The good news is that taking the initiative and seeking responsibility are traits that will help engineering and technology professionals become good project managers. The bad news is that these traits can result in project managers taking on too many responsibilities. When this happens they end up suffering the effects of overload. For people who take the initiative and seek responsibility, it is easy to fall into the trap of taking on too much. This is a common problem among project managers. When this happens, the following strategies can help get things back in balance:

- *Make a list of all current and pending obligations.* Once the list is completed, ask the following questions for each item on the list: (1) What will happen if this is not done? and (2) Is there someone else who could do this? Invariably there will be things on the list that really do not have to be done. These items sounded like good ideas originally, but in retrospect they do not really need to be done or, at least, not now. In addition, going through this exercise will often reveal that there are items on the list that could and should be delegated. Asking the two questions suggested will usually reveal ways to pare down the list. Once the list has been pared down to the items that must done, the remaining items should be prioritized. This will ensure that project managers are putting their efforts into the most important items on the list.

- ***Take stock of after-work activities.*** Taking the initiative and seeking responsibility are two traits common to people who become project managers. These traits apply as much after work as they do during the work day. Consequently, project managers often get involved in professional organizations, civic clubs, and other outside activities. Participation in organizations and activities outside of work is an excellent way to grow as a leader. However, it is easy to fall into the trap of taking on too many outside responsibilities. Balance is the key. For project managers, outside activities are like food. A certain amount of the right types are essential, but too much—even of the right types—can be harmful. When project managers find themselves overextended and need to cut back on their obligations, outside activities can be a good place to start. Project managers who overburden themselves with too many outside activities run the risk of distracting themselves from obligations at work.

Limiting Unscheduled Visitors

It is important for project managers to maintain an open-door policy for their team members. On the other hand, unscheduled visitors can take up a lot of time. Consequently, project managers must find the right balance between maintaining an open-door policy and requiring visitors to have appointments. Helping project managers strike the proper balance is the purpose of the five-minute rule that was explained in Chapter Fourteen. To review, with this rule a member of the project team can have five minutes of the project manager's time at any time unless there are circumstances that make this impossible. Within the allotted five minutes team members are expected to state their problem or concern and recommend a well-considered solution. If the team member has thoroughly prepared before asking for a five-minute audience, five minutes is typically more than enough time.

If an issue cannot be handled in five minutes, team members should be required to make an appointment so the necessary time can be blocked out on the project manager's calendar. The five-minute rule, together with requiring appointments for issues that will take more than five minutes, will help alleviate the problem of people just dropping in unannounced to chat. Unfortunately, these two strategies will not eliminate the problem completely. After all, some of the people who just drop in to chat might be superiors in the organization or a customer. Consequently, it is important for project managers to understand how to minimize the amount of time taken up by drop-in visitors. The following strategies will help (Figure 15.4):

- ***Do not allow drop-in visitors during peak times.*** Some days are busier than others and some times of the day are busier than other times. During these peak times, it is best to ask drop-in visitors to come back at another time when they can receive undivided attention, unless the following situations occur: (1) they are bringing critical information that needs to be conveyed immediately, (2) they are warning of an emergency, or (3) they are someone who must be seen right away no matter what else is on the schedule.
- ***Train administrative support personnel to come to the rescue.*** Project managers can minimize the intrusions of drop-in visitors by working out an arrangement with

---

**Strategies for Reducing Time Taken By
UNSCHEDULED VISITORS**

• Do not allow drop-in visitors during peak times.

• Train administrative support personnel to come to the rescue.

• Remain standing.

FIGURE 15.4   Drop-in visitors can waste much of the project manager's time.

administrative support personnel to come to the rescue after a prearranged amount of time, for example, say five minutes. It works like this. Whenever a drop-in visitor has been in the office for five minutes or so, the secretary buzzes or looks in and says "it's time to place that important call." This will tactfully let the drop-in visitor know that he needs to make an exit.

• ***Remain standing.*** One way to convey the message that "I am busy" without having to actually say it is to remain standing when an unannounced visitor walks into the office. Once visitors sit down and get comfortable, it can be difficult to uproot them. Continuing to stand, will tactfully convey the message that, "I can give you a few minutes, but only a few."

Overcoming Poor Delegation of Work

Poor delegation is one the easiest time-wasting traps to fall into. Some project managers find it difficult to let go of work they are accustomed to doing themselves. For example, it can be difficult to be a top-performing engineer one day and a project manager the next. Yet, this sometimes happens. In addition, some project managers find it difficult to delegate work to team members who do it differently than they would. These two phenomena can result in poor delegation, a major time waster. Tasks that are not project management tasks should be delegated. If team members cannot perform the tasks delegated to them satisfactorily, the problem is one of training or staffing not delegation. Training and staffing problems cannot be solved by refusing to delegate work. Project managers who do work that should be done by team members simply because they refuse to delegate will not be effective. If a team member who should be doing certain work cannot do it properly, he should be trained, mentored, or replaced.

Overcoming Personal Disorganization

Project managers can waste a lot of time rummaging through disorganized stacks of paperwork looking for the folder, form, drawing, or document needed. The author once worked with an engineer who had the unfortunate habit of never returning files, documents, or drawings to the appropriate place when he was done with them. Wherever this engineer happened to be when he finished with paperwork is where he left it. Worse yet, he could never remember where he left it. As a result, this otherwise talented professional could be counted on to waste valuable time looking for "missing" paperwork. He eventually earned a reputation for being habitually disorganized—a reputation that hurt his career. Because getting people and work organized is an important part of what project managers do, this talented engineer never became a project manager. Project managers can use the following

Challenging Engineering and Technology Project

PROVIDING ENERGY FROM FUSION

One of the most enticing forms of energy for the future is fusion. Fusion is the energy source for the sun. Of course the sun benefits from factors not present on earth or, at least, not to the extent they are in the sun. These factors include the enormous levels of heat and gravitational pressure that combine to compress certain atoms into heavier nuclei. This process, in turn, releases energy. Although, the heat and gravitational pressure of the sun cannot be replicated on earth, there may be another answer. Lithium, the same metallic element in the battery of most laptop computers, might be the critical element that will enable human-engineered fusion on a scale large enough to provide energy that is economical and environmentally friendly.

The challenges to providing energy from fusion are many. The fusion reaction process produces neutrons which, in turn, convert atoms into radioactive forms. Consequently, procedures will have to be developed for: 1) confining the radioactivity produced by neutrons, and 2) preventing accidental releases of the tritium fuel used in the process. Another major challenge will be developing materials that can withstand the structural weakening the neutrons will cause during the fusion process while still retaining the ability to extract the heat needed to create energy. Just building full-scale fusion-generating facilities will require developing better super-conducting magnets and advanced vacuum systems—both major technical challenges.

Producing energy from fusion is a major challenge for engineering and technology professionals, but it is a challenge with enormous potential for providing an almost limitless source of energy for commercial, industrial, and residential applications. As the problems associated with fossil fuels continue to worsen, providing energy from fusion will become even more important. *Before proceeding with the remainder of this chapter, stop here and consider how providing energy from fusion will require all of the process and people skills of project management.*

Source: Based on *National Academy of Engineering.* http://www.engineeringchallenges.org

strategies to help decrease the amount of time they waste because of personal disorganization (Figure 15.5):

- *Periodically clean off desks and workstations.* This strategy sounds so simple that one might be tempted to ignore it. But before doing so, project managers should look at their work areas, check their in-baskets, and go through their stacks of pending work. The result of this examination of the work area is often that old and outdated paperwork is found cluttering up desks and filling up in-boxes. Finding irrelevant

Strategies for
GETTING ORGANIZED

- Periodically clean off desks and workstations.
- Restack your work in priority order.
- Categorize work folders (e.g., read folder, correspondence folder, signature folder).

FIGURE 15.5   A lack of organization will cause project managers to waste time.

material in pending-work stacks is also common. Consequently, project managers should periodically go through everything on their desks and in their work areas and get rid of anything that is no longer pertinent. When a project manager decides to get organized, one of her best organizational tools is a large trash can.

- ***Restack work in priority order.*** Go through the in-box or stack of pending work and reorganize everything in order of priority. Paperwork is often stacked in the order it comes in, especially when project managers are in a hurry and do not have time to organize it. Because this can happen so frequently, it is a good idea to occasionally stop working long enough to go through the work stack and reorganize everything by priority. It's an even better idea to screen work as it comes in, placing all work in priority order from the outset.

- ***Categorize work folders.*** Organize paperwork in folders by category. Have a *Read Folder* for paperwork that should be read, but requires no writing or other action. Have a *Correspondence Folder* which contains correspondence that requires some action. Have a *Signature Folder* for paperwork that requires a signature. Organizing work in this way will save time and increase efficiency.

Making Meetings More Efficient

In spite of their value in bringing people together for the purpose of conveying information, brainstorming, planning, and discussing issues, meetings can be major time wasters. Meetings waste time because some of them are not necessary in the first place while those that are necessary, in many cases, are inefficiently run. Project managers can minimize the amount of time wasted in meetings by: (1) making sure that all regularly scheduled meetings are actually necessary and (2) ensuring that necessary meetings are run as efficiently as possible. Strategies that will help minimize the time wasted by meetings include the following (Figure 15.6):

- ***Be aware of the causes of wasted time in meetings.*** Much of the time spent in meetings is wasted. The principal reasons for this are poor preparation, the human need for social interaction, time of participants spent in conversation unrelated to the meeting's purpose, interruptions, getting side-tracked on unrelated issues, no agenda, and no prior distribution of backup materials. In addition to these time wasters, there is also the *comfort factor*. Coffee, soda, tea, water, snacks, and social interaction can create a comfortable environment that people are reluctant to leave. Project managers

---

**Strategies for
MAKING MEETINGS EFFICIENT**

- Be aware of the causes of wasted time in meetings.
- Determine if regularly scheduled meetings are really necessary.
- Hold impromptu meetings while standing up.
- Complete the necessary preparations before meetings.
- Begin meetings on time, stick to the agenda, and take minutes.
- Follow-up meetings promptly.

FIGURE 15.6   Inefficient and unnecessary meetings are a major time waster.

should make a point of minimizing these time wasters from internal meetings, especially those meetings they chair. Comfort items should remain available to customers though. Minimizing the comfort factor, minimizing unrelated conversation, preparing thoroughly, and getting organized will help enhance the efficiency of meetings.

- ***Determine if regularly scheduled meetings are really necessary.*** Most organizations have regularly scheduled weekly, biweekly, and monthly meetings of various groups. Project managers often have regularly scheduled meetings of their teams. When these meetings were established they probably had a definite purpose, and that purpose was probably valid. In fact, it might still be valid. However, it is not uncommon to find people in organizations meeting only because they have always met on a given day at a given time. Sometimes, meetings are perpetuated out of habit rather than need. If project managers call or attend regularly scheduled meetings, they should ask the following questions about each meeting: (1) What is the purpose of the meeting? (2) Is the meeting really necessary? (3) Can the meeting be scheduled less often, and (4) Can the purpose of the meeting be satisfied some other way (e-mail updates, written reports, webcasts, etc.)?

- ***Make impromptu meetings stand-up meetings.*** Impromptu meetings that should last no more than 10 minutes can be kept on schedule by holding them in standing positions. These are typically meetings without an agenda called on the spur of the moment to quickly convey information to a select group or to get input from that group. These types of meetings are best held in a location other than a conference room. When meetings are held in a conference room, it is difficult to keep participants from pulling up chairs and getting comfortable. Holding a meeting standing up conveys the message that "this is going to be a brief meeting" without the project manager having to actually say so.

- ***Complete the necessary preparations before meetings.*** For sit-down, scheduled meetings, project managers should have an agenda that contains the following information: purpose of the meeting, starting and ending time, list of agenda items with the person responsible for each item noted, and a projected amount of time to be devoted to each item. Set a deadline for submitting agenda items and stick to it. Require all backup material to be provided at the same time as the corresponding agenda items. Distribute the agenda, backup material, and the minutes of the last meeting at least a full day before the meeting. Distributing meeting materials too far in advance is not advisable because participants will tend to put them aside and forget about them. In addition, distributing materials too far in advance limits the time participants have to submit agenda items and backup material. On the other hand, if project managers wait until during the meeting to distribute the agenda and backup materials they will waste time handing them out and waiting while participants read them. Better to ask participants to read the agenda and backup materials before the meeting so that they come prepared to participate in an efficient and effective manner. This is why the agenda and backup materials are distributed in advance. In other words, project managers should come to meetings well prepared and insist that other participants do the same.

- ***Begin meetings on time, stick to the agenda, and take minutes.*** An important rule of thumb for project managers who run meetings is: begin meetings on time. Waiting for latecomers to arrive only reinforces tardiness. Project managers who run meetings should remember the management adage that you get what you reinforce. Reinforce tardiness and you will get tardiness. If participants know that meetings are going to start on time, most will eventually discipline themselves to arrive on time. Project managers should assign someone to take minutes or bring an administrative assistant to meetings who can take minutes. In the minutes, all action and follow-up items should be typed in bold face so they stand out from the routine material. Make the minutes of the last meeting the first item on the agenda of the current meeting. In this way, the first action taken in each meeting will be following-up on assignments and commitments (those items that appear in boldface in the minutes) made during the last meeting. In running meetings, project managers should stay focused and insist that participants stick to the agenda and stay on task. The last agenda item should be "Around-the-Table Comments." This final item gives participants an opportunity to bring up issues that are not on the agenda without getting the meeting sidetracked before all agenda items have been discussed. Around-the-table comments should be limited to minor informational items that do not warrant a place on the agenda, require no discussion, and will take only a little time—matters such as announcements. Project managers who allow participants, during the around-the-table portion of the meeting, to discuss issues that should have been put on the agenda just encourage participants to ignore the agenda preparation process. When this happens, project managers lose control of their meetings and find themselves wasting time. Project managers should also ask participants to silence cellular phones and other handheld communication devices during meetings so they can focus on the purpose of the meeting. Interruptions from cellular phones can be a major distraction and time waster in meetings.

- ***Follow-up meetings promptly.*** Project managers should ensure that the minutes of meetings are distributed as soon as possible following meetings—ideally on the same day the meeting occurred. Once the minutes have been distributed, the next step is to allow an appropriate amount of time for participants to act. Once team members have had sufficient time to get started on actions items from the meeting, project managers should begin the follow-up process. It is unwise to wait until the next meeting to follow-up on progress made by team members in completing action items from the previous meetings.

The strategies recommended in this chapter will help project managers take control of one of their most valuable assets—time. The same strategies will also help project managers teach team members how to manage their time efficiently. When project managers and their team members manage time well, project teams are more likely to complete projects on time, within budget, and according to specifications.

Project Management Scenario 15.2

His meetings waste too much time

Edward Andrews, Vice President for Future Tech, Inc., is concerned about one of his project managers—Mack Day. It seems that nobody wants to serve on project teams led by Day. No one has said anything specific, but Andrews can tell that the company's personnel are reluctant to work with Day. To get to the bottom of the situation, Andrews asked one of his most trusted colleagues to stop by his office for a brief meeting. When the meeting convened, Andrews came right to the point. "I don't understand why so few people are willing to work with Mack Day. I need to know why. What's going on?" The response Andrews got from his colleague was frank and straightforward.

"Mack Day is world-class time waster. He is disorganized, refuses to delegate, micromanages his team members, and his meetings waste too much time. He will answer his cell phone in the middle of a conversation, get sidetracked on unrelated issues, and is constantly frantically rushing around trying to find paperwork he has misplaced. I don't see how Mack manages to get dressed in the morning. I wouldn't be surprised if he shows up one day without his shoes."

Discussion Questions

This case illustrates one of the problems that can occur when project managers are poor time managers: Nobody wants to work with them. Have you ever known a person who was so disorganized that nobody wanted to work with him or her? If so, describe the situation. If Mack Day was your colleague and he asked for advice about how to be a better time manager, what would you tell him?

SUMMARY

Project managers who fail to manage their time effectively are not likely to complete projects on time, within budget, and according to specifications. Poor time management can cause a number of problems, including wasted time, added stress, lost credibility, missed appointments, poor follow through, inattention to detail, ineffective execution, and poor stewardship.

Crisis situations rob project managers of valuable time. Crisis situations may be reduced if project managers will end each day by planning the next, scheduling loosely, refusing to get bogged down in the personal problems of team members, and remembering that most tasks take longer to complete than planned. The telephone can be a time waster unless project managers learn to use e-mail instead of the telephone when appropriate; categorize calls as important, routine, and unimportant; use a cellular phone to return calls between meetings and during breaks; block out time on the calendar for returning calls; and limit unrelated dialogue during calls.

Project managers can create problems for themselves by taking on too much. When this happens, the key is to cut out all nonessential responsibilities. This can be done by making a list of all current and pending obligations and deciding which items on the list are not essential. It will also help to take stock of after-work activities to determine if they are essential. Unscheduled visitors can rob project managers of valuable time.

The interruptions of unscheduled visitors can be limited by refusing to meet with them during peak work times (unless the visitor is a superior, a customer, or a team member with an emergency). It will also help to train administrative assistants to come to the rescue and to remain standing when unscheduled visitors pop in.

Poor delegation and personal disorganization can rob project managers of valuable time. Project managers who fail to delegate because they have no confidence in team members should either arrange training for these team members or replace them. Personal disorganization can be overcome by cleaning out one's desk every six months, stacking the work in one's in-basket in order of priority, and using categorized work files: read folder, signature folder, and correspondence folder.

One of the most effective ways for project managers to save time is to eliminate unnecessary meetings and to make necessary meetings more efficient. Strategies that will help project managers minimize the amount of time wasted by meetings include being aware of the causes of wasted time in meetings, determining if regularly scheduled meetings are actually necessary and if they can be scheduled less frequently, holding impromptu stand-up meetings, preparing well for meetings and requiring participants to prepare, beginning meetings on time and sticking to the agenda, and following up promptly after meetings.

KEY TERMS AND CONCEPTS

Time management	Drop-in visitors
Poor time management	Remain standing
Crisis situations	Poor delegation
Schedule loosely	Personal disorganization
Unrelated dialogue	Unnecessary meetings
Nonessential responsibilities	Impromptu meetings
Pending obligations	Stick to the agenda
After-work activities	
Unscheduled visitors	

REVIEW QUESTIONS

1. How can project managers reduce the amount of time in their days wasted by crisis situations?
2. Explain the concept of scheduling loosely.
3. Why should project managers avoid the temptation to play amateur psychologist with their team members?
4. How can project managers make the telephone a more efficient work tool?
5. What is the problem with unrelated dialogue during telephone calls?
6. When project managers begin to feel overloaded, how should they approach the problem?
7. How can project managers limit the time wasted by unscheduled visitors?
8. If you knew a project manager who wasted time because of personal disorganization, what advice would you give him or her about how to get organized?
9. Describe how you would go about reducing the time wasted in an organization by meetings.
10. Why is it important to begin meetings on time?

APPLICATION ACTIVITIES

The following activities may be completed by individual students or by students working in groups.

1. Make a list of upcoming activities in your life and estimate how long it will take you to complete each activity. Record how long each activity actually takes. Discuss your list and the results of your experiment with other students.
2. Record how much time you spend on the telephone for a week. During each call, estimate how much time was devoted to the subject of the call and how much was unrelated. Record your results and discuss them with other students.
3. Make a list of all of your current and pending obligations and responsibilities. Are there any on the list that could be eliminated without causing problems. Discuss your list and the results with other students.
4. Identify a project manager in an organization who is willing to be interviewed. Ask this individual to list all regularly scheduled meetings he or she must attend and how often. Then ask if the meetings are essential or if they could meet less regularly without causing problems. Discuss the results of your interview with other students.

CHAPTER 16

Project Managers and Change

Change is as much a fact of life in engineering and technology firms as it is in any other kind of organization. Change in engineering and technology projects occurs primarily for two reasons: (1) in response to change orders and (2) in response to change initiatives undertaken to make the firm more competitive by improving productivity, quality, safety, or some other critical factor. Change orders are an ever-present part of projects and usually result from a change instigated by the customer, but they can also be initiated to compensate for errors made in the design or quality of the work.

Change initiatives are sometimes undertaken by engineering and technology firms as part of the continual improvement process. For example, a firm might decide to improve quality by adopting the concept of lean six-sigma, reach new markets by adding a metal plating capability, improve human performance by providing leadership training for all supervisors, or any number of other initiatives. Each of these internal improvement projects would require change. Improving anything requires changing it in some way, and change must be managed if it is going to bring the desired result.

To manage change effectively project managers must be prepared to deal with an unchanging fact of life. People get comfortable with how things are and, as a result, are prone to resist change. Because of this innate human tendency to resist change, project managers need to be prepared to deal with a concept called comfort-induced inertia.

Engineering and technology students will recall that inertia is a concept from physics in which a body at rest will tend to stay at rest until sufficient force is applied to move it. People working on projects can be like the body at rest in physics in that their tendency is to maintain the *status quo* until sufficient *force* is applied to break the inertia. "Force" in this case refers to leadership provided by the project manager. The concept of comfort-induced inertia, as it applies to project teams, can be summarized as follows: A project team's members will prefer to stay in their comfort zones and maintain the *status quo* until sufficient leadership is applied to convince them to embrace a proposed change.

CHANGE MANAGEMENT MODEL FOR PROJECT MANAGERS

In a competitive environment, engineering and technology firms and, in turn, project managers must be attuned to finding ways to continually improve the performance of people and processes, productivity, quality, safety, and all of the other factors that add value to a firm's work. They must also be prepared to implement change orders on the projects they manage. Managing change effectively is an important responsibility of project managers. Therefore, project managers who find themselves in the position of being responsible for implementing a significant change or even just a part of one will find the change-management model in Figure 16.1 helpful. This model can be used regardless of the cause of the change in question.

This model is designed to be used to implement changes that are caused by change orders as well as those that are internal improvement initiatives. It works best on major changes that affect factors that are important to stakeholders inside and outside of the engineering and technology firm such as time, money, quality, safety, long standing procedures, and even personal preferences. Each step in the model is important. Consequently, no step should be skipped.

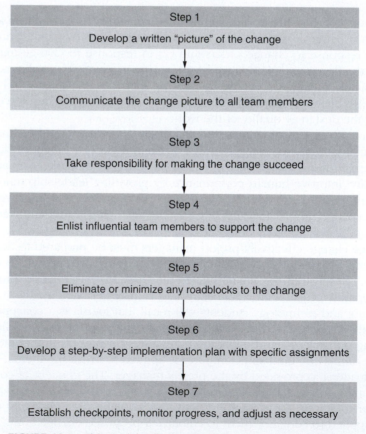

FIGURE 16.1 This model will allow change to be made systematically.

Project Management Scenario 16.1

Mary changes things just for the sake of change

When Sandra Barker told her supervisor she would rather not serve on a project team managed by Mary Andrews, the supervisor was shocked. "I thought you liked Mary. In fact, I thought she was your best friend." "She is. That's why I know her so well, and knowing her so well is why I don't want to be on her project team." Barker went on to explain that Andrews become bored easily and that as a result she likes to change things. "She changes the way she drives to work every week, changes the arrangement of the furniture in her apartment at least once a month, and changes clothes at lunch. In fact, Mary changes things just for the sake of change." Barker claimed that the last time she served on a team led by Mary Andrews her fellow team members nearly revolted over the constant unnecessary changes.

Discussion Questions

In this scenario, Mary Andrews seems to make changes that affect her team members for no reason other than boredom and a desire for variety. Have you ever known someone who liked to make changes for no reason but the sake of change? If so, did this individual's restlessness cause problems? What advice might Sandra Barker give her friend, Mary Andrews, about when and why to make changes in a project team?

Step 1: Develop a Written "Picture" of the Change

What people like least about change is the unknown. In fact, it is typically fear of the unknown that makes people uncomfortable with change. This, in turn, is one of the main reasons they resist change. Even when people do not like the conditions in which they work, they still find a measure of comfort in the concept of familiarity. In most situations, people will gravitate toward the known out of a fear of the unknown. Because of their comfort with the familiar, members of project teams often adopt an attitude toward change that is best summarized by the adage, *the devil you know is better than one you don't*. Consequently, when making changes it is important for project managers to begin by eliminating the fear of the unknown. Thus, eliminating fear is the first step in the change management model.

When a change must be made, project managers should remember that their team members know how things are now. What they do not know is how things will be after the proposed change, and this is what will concern them. When a major change order is in the works or a major change initiative is being discussed, rumors will begin to circulate unless the project manager takes the initiative to get out in front of them. To eliminate fear of the unknown, project managers should replace the unknown with the known. The unknown can be transformed into the known by developing a compelling, informative change picture. A change picture is a comprehensive but brief written explanation of how things will be after the change.

The key to making a change picture compelling is for project managers to view the change from the perspective of those affected by it—their team members—and develop the change picture accordingly. People take change personally and react to it on an emotional level. Consequently, the first thing they will want to know about the proposed change is how it will affect them. Will the change mean more work or less? Will the change require workers to do things they do not yet know how to do? Will the change require workers to scrap their comfortable procedures and learn new ones? Will the change threaten their job security? Project team members will have these types of questions, even if they do not verbalize them. Whether the news is good or bad, a well-written change picture will answer these types of questions as well as others by providing the following information: what, when, where, who, why, and how (five Ws and one H). Even if the news is will not be welcomed by stakeholders, it is better for them to know the facts than to depend on rumors because rumors, by their very nature, tend to be exaggerated.

A good change picture explains the nature of the change (*what*), *when* the change will occur, *where* the change will occur, *who* the change will affect, *why* the change is being made, and specifically *how* it will affect team members. This final consideration—how the change will affect team members—is important because it is the aspect of the change team members will be the most concerned about. A good change picture will explain the "how" question from the perspective of those affected by it. For example, if there will be changes to deadlines, more work added to the project, cutbacks that could result in layoffs, changes in the normal work hours, altered working conditions, or anything else that will affect team members, the change picture should explain the change in terms that are open, honest, candid, easily understood, and from the perspective of team members. Project managers who keep their team members guessing about how a change will affect them run the risk of losing their team members' trust and damaging their morale. It cannot be stressed enough that facts about changes—no matter how unwelcome the changes—are better than rumors.

Step 2: Communicate the Change Picture to All Team Members

Once the change picture has been written it must be communicated to team members. Unless the team is scattered among different locations and connected only electronically and by telephone, the change picture should be provided to team members in face-to-face meetings that allow for maximum give and take. However, no matter what communication methods are used, it is imperative that there be a convenient feedback mechanism so that stakeholders can ask questions, express concerns, point out problems, or just vent. This is why face-to-face meetings are the best approach for communicating the change picture.

An important part of communicating the change picture to team members involves conducting what the author calls a *roadblock analysis*. The roadblock analysis involves asking team members to identify any and all roadblocks that might derail the change initiative. A roadblock analysis can be accomplished by e-mail or even telephone, but the best approach is face-to-face. By asking for the help of team members who are closer to the situation than they are, project managers can identify roadblocks that might impede or even undermine the change implementation process. Once they have been identified, roadblocks can be removed or, at least, minimized before they derail the implementation process. Removing or minimizing roadblocks is explained in a later step. This step requires only that they be identified.

Step 3: Take Responsibility for Making the Change Succeed

Change in projects does not just happen—it requires commitment, persistence, and a lot of work, both mental and physical, from team members. More than anything, it requires team members to step forward and take responsibility for doing their part to make the change succeed. Often, the first response of team members to change is to resist it. Some team members will react to change by openly resisting it, but many will opt for a more subtle, more passive form of resistance.

One of the more common forms of passive resistance is the wait-and-see attitude. Team members who adopt a wait-and-see attitude do not necessarily work against the change. On the other hand, they do not work for it either. Wait-and-see team members might say all the right things about supporting the change, but in reality they put no effort into making it succeed. This is the "passive" aspect of their resistance. Passive resisters often try to play on both sides of the fence until they know whether the change is going to succeed or fail. They like to be able to say "I told you so" if the change initiative fails and "I supported it from the beginning" if the initiative succeeds.

The wait-and-see crowd can be even more detrimental to change initiatives than those who openly oppose them. This is because making changes in the middle of a project is like pushing a boulder up a hill. The team needs all of its members pushing together in a coordinated and concerted effort. Those who openly oppose the change can be neutralized by isolating them from the implementation process. But those who act like they are helping when they are not, can undermine the process. When trying to push a boulder up a hill, a lot of people are needed who will push with all their might.

Team members who act like they are pushing when they really are not make the task more difficult for those who are. This is why wait-and-see team members do not have to openly work against changes they oppose. All they have to do is sit back and let them fail. However, when the team members who do their part manage to get the boulder over the crest of the hill the task becomes easier from that point on. When this happens, those who contributed nothing to getting the boulder up the hill typically join the parade of team members following it down the other side, all the while acting as if they had pushed with all their might. On the other hand, if the boulder bogs down and fails to make it up the hill, members of the wait-and-see crowd will be quick to join the naysayers who openly opposed the change.

Changes in projects—whether they are caused by change orders or change initiatives—will succeed only if the team members responsible for implementing them are willing to: (1) take responsibility for pushing the boulder up the hill and (2) commit to doing what is necessary to push the boulder over the crest. Once the inertia has been broken and the boulder is moving uphill, responsible team members cannot rest until it is over the crest and momentum is on the side of success.

Step 4: Enlist Influential Team Members to Support the Change

In every team, some members are more influential than others. The source of their influence might be seniority, talent, popularity, strength of personality, a combination of these, or a variety of other factors. Regardless of why they are influential, other team members look to these special few for direction and approval. These influential team members can contribute greatly to the success or failure of a change. This step is more important when implementing improvement initiatives than it is when implementing change orders, since there is a contract-driven urgency to change orders. This contract-driven motivation is missing when implementing change initiatives. None-the-less, this step should not be skipped, even when the change in question is caused by a change order.

It is always best to enlist the support of influential team members when implementing changes. For example, assume that a major change order has been approved that is going to require a prototype to be broken down and rebuilt from the ground up. It will be important to enlist the support of the manufacturing personnel who will have to complete the rework process. Nobody likes to redo completed work, but the manufacturing personnel will be more likely to do what has to be done if they see their influential team members supporting the change. Enlisting influential team members can be a challenge, especially if they are opposed to the change or fall into the wait-and-see category. Consequently, before attempting to enlist influential team members it is important for project managers to have face-to-face, one-on-one conversations with them.

During these conversations, project managers should determine where the influential team member in question stands concerning the change. Is he for it, against it, or just waiting to see what will happen? If he is against the change, does he intend to throw up roadblocks or just sit back and to see what will happen? If influential personnel are against the change or are in wait-and-see mode, project managers have two options: (1) isolate them from the change implementation process to limit their negative influence or (2) give them responsibility for some aspect of the implementation plan. Sometimes, the fastest way to get an influential individual committed to a change is to give him or her responsibility for some aspect of it.

When an influential team member is given responsibility for some aspect of the change implementation, it is important to apply appropriate reinforcement methods. Typically, this means applying both the carrot and the stick. The *carrot* consists of incentives the individual will receive when the change succeeds. The *stick* consists of the negative consequences that will occur if it does not. Assurance of full cooperation should be gained before putting an influential team member to work on behalf of the change. To do otherwise is likely to just undermine the change implementation process.

Step 5: Eliminate or Minimize any Roadblocks to the Change

Part of an earlier step in this model involved conducting a roadblock analysis to identify any obstacles that might impede a successful implementation of the change initiative. In this step, the obstacles identified are either eliminated or minimized. In order to explain the change management model in a step-by-step manner, it is necessary to put this step here. However, in reality, removing or minimizing roadblocks—this step—begins as soon as they are identified. This step can be undertaken in parallel with those that have already been explained.

More often than not the obstacles identified are internal impediments decision makers have not thought of. With almost any project, there will be unforeseen difficulties that result from unintended consequences of well-intended actions or decisions. There will also be unforeseen obstacles. Dealing with these unforeseen circumstances and factors is an important part of project management. For example, the author once worked on a manufacturing project that required extensive use of the company's most advanced machining center—the only machining center in the company that could hold the demanding tolerances specified for the job in question. Unfortunately, the machining center in question was still tied up by another job that should have been completed at the time but was running well behind schedule. The team assigned to the new project did not have access to the machining center needed to complete it.

The project manager discussed and debated several options with higher management (e.g., purchase another machining center, lease the needed machining center, solicit bids from subcontractors to do the work). Purchasing or leasing the needed machining center were not viable options because shipping, delivery, and set up would take to long. In order to meet the demanding schedule specified in the contract for the new project, work had to begin right away. The only subcontractors capable of doing the work on the new project had been competitors in the bidding process. Winning the bid over them and then asking one of them to do the work did not seem to be a good idea. The customer might simply cancel our firm's contract and delegate it to the competitor that could do the work.

The best option turned out to be adding another shift for the duration of the new project. The machining center was available from 6:00 PM every day until 6:00 AM the next day. The project manager's challenge became finding a team of machinists who were willing to work the late shift for the duration of the new project. The change was a major disruption in the lives of those who agreed to work the late shift. Consequently, financial incentives had to be offered and some overtime had to be paid. But the project was completed on time, within budget (just barely), and according to specifications.

Roadblocks are going to pop up during the course of projects. This is a fact of life that project managers must be prepared to deal with. The key is to understand that roadblocks do not change the destination of a project, just the route taken to get there. Convincing team members to embrace the new route and make sure it succeeds is a major change-management responsibility of the project manager.

Step 6: Develop a Step-by-Step Implementation Plan with Specific Assignments

Once the roadblocks to a successful implementation have been removed or minimized, a step-by-step implementation plan is developed. The plan lists every action step that must be taken to get the change initiative successfully implemented. Every action that is to be taken is then assigned to a specific individual—not a group or a team, but an individual. This is important. Although changes in projects often require several members of the team to be involved in their implementation, specific individuals be given responsibility for the assigned tasks. These individuals, then, take the lead in making sure that the assigned work gets done. Avoiding confusion concerning who is supposed to do what is essential to the successful implementation of a change initiatives and change orders. Saying that everybody is responsible is the same as saying nobody is responsible.

Step 7: Establish Checkpoints, Monitor Progress, and Adjust as Necessary

One of the reasons changes in projects fail is that project managers ignore the need to monitor progress after implementation has begun. Assigning responsibility for specific tasks—as recommended in the previous step—is important, but even doing this is no guarantee of a successful implementation. To ensure a success, project managers must stay on top of the implementation. This means establishing a deadline for completion of every action step that is assigned to a specific team member, establishing incremental progress points that precede these deadlines (e.g., deadlines for 25%, 50%, 75% completion), monitoring to ensure that satisfactory progress is being made, and making adjustments when unanticipated problems occur.

When implementing a change initiative, it is common for problems to arise that even the roadblock analysis did not anticipate. When this happens, project managers need to know about it so they can take immediate corrective action. Taking immediate corrective action involves adjusting as necessary to keep the implementation on track and the momentum working on the side of a successful implementation.

The model for implementing changes in projects presented in this chapter will help project managers become effective change agents and change managers. Project managers who are good change managers will, in turn, be better able to ensure that their projects are completed on time, within budget, and according to specifications.

Project Management Scenario 16.2

I know how to manage projects, but I need to know how to manage change

Juan Padea has been a project manager for more than three years. During this time, his teams have always performed well. In fact, Padea is known throughout his firm for bringing projects in on time, under budget, and according to specifications. But Padea's current project is presenting him with a challenge he has not had to deal with on previous projects: a major change resulting from a contract amendment requested by the customer. The owner has decided he wants two additional floors added to the hotel that Padea's firm is building.

The customer has agreed to pay for the change order. This is the good news. The bad news is that he wants to retain the original date for project completion. Word has spread about the contract amendment and Padea's team members are nervous. His team members want to know what the change will mean to them. In fact, Padea's team members are spending so much time worrying about the impending change order that they are not attending to the work at hand. Suddenly the project is falling behind schedule even without the added work in the change order. When a colleague asked Padea how he planned to handle the change order, he replied: "I'm not sure. I know how to manage projects, but I need to know how to manage change."

Discussion Questions

In this scenario, Juan Padea faces a problem that often challenges the leadership skills of project managers: having to make changes in the middle of a project that is well underway. Have you ever started on a project in school or any other setting and had to make a major change in midstream? If so, how did you respond? Was the situation frustrating? If Juan Padea asked you for advice concerning how to manage the change his team is facing, what would you tell him?

SUMMARY

Change is a fact of life in engineering and technology firms. This is why the change order process exists. Even without change orders, continual improvement means continual change because doing something better than it is currently being done necessarily means doing it differently. To manage change effectively, project managers must be prepared to deal with a phenomenon known as *comfort-induced inertia*: the tendency of people to maintain the *status quo* until sufficient leadership is applied to convince them to change.

An effective change management model has the following steps: (1) develop a written "picture" of the change initiative, (2) communicate the change picture to all team members and conduct a roadblock analysis, (3) take responsibility for making the change initiative succeed, (4) enlist influential team members to support the change initiative, (5) eliminate or minimize any roadblocks to the change initiative, (6) develop a step-by-step implementation plan with specific assignments, and (7) establish checkpoints, monitor progress, and adjust as necessary.

KEY TERMS AND CONCEPTS

Comfort-induced inertia
Change management model
Change picture
Roadblock analysis
Change initiative

Influential team members
Carrot
Stick
Implementation plan
Monitor progress

REVIEW QUESTIONS

1. Why is it important for project managers to be effective change managers?
2. Explain the concept of organizational inertia. How can it be overcome?
3. Explain what is meant by developing a written "picture" of a change initiative.
4. What information should be contained in a change picture?
5. Ideally, how should a project manager go about communicating a change picture to team members?
6. What is a roadblock analysis?
7. Explain how team members can passively resist a change initiative.
8. How can a project manager enlist influential team members to ensure that a change initiative succeeds?
9. Explain the carrot-and-stick approach for dealing with influential team members project managers want to enlist on the side of a change order or initiative.
10. Think of an example of a change that you would like to make. Now identify any roadblocks that might keep you from making the change. How would you eliminate or, at least, minimize the roadblocks?
11. What is the significance of making specific assignments when implementing a change order or initiative?
12. Explain why it is important to establish checkpoints and monitor progress when implementing change initiatives.

APPLICATION ACTIVITIES

The following activities may be completed by individual students or by students working in groups.

1. Identify an engineering and technology firm in your region that will cooperate in completing this project. Ask the firm to see a major change order that occurred during a project. Then ask for an explanation of how the change order was handled. How could the seven-step model for implementing change presented in this chapter have been used to implement the change order?

2. Assume that you work for a large engineering and technology firm that plans to purchase new CAD software. The new software is much more advanced than the current CAD software used by your firm, but its operational procedures are much different. The software will require a corresponding hardware upgrade. Using the change management model explained in this chapter, explain what will need to be done in each step of the model. Do not forget to conduct a roadblock analysis.

CHAPTER 17

Project Managers and Diversity

Americans can trace their lineage to virtually every country, race, and culture in the world. In fact, America is one of the world's most diverse countries, and this diversity is reflected in the workplace. Because of this, project managers can expect to lead diverse teams. The term *diversity* as applied in the current context refers to human differences of all kinds. Dealing with these differences in ways that make the team stronger is the responsibility of the project manager.

There are many ways in which people can be different, but the ways that always seem to command the most attention are race, culture, national heritage, gender, religion, politics, level of education, worldview, and personality. Because people in project teams can be different from each other in all of these ways as well as many other ways, it is important for project managers to learn how to lead a diverse team and to teach team members, through words and by their example, how to work well in a diverse environment.

DIVERSITY DEFINED

Try this experiment. Ask several friends or fellow students what comes to mind when they hear the term *diversity*. Do not be surprised to find that people tend to associate the term with racial differences. This perception is partially correct, but only partially. Diversity includes racial differences but it is a much broader concept than just racial differences. As it applies in the context of project management, diversity encompasses all of the many ways in which people can be different.

What follows is a partial list of ways that people can be different: race, mental ability, physical ability, physical appearance, age, marital status, geographic status, religion, denominations within religions, ethnicity, nationality, worldview, education level, values, political beliefs, interests, personality, cultural background, height, weight, career status, white collar, blue collar, and personal preferences (food, clothing, music, hobbies, etc.) as shown in

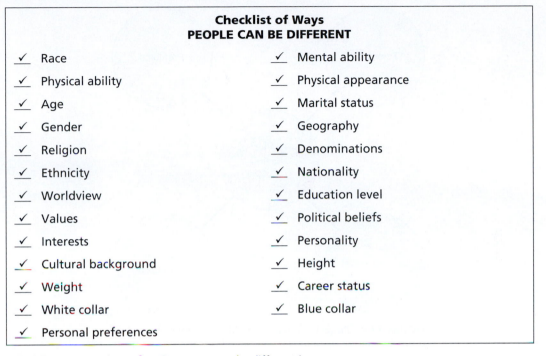

**Checklist of Ways
PEOPLE CAN BE DIFFERENT**

✓ Race   ✓ Mental ability

✓ Physical ability   ✓ Physical appearance

✓ Age   ✓ Marital status

✓ Gender   ✓ Geography

✓ Religion   ✓ Denominations

✓ Ethnicity   ✓ Nationality

✓ Worldview   ✓ Education level

✓ Values   ✓ Political beliefs

✓ Interests   ✓ Personality

✓ Cultural background   ✓ Height

✓ Weight   ✓ Career status

✓ White collar   ✓ Blue collar

✓ Personal preferences

FIGURE 17.1   Members of project teams can be different in many ways.

Figure 17.1. This is a long list, but it is not complete. There are many more ways in which people can be different.

Obviously, people can be different in a lot of ways. In fact, members of the same family—even identical twins—can be vastly different when it comes to their physical abilities, intelligence, personal interests, religious beliefs, level of education, personality, work skills, appearance, attitudes, ambition, political beliefs, and worldviews. If there can be this many differences between family members, imagine all the ways in which people in project teams can be different. It is a certainty that project managers are going to have to lead team members who are different from them in how they talk, dress, eat, interact, socialize, believe, think, and approach their work. Being able to effectively lead diverse teams is an essential people skill for project managers.

DIVERSITY-RELATED CONCEPTS

In conversations about diversity, certain terms will surface over and over again. Project managers should be familiar with these terms themselves and help their team members become familiar with them. Important diversity-related concepts include the following (Figure 17.2):

- Prejudice
- Stereotyping/labeling
- Discrimination
- Tolerance

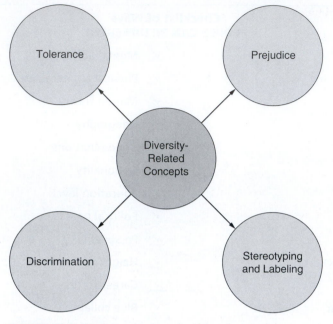

FIGURE 17.2 Project managers should understand the diversity-related concepts.

Prejudice

Prejudice is a predisposition to adopt negative perceptions about groups of people. For example, one might be prejudiced against conservatives, liberals, northerners, southerners, easterners, westerners, men, women, whites, blacks, accountants, engineers, Democrats, Republicans, youth, elderly, or any other identifiable group of people. People who are prejudiced have learned to harbor negative perceptions toward a distinct group or groups of people. These negative perceptions can result in ill feelings toward members of the group in question. For those who are prejudiced against a given group of people, all members of that group are subject to the same negative perceptions. For example, an individual who is prejudiced against teenagers might claim that all teenagers have slouchy posture and bad attitudes and that they spend too much time hanging out at shopping malls. Although this description might be accurate for some teenagers, it is not accurate for all teenagers. This partially-accurate-but-mostly-wrong phenomenon is true of any group that is the target of prejudiced attitudes.

People who are prejudiced tend to divide the world into two distinct groups: "us" and "them." Their unwritten, unspoken but deeply felt philosophy can be summarized in these words: *people who are not like us are wrong or they are bad*. Those who appear to share their values, cultural mores, worldviews, and other pertinent characteristics are part of the "us" group. Everyone else is lumped into the group known as "them." Prejudiced people do not want to associate with those they are prejudiced against.

Prejudice can manifest itself in project teams in a number of different ways—all negative. People in project teams may act out their prejudices through discrimination, stereotyping, and labeling. They may also act out their prejudices in less overt, but equally negative ways. For example, a project manager who is prejudiced against a given group might refuse to admit to himself that anyone in that group ever had a good idea, did a good job, was worthy of promotion, or could sufficiently satisfy any performance measure. This individual might never utter a prejudiced word, but his attitude toward the group in question will be reflected in his failure to give members of the group a chance to excel. With prejudiced people, members of the "us" group receive the benefit of the doubt in all situations. Members of the "them" group receive only the doubt.

Prejudice can blind project managers and members of project teams to the fact that people who they are prejudiced against are capable of excellent performance. As a result, prejudiced project managers might not want certain people to serve on their project teams, might not give them a chance to prove themselves capable, or might not invest the time and effort in them that they invest in others. Not only is this wrong from the perspective of equity, it is unwise from the perspective of effective project management. Prejudice can blind a project manager to the talent and potential of others.

A team member's performance on a project is a function of individual talent, attitude, motivation, and other job-related factors. It is not a function of race, sex, politics, cultural heritage, worldview, or other diversity-related factors. Consequently, when project managers allow prejudice to influence their actions and decisions, they limit their ability to get the best possible performance from their teams. In addition, they often pass on their prejudice to team members, corrupting their views concerning diversity. Remember, prejudice is a learned behavior.

Stereotyping/Labeling

Stereotyping is a negative by-product of prejudice. It is the act of generalizing certain characteristics to all members of a given identifiable group. Consider the example of so-called *blond jokes*. What blond jokes have in common is the assertion that all blonds are dumb. In order to "get" a blond joke, one must accept this assertion. Of course, such an assertion is not just illogical but absurd. One need only think of all the people he or she has known in life. Many would be blonds—male and female. Some blonds are brilliant, some are smart, some are average, and some are below average when it comes to native intelligence. This statement can be made with an acceptable degree of certainty because the same thing can be said about all people, regardless of hair color or other physical attributes. Making or just accepting assertions such as blonds are dumb is engaging in stereotyping.

It was stereotyping that for years denied even the best African-American college quarterbacks opportunities to lead football teams at the professional level. For decades outstanding college quarterbacks who happened to be black were automatically converted to receivers and defensive backs when drafted by professional football teams. The stereotype behind this unfortunate practice was that blacks were not smart enough to lead teams at the professional level. Now that African-American quarterbacks are have excelled in leading professional football teams, people who are too young to have experienced this stereotype have trouble believing it ever existed. But, of course, it did. The same kind of stereotype once applied to football coaches at the college and professional levels. Now that some of the most successful coaches who ever walked the sidelines are black, the stereotype has been discarded and forgotten. However, what should not be forgotten is how wrong the stereotype was because it shows how illogical and damaging stereotyping can be.

Stereotyping results in lose-lose-lose situations in project management. Victims of stereotyping lose because they are denied opportunities to prove themselves and to contribute to the success of the team. The team loses because it is denied the talent of the stereotyped individual. The individual who engages in stereotyping loses because he is responsible for the first two losing propositions. This, in turn, renders him less effective at his job be it project manager or CEO.

Labeling is an extension of stereotyping in that it involves attributing a certain characteristic to a distinct group of people and then using that characteristic to label all people in that group. For example, people who are talented in the fields of engineering, science, math, or computers are often labeled as "nerds" or "geeks." These labels are supposed to conjure up an image of a socially inept person who can solve quadratic equations but cannot hold a decent conversation on any topic outside of her field. While there may be some academically gifted individuals who are socially inept, to lump all scholars into a homogeneous group in this way is not just illogical, it is narrow minded.

Prejudiced behavior such as stereotyping and labeling harm both the victims and the perpetrators. Just because some people in an identifiable group exhibit certain characteristics hardly means that all people in that group share those characteristics. After all, some people in almost any group will exhibit certain identifiable characteristics. In fact, the prejudiced person often exhibits some of the very characteristics he is prejudiced against. The mistake made by prejudiced people is attributing those identifiable characteristics to all members of the group in question.

To understand how illogical the practices of stereotyping and labeling can be, take the prejudicial practice of referring to all southerners as rednecks. Labeling someone a *redneck* is supposed to conjure up an image of an uneducated, uncouth, tobacco spitting oaf. Labeling all southerners as rednecks is a stereotype that is easily disproven. Some of America's most brilliant scholars, poets, musicians, artists, scientists, and political leaders have been men and women from the South. In fact, several southerners have been president of the United States, including George Washington, who is known as "the father of our country" and Thomas Jefferson, the author of America's Declaration of Independence. An objective observation by anyone who is not prejudiced will reveal the fallacy of stereotyping and labeling.

Project managers who allow stereotyping and labeling to influence their decisions do their victims, their teams, themselves, and their firms a great disservice. Try to imagine a baseball coach leaving his best hitter on the team sitting on the bench because of stereotypical thinking. In essence, this is what project managers do when they allow prejudiced practices such as stereotyping and labeling to influence their decisions. Project managers who excel make their decisions on the basis of such factors as productivity, performance, and quality. In other words, their decisions are based on who will contribute the most to getting projects completed on time, within budget, and according to specifications. This focus on performance rather than prejudice is essential to achieving excellence as a project manager.

Discrimination

Project managers must constantly discriminate. They discriminate between effective and ineffective work practices, high and low-performing team members, efficient and wasteful processes, good and bad quality, and other factors that affect performance. Discrimination is not an inherently bad concept. For example, someone with discriminating taste is a person who recognizes good food. Someone with a discriminating mind is a person who is able to separate fact from opinion. Obviously, there are types of discrimination that are appropriate. However, discrimination—as it applies in the context of diversity—is not one of them.

Diversity-related discrimination is a negative concept. In this context, discrimination means putting prejudice into action. It means allowing diversity-related factors such as race, gender, culture, and worldview to influence one's actions and decisions, a practice that is illegal, unethical, and counterproductive. Diversity-related discrimination can quickly and effectively undermine a project team's performance and the morale of its team members. Wherever it exists, discrimination is an obstacle to the achievement of excellence because it undermines individual, team, and organizational performance as well as team morale.

DISCRIMINATION AND TEAMWORK. Project teams are like football teams. In football, every time the ball is snapped eleven people on the offense have their individual assignments that must be carried out properly if the ball is to be advanced. In addition to carrying out their individual assignments, each team member is responsible for cooperating with and supporting his team mates.

Take, for example, when the quarterback throws a pass. Before the ball is thrown, the offensive line is responsible for protecting the quarterback from the rush of the opponent's defensive line and blitzing linebackers. In order to do this well, the offensive linemen must cooperate with each other in mutually supportive ways. Then, if the ball is caught, these same linemen immediately switch their support to the receiver and cooperate in trying to help him score by providing downfield blocks. Consider what would happen if the offensive linemen decided they would not block for the quarterback because he is an easterner or would not hustle downfield to block for the receiver because he is a westerner. Teamwork would be undermined and an attitude of us-against-them would set in. The same thing will happen in project teams where diversity-related discrimination is present.

DISCRIMINATION AND TEAM PERFORMANCE. In order for a project team to excel, the project manager must lead in ways that encourage, support, and facilitate peak performance. Few things will undermine a project manager's efforts to promote peak performance in her team faster than diversity-related discrimination. This is because team members who feel discriminated against will quickly lose their motivation. Further, the energy team members should be putting into their work will be used up by their efforts to fight against discrimination.

Discrimination affects its victims on a deep and personal level. Because this is the case, team members cannot give their best efforts when they are being discriminated against. The more energy team members invest in fighting discrimination, the less they will have to invest in achieving peak performance. Members of project teams can spend their time trying to excel or they can spend it fighting discrimination, but they cannot do both. Members of project teams do not have the time or the energy to do both.

DISCRIMINATION AND MORALE. Of the various factors that can affect the performance of project teams, few are more powerful than morale. Morale refers to the spirit of team members—how they feel about themselves, the team, and the project manager. Team members with high morale exhibit such characteristics as trust, loyalty, pride, and faith in the team as well as commitment to the mission. High morale is synonymous with team unity. On the other hand, team members with low morale are not likely to excel. In fact, the opposite is more likely to happen. Typically, low morale results stress, frustration, and eventually, poor performance.

Obviously maintaining the highest possible morale should be a goal of project managers. Morale is a function of various factors, one of the most important being the perceptions of team members concerning how they are treated by the project manager and their team mates. Members of project teams want to be appreciated, respected, and recognized for their contributions to the team. Team members with low morale typically feel as if the project manager does not care about them and does not appreciate, respect, or properly recognize them. It should come as no surprise then that discrimination results in low morale.

Discrimination against certain team members by the project manager is certainly evidence that he does not care about his victims, but discrimination goes well beyond just not caring. In fact, it is an oppressive practice that involves denying opportunities to some team members while providing them to others without considering talent, motivation, commitment, or other performance-related factors. Such a practice is not just illegal and unethical, it is counterproductive. Team members who believe they are being discriminated against are likely to suffer from low morale, and low morale is the enemy of team performance. Worse yet, low morale is contagious. Like chickenpox it can spread through contact. Team members suffering from low morale will soon spread their *disease* to other team members. The inevitable result of low morale is poor performance.

Project Management Scenario 17.1

I think you are prejudiced

Mack Murphy is known as a project manager who runs a tight ship. He always develops a detailed schedule and monitors closely to make sure that his team members stay on schedule. Murphy monitors his team's budget down to the penny. He is also very particular when it comes to choosing team members for his project teams. Murphy has a well-deserved reputation for getting projects done on time, within budget, and according to specifications. But, there is a disturbing side to Mack Murphy that has become apparent only in recent months as his company's ratio of female engineering and technology professionals has increased. Murphy refuses to have women on his project teams.

When asked about the issue, Murphy always responds, "I just pick the best people for my projects. It doesn't matter to me whether they are men or women. I just want people who will give me their best." This explanation worked for a while, but when he refused to select a high-performing women for his team—a women who is well-thought-of by other project managers in the company—eyebrows were raised. This time when asked about the issue, his standard answer did not ring true. Murphy's supervisors said, "Mack, I think you are prejudiced."

Discussion Questions

In this scenario, Mack Murphy—a high-performing project manager—does not seem to want to have women on his project teams. Have you ever worked with someone who appeared to be prejudiced against people in some identifiable group? If so, discuss the situation. Did it cause problems? Have you ever been the victim of prejudice yourself? If so, discuss any aspect of the situation you are comfortable talking about? If you were Mack Murphy's supervisor, what would you tell him about the potential negative effects of prejudice?

Tolerance

The diversity-related concepts explained so far—prejudice, stereotyping, labeling, and discrimination—have all been negative. The concept explained in this section—tolerance—is positive. At least for as far as it goes. Tolerance is a willingness to interact positively with people in spite of differences and to remain open to differing points of view. People who are tolerant do not automatically reject another person, opinion, or perspective on the basis of diversity-related factors. As a result, tolerant people are more likely to make decisions on the basis of performance than on diversity-related factors.

Tolerance is a step in the right direction in that it moves people away from prejudice, stereotyping, labeling, and discrimination. But, there are two potential problems with tolerance that project managers should understand and avoid: (1) in project teams that are striving for excellence simply tolerating differences is not enough and (2) well-intentioned people can misinterpret the concept and end up being too tolerant in that they tolerate behaviors that are unacceptable because of diversity-related considerations. Project managers should understand these two potential problems and take appropriate steps to prevent their occurrence.

TOLERANCE IS NOT ENOUGH.   It is often said that diversity is an asset to organizations. In fact, some organizations adopt slogans such as "…diversity is our greatest asset." A more accurate rendering of this philosophical ideal would be that diversity can be an organization's greatest asset. It is a concept with enormous potential for good, but potential seldom becomes reality without a concerted effort to make it so. The hard truth about diversity is that it becomes an organizational asset only when it is handled well. Project managers who do a poor job of handling diversity will not just miss out on its potential benefits they will introduce diversity-based conflict into their teams. When it comes to making diversity an asset, tolerance is a step in the right direction but it is not enough.

In actual practice, what passes for tolerance too often amounts to people just putting up with each other because they think they are supposed to, because they feel pressure to do so, or simply for the sake of harmony. But in project teams the goal is performance not harmony. Further, putting up with others is not the same thing as embracing them. Too often the unstated message behind tolerance is: *I will put up with your differences, but only because I have to.* In practice there is often a begrudging aspect to tolerance. Begrudgingly putting up with others is better than overtly or covertly exhibiting prejudice, but it is not enough to help a project team achieve excellence. Project managers and project teams that want to enjoy the potential benefits of diversity must go beyond just tolerating it to embracing it. This is an important point for project managers to understand and help their team members understand.

Embracing diversity is tolerance taken to a higher level. People who embrace diversity do not just put up with others who are different they seek them out and ask for their opinions, perspectives, and input. People who embrace diversity understand that human differences—race, gender, background, experiences, culture, education—can produce differences in perspectives, points of view, and opinion. Further, they understand that one of the best ways to solve problems and improve decisions is to get people with different perspectives, backgrounds, and experiences involved in dealing with them.

People who are uniform in their perspectives, opinions, and points of view will tend to see problems in the same way. Give them a problem and they are all likely to see the same cause and same solution. But people with diverse perspectives are more likely to achieve a 360 degree view of the problem. They will see causes others would miss and, as a result, be in positions to suggest solutions that others might overlook. Uniformity tends to narrow the possibilities when trying to solve problems or find ways to improve performance.

The opinions of project managers and their team members are informed by their individual backgrounds and experiences. This is why different people can look at the same problem and see different causes and solutions. One person, no matter how talented, can see a problem from just one perspective—her own. But a diverse group of people will view the problem from a variety of perspectives. This diversity of perspectives can lead to better decisions, better ideas, and better solutions if it is embraced and put to good use. Better decisions, ideas, and solutions, in turn, lead to better performance. Project managers learn to embrace diversity and help their team members do the same by applying the following strategies:

- Talking with team members about the concept and making sure they understand how it can benefit the team's performance
- Ensuring diversity when selecting the members of project teams
- Seeking out a broad base of opinions before making decisions and playing devil's advocate to encourage differing points of view
- Encouraging team members to break out of their comfort zones and seek out team mates who are different from them

Challenging Engineering and Technology Project

IMPROVING MEDICAL INFORMATION SYSTEMS

Much progress has been made in developing basic medical information systems for maintaining patient records and insurance data. Progress has also been made in developing systems for sharing medical information between and among different stakeholders (e.g. physicians, hospitals, insurance companies). However, much work remains to be done in the following areas on medical information systems: 1) maximizing the effectiveness and usefulness of these systems, 2) ensuring the confidentiality of medical information in these systems, and 3) guarding against the misuse of medical information in these systems.

The challenges in improving medical information systems are many. Three of the largest challenges are: 1) technological incompatibility between different hardware and software platforms, 2) difficulties in sharing information over regional, national, or global networks due to differences in computer networks and data recording rules, and 3) providing "just-in-time" information specific to individual patients to assist physicians in making optimum decisions about diagnoses and treatments. Meeting these challenges will require creative thinking, innovation, and persistence on the part of engineering and technology professionals. It will also require effective project management. Hardware design, software development, systems integration, and user training are all part of the challenge facing engineering and technology professionals in improving medical information systems. *Before proceeding with this chapter, stop here and consider how improving medical information systems will require all of the process and people skills of project management.*

Source: Based on *National Academy of Engineering.* http://www.engineeringchallenges.org

- Recognizing and rewarding team members strictly on the basis of performance
- Setting an example of embracing diversity

TOLERATING THE UNACCEPTABLE IS A MISTAKE. Tolerance is a practice that can enhance a project team's performance, especially when it is taken to the level of embracing diversity. When promoting tolerance among team members, project managers should keep its potential impact on performance foremost in their minds. This is important because when it comes to the practical application of tolerance, some project managers do not understand the concept. It is not uncommon for project managers to be so anxious to prove they are tolerant that they tolerate behavior that is unacceptable. One of the things tolerant project managers should not tolerate is behavior that undermines performance. Not only does this misconstrue the concept, but also it can quickly undermine the project manager's credibility, and the project team's performance.

The race, gender, background, culture, or education level of team members who are not productive should never be used as an excuse for ignoring poor performance. Unproductive work practices and counterproductive behavior in a project team are wrong, and should not be tolerated. Tolerance does not mean giving team members who refuse to do their jobs a free pass because of diversity-related factors. Tolerance practiced properly means embracing people who are different, seeking out their perspectives for the purpose of finding better solutions, and basing decisions on performance rather than diversity-related factors.

PREJUDICE IS LEARNED BEHAVIOR THAT CAN BE UNLEARNED

Prejudice in project teams is both unethical and counterproductive. In spite of this, there are still project managers and team members who struggle with embracing or even tolerating diversity. People who are uncomfortable working in a diverse environment are not necessarily bad people, but they have learned some bad habits. This concept of learned behavior is important for project managers to understand because it applies to them personally as well as to their team members.

People are not born prejudiced, but some learn to be. To test this theory, consider the example of young children. Left to themselves, little children will happily play with each other without concern for race, gender, or other differences. They are just happy to have play mates. It is only as they grow older that they learn to adopt negative attitudes toward people who are different from them. Over time these negative attitudes can harden into prejudice. It is important for project managers and their team members to understand that prejudice is a learned behavior because what is learned can be unlearned. Further, if people can learn negative behaviors they can learn positive behaviors. Once they learn positive behaviors, they can substitute them for the negative and, in the process, overcome them.

OVERCOMING PREJUDICE AND EMBRACING DIVERSITY

As has already been explained, prejudice, stereotyping, labeling, and discrimination are learned behaviors. They are not genetic. This is actually good news because what can be learned can be unlearned, just ask a former smoker. The comparison with smoking is appropriate because giving up long-held prejudices can be as hard as giving up smoking or any other bad habit. Doing so requires commitment, persistence, and a willingness to change. Further, it requires replacing a bad habit with something better.

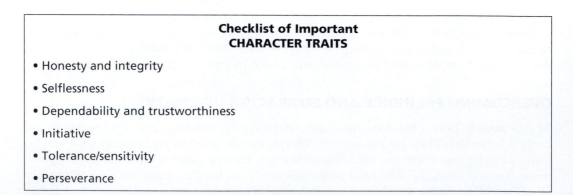

**Strategies for
EMBRACING DIVERSITY**

• Focus on character not race, gender, culture, or other differences

• Look for common ground with others

• Focus on what really matters

• Relate to people as individuals

FIGURE 17.3   Project managers should learn to embrace diversity and make it an asset.

Effective strategies for unlearning prejudice and embracing diversity include: (1) focus on the character of people rather than race, gender, culture, or other differences; (2) look for common ground with others; (3) focus on what really matters; and (4) relate to people as individuals (see Figure 17.3). These strategies can be used by project managers for overcoming their own prejudices and for helping team members overcome theirs. Of course, stating these four strategies is easier than applying them. Remember, breaking bad habits requires persistence, patience, and concerted effort. There will be ups and downs. When trying to break bad habits, taking two steps forward and one back is normal—it is to be expected. However, once started down the right path, project managers should keep going until they have replaced learned prejudice with a willingness to embrace diversity.

Focus on Character Rather Than Diversity-Based Differences

The various ways people can be different—race, gender, culture, age, background, politics, nationality—are the wrong things to consider when forming opinions of people. These factors occur as a result of birth, not merit. A better approach is to focus on the traits that really make people who they are, those known as *character traits*—traits that are determined by individual choice, effort, and merit rather than birth. Character traits that are especially important for project managers and their team members include the following (Figure 17.4):

• Honesty and integrity
• Selflessness

**Checklist of Important
CHARACTER TRAITS**

• Honesty and integrity

• Selflessness

• Dependability and trustworthiness

• Initiative

• Tolerance/sensitivity

• Perseverance

FIGURE 17.4   Project managers should learn to focus on character rather than differences.

- Dependability and trustworthiness
- Initiative
- Tolerance/sensitivity
- Perseverance

Race, gender, culture, and the other ways people can be outwardly different are not character traits. They are determined by birth, and do not make people who they are. Character traits, on the other hand, have the element of choice. People can choose to be honest, selfless, dependable, and so on. There are people of every race, both genders, and all cultures who are honest, dependable, and selfless, and, of course, the obverse is also true. It is character, not diversity factors, that makes people who they are.

Look for Common Ground with Others

People who are willing to look beyond differences in others will typically find that there is more common ground between them than differences. This is the case even when the people in question appear at first glance to have little or nothing in common. Even people from different countries who speak different languages, are of different races, and have different cultural backgrounds have more similarities than differences. People tend to share the same desires, hopes, fears, and needs regardless of their diversity-related differences. The key to finding common ground with other people is to look past diversity-related differences and get to know them well enough to see recognize common ground.

Focus on What Really Matters

Project managers who are trying to lead teams in completing projects on time, within budget, and according to specifications soon learn what is really important about the members of project teams. What project managers need from their team members above all else is peak performance. Once this fact is understood, the door to a whole new perspective can open up for even the most prejudiced project manager. This new perspective is gained when an individual comes to realize and is willing to acknowledge that what really matters are not diversity factors such as race, gender, age, and culture, but performance factors such as talent, motivation, attitude, and teamwork.

This is precisely the perspective needed by project managers who work for engineering and technology firms that operate in a globally competitive environment. People who make project teams perform better should be the ones selected to serve on those teams regardless of such factors as race, gender, age, culture, politics, or any of the other diversity factors that make people different. In a competitive environment, project managers who make decisions on the basis of diversity factors that are unrelated to performance will eventually fail.

Relate to People as Individuals

One of the things that can make overcoming prejudice difficult is that prejudiced people can always find someone of another race, gender, age, or culture who fits their stereotypes; someone they can use to validate their prejudice. There usually are *some* people in any group who will actually display the characteristics that prejudiced people attribute to the whole group. This is one of the ways that people develop their prejudices—by attributing characteristics or behaviors they see in a few members of a group to everyone in that group.

An inescapable fact is that whenever people attribute any characteristic to everyone in a given group, they are automatically wrong. People, no matter how they might be grouped, are individuals. As was explained earlier in this chapter, even identical twins can be vastly different in a variety of ways. Consequently, the only dependable, sensible, fair, and equitable way to relate to people is as individuals rather than as members of racial, gender, age, cultural, or political groups. Project managers who understand this have an advantage over those who do not.

In engineering and technology firms there are people of different races, ages, genders, and cultures who are positive, talented, motivated, team players who perform at peak levels. Conversely, there are also those who do not measure up. In either case, their work-related behavior is a product of their character, not their race, age, gender, cultural background, or other diversity-related factors. Project managers who make an effort to relate to others as individuals will find it difficult to maintain the prejudices they have learned. This, in turn, will make them better project managers.

Project Management Scenario 17.2

You need to overcome the prejudice you've learned

Before going away to college, Elizabeth Martin had led an insulated life. Growing up, the people in her neighborhood, the students in her schools, and the friends she spent most of her time with were just like her. Martin's parents and the parents of her friends were all successful white-collar professionals who were able to give their children the material advantages of wealth. They talked alike, dressed alike, drove the same kinds of cars, liked the same kinds of food, were members of the same clubs, and voted for the same political candidates. There was little or no diversity in Elizabeth Martin's life from birth through high school. However, this situation changed radically when she went away to college.

Rather than enroll at the same elite private institution most of the young people from her neighborhood attended, Martin chose to attend her grandfather's alma mater—a public university located in a major city. As a result, overnight Martin went from a life of uniformity to a life of diversity. The starkness of the change in her life was shocking. Martin felt like she had been dropped onto an alien planet. She had never interacted with people of other races, cultures, worldviews, and perspectives. The change was overwhelming.

At first, Martin had difficulty adjusting. In fact, she became so distraught that her grades suffered and she contemplated dropping out and going home. After seeking the help of a member of the university's counseling staff, Martin decided to make some adjustments and persevere. As Martin walked out of the counselor's office, his words echoed in her thoughts: "You need to overcome the prejudice you've learned."

Discussion Questions

In this scenario, Elizabeth Martin grew up in a corner of the world where people were almost uniformly alike. Consequently, when she went away to college the diversity she encountered was so alien to her as to be overwhelming. To her credit, Martin decided to stay in college and work on overcoming her learned prejudice. Have you ever known someone who was uncomfortable in a diverse setting? If so, discuss the situation and its outcome. If Elizabeth Martin was your friend and asked for advice about overcoming the prejudice she learned growing up, what would you tell her?

SUMMARY

America is one of the most diverse countries in the world and this diversity is reflected in the workplace. Consequently, project managers can expect to lead diverse teams. The term diversity as applied in the current context refers to all of the ways that people who serve on project teams can be different. Dealing with these differences in ways that make the team stronger is the responsibility of the project manager. Diversity is a concept that encompasses all of the ways that people can be different. Important diversity-related concepts include prejudice, stereotyping, labeling, discrimination, and tolerance.

Prejudice is a predisposition to adopt negative perceptions about groups of people. People who are prejudiced tend to divide the world into two groups: us and them. People in groups often act out their prejudice in negative and even harmful ways. Stereotyping is a by-product of prejudice. It is the act of generalizing certain characteristics of an individual to all members of a given group. Labeling is an extension of stereotyping in that it involves attributing a certain characteristic to a distinct group of people and then using that characteristic to label all people in that group.

Diversity-related discrimination is a negative concept. It means allowing diversity-related factors such as race, culture, gender, and worldview to influence one's actions and decisions. Discrimination can quickly undermine the effectiveness of a project team and the credibility of project managers. Tolerance is a willingness to interact positively with people in spite of differences based on such factors as race, gender, culture, worldview, political beliefs, and so on. Tolerance is a step in the right direction, but as a concept it is not without problems. The problems with tolerance in project teams are that it is not enough, and it is sometimes misapplied by well-meaning project managers who do not understand what it means. Just tolerating differences in people is not enough to mold them into an effective team. Project managers and their team members need to embrace their differences. On the other hand, well-meaning project managers should never use tolerance as an excuse for ignoring behavior that is counterproductive.

Prejudice is learned behavior that can be unlearned. People are not born prejudiced—they learn to be as they age. Behavior that is learned can be unlearned and replaced with more appropriate behaviors. The key to overcoming learned prejudice and then embracing diversity in people include the following: (1) focus on character rather than differences, (2) look for common ground with others, (3) focus on what really matters rather than differences, and (4) relate to people as individuals.

KEY TERMS AND CONCEPTS

Diversity
Prejudice
Stereotyping
Labeling
Discrimination
Tolerance
Discrimination and teamwork
Discrimination and team performance

Discrimination and morale
Embracing diversity
Tolerating the unacceptable
Learned behavior
Embracing diversity
Diversity-based differences
Character
Common ground

REVIEW QUESTIONS

1. Define the term "diversity" as it relates to project management.
2. Why is it important for project managers to be able to lead diverse teams?
3. Define the term "prejudice."
4. Explain the concept of stereotyping. Give an example.
5. What is meant by labeling? Give an example.
6. Define the term "tolerance."
7. Explain the potential problems associated with tolerance.
8. Explain how discrimination can undermine teamwork.
9. Describe the effect discrimination can have on morale in project teams.
10. Explain what is meant by the phrase "prejudice is learned behavior…"
11. What is meant by embracing diversity? Explain how one goes about embracing diversity.

APPLICATION ACTIVITIES

The following activities may be completed by individual students or by students working in groups:

1. Have you ever experienced or observed an act of prejudice, stereotyping, labeling, or discrimination? Write down or discuss the situation. What happened? Were you the victim or just an observer? If you were the victim, how did the act of prejudice make you feel?
2. Have you ever experienced or observed tolerance being taken too far (i.e., someone tolerating behavior that was counterproductive)? Write down or discuss the situation. What happened? Why do you think the individual who used tolerance as an excuse for overlooking counterproductive behavior did so?
3. Do the research necessary to create a list of stereotypes that existed at one time but either no longer exist or are dying out. Discuss your list with other students.
4. Identify a project manager in an engineering or technology firm and discuss the issue of diversity with that person. Determine if he or she has ever experienced diversity-related problems that affected the morale or performance of the team. If so, how was the situation handled?

Project Managers and Adversity

The projects that engineering and technology professionals work on are like baseball games in that they consist of more than just one inning. In a baseball game, each batter comes to the plate several times and even the most talented player strikes out occasionally. In addition, players occasionally have to hit the dirt when the opposing pitcher brushes them back with a fastball thrown high and inside. But the best players do not let these temporary instances of adversity keep them down or make them quit. Instead, they pick themselves up, brush themselves off, and step right back up to the plate. They persevere no matter how difficult the game becomes and no matter how many times they strike out.

The same thing is true of the best project managers. There are going to be problems over the course of a project. Projects will not always run smoothly or as originally planned. Project managers and their team members will face adversity. This is a fact of life that those who aspire to be project managers should understand. Consequently, a willingness to persevere in times of adversity is critical to success as a project manager. It is one more factor that separates the best project managers from the mediocre.

One of the reasons champions become champions is because they refuse to give up and they never quit. On the other hand, even champions are human. Like all people, during times of adversity, they can become discouraged. But what separates champions from mediocre performers is their steadfast refusal to give in to discouragement. One of the characteristics of the most successful project managers is that they try to view every set back as an opportunity for a comeback. When knocked down, they get back up and try again.

In a competitive environment, engineering and technology firms need project managers who are willing to persevere in spite of obstacles, setbacks, difficulties, and other kinds of adversity. Project managers, in turn, need team members who will do the same. Having project managers who set a consistent example of persevering during times of adversity is essential to project teams that are striving for excellence. Following the project manager's example is how team members learn to persevere. Members of project teams who witness their project manager maintaining a positive can-do attitude during difficult times will have a worthy example to emulate.

Every engineering and technology firm experiences adversity. This, in turn, means that every project team will experience adversity. Consequently, having project managers who can stay positive and persevere in spite of the difficulties is essential to getting projects completed on time, within budget, and according to specifications. Contracts will be revised, important team members will resign, deadlines will change, hardware will break down, software will develop bugs or viruses, corporate mergers and buyouts will occur, shortages of critical materials will happen unexpectedly, work stoppages will occur, the list goes on and on.

There is never a scarcity of adversity in firm's that are trying to compete in today's globally competitive environment, and adversity that affects the organization soon affects its project teams. For this reason, engineering and technology firms need project managers who understand that adversity comes with a gift in its hand. That gift is the opportunity to: (1) grow stronger by persevering and working through the adversity and (2) improve by learning from the experience and being better prepared next time.

DO NOT GIVE UP AND NEVER QUIT

Excellence in project teams is not just a matter of talent. There are plenty of talented people in engineering and technology firms who never achieve excellence because when the work becomes difficult—when they are faced with adversity—they become overwhelmed and just give up. This is unfortunate because the best results are achieved by those who are willing to keep trying—those who refuse to give up and quit.

A good example of the benefits of perseverance is America's Olympic women's softball team. The team had to work long and hard just to get the sport accepted as an Olympic event. Turned down numerous times, team leaders became frustrated, discouraged, and even angry, but they persevered and refused to quit. As a result, they were eventually successful and softball became an Olympic event. Just getting the sport accepted by the International Olympics Committee was a hard-won victory. But America's team did not stop there. Having won the battle for Olympic recognition the softball team went on to win three consecutive gold medals in its sport.

Examples abound of perseverance paying off. Think of the boxer who loses the first nine rounds of the fight, but perseveres and wins by a knockout in the 10th round. Think of all the times baseball games have been won when the last batter in the bottom of the ninth inning knocks a home run. Think of football games won by a field goal as the clocks winds down in overtime. Think of great basketball games won at the buzzer by a desperation shot tossed up from midcourt. In all of these examples, victory went to the team that persevered, that refused to give up and quit. Sometimes victory is just one step beyond the last step you think you can take.

These examples come from the world of sports because that is where the best-known and most dramatic instances of victory through perseverance can be found. But the never-quit-never-give-up attitude applies just as directly in project teams as in sports. Consequently, it is just as important for project managers and their team members to learn to persevere as it is for athletes. There will be times when changes will throw the project team's work into disarray, when so much work will pile up that team members begin to doubt they will ever get it done, and when long hours will result in fatigue and frustration. It is during times such as these that the willingness to persevere distinguishes project teams that excel from those that never get beyond mediocrity.

Perseverance Strategies

Perseverance means staying the course and continuing to try when inclined to give up and quit. Persevering is more of a mental than a physical exercise. In fact, perseverance can be thought of as the mental equivalent of physical stamina. Achieving excellence in project teams requires that team members do a lot of things well, often difficult things. The road to excellence is never easy, nor is it final. Once achieved, excellence must be achieved again and again on each successive project. Market forces, competition, customer demands, supplier problems, the economy, close schedules, tight budgets, demanding specifications, technological developments, and a variety of other factors make completing projects on time, within budget, and according to specifications a never-ending challenge for project managers and their team members.

Consequently, engineering and technology firms need project managers who can bear up under the never-ending pressure of leading project teams and keep going in spite of anxiety, frustration, fear, uncertainty, and fatigue, both mental and physical. The following strategies will help project managers and their team members keep the proper perspective when facing adversity:

- When facing an intractable problem and it seems that everything has been tried without results, remember the lesson of the great inventor, Thomas Alva Edison. In trying to invent such useful products as the storage battery and a durable filament for the light bulb, he failed repeatedly. It is said that it took him almost 25,000 attempts to finally succeed with just these two inventions. But when others might have quit, Edison refused to give up. Instead, he persevered, and finally succeeded. The world can be thankful he did.
- When a team member fails at something important and he is discouraged, remind him that every time he tries something and fails, he is better prepared to succeed the next time. A failed attempt does not become a failure unless the individual involved quits. A failed attempt is just another opportunity to try again better prepared the next time.
- When a team member is unsure of her ability to complete a given assignment or meet a given challenge—when the consequences of failure are frightening—encourage her to focus on what will happen when she succeeds rather than what might happen if she fails. Tell her to stay focused on the ball that is in play, not on the scoreboard. Tell her to play to win rather than to avoid losing.

Project Management Scenario 18.1

My team members give up too easily

Juan Lopez had to struggle to work his way out of poverty. Even as a child he worked to help his mother support their family after his father died in a tragic accident. Then he worked himself through college, secured a good job, and worked hard to climb the career ladder. Consequently, Lopez knows about persevering during hard times. As a result, he has high expectations of his team members when the work on a project becomes difficult. Lopez expects his team members to persevere when they face adversity. Unfortunately, his team members lack his experience at dealing with difficult times.

Lopez's team is going through tough times right now and he is not happy with how his team members are responding. They seem to have developed a defeatist attitude toward their work. They just seem to be going through the motions without really trying. While discussing this situation with a colleague, Lopez shrugged in frustration and said: "My team members give up too easily. Do you have any suggestions for how I can get them to toughen up mentally?"

Discussion Questions

In this scenario, the project manager is an individual who knows all about dealing with hard times. His whole life has been characterized by the need to persevere during difficult times. Consequently, he does not understand people who are unwilling to persevere. Have you ever been in a situation in which the people involved wanted to quit in the face of adversity? If so, explain the circumstances and what eventually happened. If you were Juan Lopez's colleague in this scenario, what advice would you give him?

FACE ADVERSITY AND OVERCOME IT

There is a tendency to think that success comes easily for those who make a name for themselves in a given profession. In reality, this is hardly ever the case. Further, the concept of the overnight success is typically a myth. Few people who succeed do so overnight. Most so-called overnight successes are people who struggled long and hard to finally succeed. This fact applies to project managers too. Successful people are rarely strangers to adversity. An excellent example of facing adversity and overcoming it is Franklin Delano Roosevelt, President of the United States, leading up to and during World War II.

Roosevelt was elected President when the United States was firmly in the grip of the Great Depression. Unemployment was at its highest level in America's history, the nation's banking system had crashed, small businesses were closing daily, people were losing their homes because they could not pay their mortgages, the Midwestern farming states had become a vast dust bowl no longer suited for farming, and many people woke up every morning wondering what, if anything, they would have to eat that day.

Into this bleak picture stepped Franklin Delano Roosevelt, the former governor of New York and a man of great optimism. Soon after taking office he began to use radio broadcasts he called "fireside chats" to reassure Americans that the economy would pick up again and their lives would get better. He spoke to the country in such calm and optimistic terms that Americans began to gain a sense of hope. Then, before any of the President's economic recovery programs had time to produce results, the Japanese attacked Pearl Harbor and the United States found itself embroiled in World War II.

Faced with an even bigger crisis than the Great Depression, President Roosevelt again calmed the anxiety of Americans. In a nationally broadcast address he set a confident and hopeful tone when he said: "All we have to fear is fear itself." Roosevelt used the same calm optimism to face the adversity of World War II that he had used to face the adversity of the Great Depression. The President's optimism during times of unprecedented adversity is a story unto itself, but what is even more instructive about his example is that he held the country together while suffering from a severe case of polio. President Roosevelt—a man who portrayed himself as a robust, vibrant leader—could not even walk without assistance. But he hid this fact so well that many Americans were not even aware of it until after his death.

In spite of having to struggle daily against the increasingly debilitating and painful effects of a crippling disease, Franklin Delano Roosevelt remained calm and optimistic in the face of tremendous adversity. He could have simplified his life and eased his daily pain by agreeing to use in public the wheel chair he used only in private. But the attitudes of Americans toward people in wheelchairs were less enlightened in those days, not to mention the attitudes of America's enemies. Roosevelt knew that the times and circumstances demanded an image of strength. Consequently, he endured incredible pain and inconvenience to provide that example.

Not only did the president fight courageously every day to win the war and revitalize the nation's economy, but he did it while supporting himself, when in public, with heavy metal braces worn under his trousers that bit painfully into his frail and paralyzed legs. The leg braces and the façade of healthful vigor he felt it necessary to maintain only added to the adversity this courageous man faced every day. But in spite of the pain and inconvenience, Roosevelt maintained the necessary façade from the moment he was elected President of the United States until the day his heart finally gave out during a rest-and-recuperation visit to his "Little White House" in Warm Springs, Georgia.

One might wonder why a wealthy President from Hyde Park, New York, would have a vacation retreat in the rural, poverty-stricken town of Warm Springs, Georgia. The reason reveals even more about the incredible courage and perseverance of this unique individual. Using his own money, Roosevelt had founded a treatment center at Warm Springs for polio victims. The naturally occurring warm springs that bubbled up in this tiny, out-of-the-way Georgia town had a stimulating effect on the withered limbs of polio victims, especially children. Although serving as President of the United States during some of America's darkest hours left him no time to undergo the physical therapy offered at his own treatment center, Roosevelt wanted to make sure that future generations of polio victims had a chance to receive the treatments. Thanks to Roosevelt they did.

Project managers facing periods of adversity would do well to remember the example of President Franklin Delano Roosevelt and set a similar example for employees. They should stay positive and optimistic, be calm and reassuring, and focus on solutions rather than problems. No matter how difficult the problems they must face become, project managers should provide a consistent, calm, and resolute example of perseverance that sends the message, "We can get through this." Of course, this kind of example is important at all times, but it is even more important during times of adversity. During times of adversity, project managers are just like emergency room doctors. They have to maintain their calm and their focus no matter what is going on around them because others will take their cues from them.

ACCEPT ADVERSITY AS A NORMAL PART OF LIFE

Completing engineering and technology projects on time, within budget, and according to specifications requires excellence and peak performance from project teams. If achieving excellence and peak performance in project teams were easy, all teams and all project managers would achieve these things. Unfortunately, achieving excellence and peak performance are seldom easy. Rather, they are difficult to achieve and even more difficult to maintain over time.

There are and will always be plenty of obstacles for teams trying to complete projects on time, within budget, and according to specifications—obstacles that inhibit performance. It is when dealing with these inevitable obstacles that a willingness to persevere will separate one project team from another. This is why it is so important that, during hard times, project managers set an example for their team members that says "Adversity is normal—just keep going."

In a competitive environment, adversity is often the rule rather than the exception. Consequently, the project managers and team members most likely to thrive in such an environment are those who can maintain their focus, positive attitude, and solution orientation in spite of the difficulties. People who approach adversity as if it is a one-time event to be survived rather than as the normal state of things to be dealt with on a daily basis will eventually feel overwhelmed and give up. This is important because in a competitive environment, adversity can be the normal state of things. Even in the best of times there will be periods of adversity. This means that project managers and their team members must learn to deal with adversity in a positive way as a normal part of their jobs.

Some strategies that will help project managers and their team members persevere when facing hard times are as follows:

- Understand that adversity is a normal part of work and life.
- Refuse to get caught up in the here and now of the situation—look down the road past the difficulties.
- Focus on solutions rather than the problems.
- Develop a course of action for getting past the difficulties faced and implement it.
- Prepare physically and mentally for the next crisis and accept that there will be one.
- Stay positive and take adversity in stride.

Not only should project managers internalize these rules of thumb, but they should also teach them to their team members through both conversation and example.

ACCEPT THAT LIFE CAN BE UNFAIR

One of the factors that can make it difficult for people to deal with adversity in a positive manner is the seeming unfairness of it. People facing adversity often wonder, "Why is this happening to me? It's not fair." While it might be true that the circumstances in question are unfair, getting caught up in the unfairness of adversity is a mistake. Focusing on the unfairness of a situation can cause people to become discouraged and give up. This is a point that project managers should understand and help their team members understand.

During times of adversity, it is important for project managers to help their team members understand that they are not alone. A bad situation can seem less unfair when others are facing similar circumstances. If the project team is going through hard times, all team members are probably affected. Perhaps in different ways, but they are all affected nonetheless. Making this point to team members can help them see that they are not alone in dealing with adversity.

Watch people in the aftermath of a natural disaster such as an earthquake, tornado, or hurricane. Amazingly, people who have lost everything can be seen working to help others restore a semblance of normalcy to their wrecked lives. Knowing that they are not alone in their trials and that there are others going through the same difficulties bolsters people and gives them hope. But when people feel isolated in their misery, the unfairness of the situation can give rise to frustration, hopelessness, and despair. People who feel this way are inclined to give up and quit.

This is why it is important for project managers to let their team members know that work and life are not always fair, and they should not expect them to be. Further, no matter how hard well-intended people try, work and life are not likely to ever be completely fair. Project managers should endeavor to be fair and equitable with their team members, colleagues, and customers. Engineering and technology firms should adopt policies, procedures, and practices that ensure as much fairness as possible for their personnel. But project managers should never lose sight of the fact that bad things do happen to good people, to good teams, and to good firms. This fact should also be understood by team members. When the concept of unfairness is understood and accepted, team members will be better prepared to deal with the adversity they will inevitably face.

HELP TEAM MEMBERS WHO ARE FACING ADVERSITY

When facing adversity, it is easy to become self-focused. After all, the fear, frustration fatigue, and uncertainty associated with adversity are felt on a deeply personal level. On the other hand, people who respond in a self-focused way to adversity are not likely to get through it. Experience shows over and over that one of the most effective ways to deal with adversity is to help someone else who is facing difficulty. This is a strategy that project managers should adopt themselves and teach to their team members.

In any organization, if one project team is facing hard times there are probably others going through the same or similar troubles. Reaching out to other project managers who are facing budget, schedule, or personnel problems may be the best way there is for project managers to avoid becoming self-focused and giving in to gloom and defeatism. The same is true for team members.

This concept of helping oneself by helping others should be thoroughly understood by project managers who are trying to encourage perseverance among their team members. When a project team goes through hard times, getting its personnel to focus on helping each other is an excellent way to lessen everybody's burden and to teach team members how to deal with adversity. Looming deadlines, demanding executives, tight budgets, personnel issues, and other adversity-inducing events are a normal part of life for project managers and project teams. Consequently, helping team members develop the willingness and ability to persevere during hard times is an essential ingredient in the formula for excellence in project teams. Project managers who wish to develop peak-performing team members should teach them, especially by example, how to persevere in times of adversity.

DEALING WITH MICROMANAGERS WHO CREATE PROBLEMS

An adversity causing challenge that project managers must be prepared to occasionally deal with is the micromanager. A micromanager is a superior in the organization who delegates the job of project management to an individual but then refuses to give him the latitude to do the job. Micromanagers stay involved in the everyday details of the project manager's job to an extent that is inappropriate and even inhibiting. They constantly look over the project managers' shoulders and question every detail of what they do.

Not only are micromanagers annoying, but they can also create problems by undermining the project manager's credibility with team members. When a superior in the organization micromanages a project manager, team members begin to wonder "who is in charge here and who do we report to?" If not handled properly, a micromanager can render a project manager irrelevant, create stress in the team, and damage morale to the point that the team's work suffers.

The first step in dealing with micromanagers is to understand why they micromanage. There are a variety of reasons, and not all of them apply to every micromanager. However, when project managers find themselves dealing with micromanagers, the factors explained below will help in establishing an understanding of them. Once it is understood why an individual micromanages, it is easier to know how to deal with that person.

What follows are common reasons why some people feel compelled to micromanage. People are compelled to micromanage because of the following reasons: (1) they think no one can do the job correctly but them, (2) they struggle with letting go of the type of work they used to do before being promoted, (3) they do not understand the concept of delegation, (4) they have no confidence in the project manager, and 5) they are insecure and fear that the team will perform poorly which will reflect badly on them. All of these factors apply to some micromanagers, but more commonly just one or two apply.

Strategies for Overcoming Micromanagement

In general terms, project managers who work for a micromanager should be patient, refuse to take the intrusiveness personally, and understand that the micromanager is an individual who is dealing with issues that might be deep-seated and personal. It can take time to cure a micromanager of his propensities. More specifically, project managers who work for a micromanager—who thinks only he or she can do the job correctly—should patiently demonstrate that they can do the job. Keep the individual informed of the team's progress on a regular basis. Show that the project is moving forward on time, within budget, and according to specifications. When a problem arises, project managers should explain the situation to the micromanager and describe what they plan to do about it.

Project managers who work for a micromanager who struggles with letting go of the kind of work he or she once did before being promoted should: (1) make a point of involving him or her in ways that will satisfy his or her need to stay in touch with the work (e.g., periodic meetings, requests for advice, frequent e-mail updates) and (2) involve him or her in the problem-solving process when unexpected difficulties arise. When this kind of micromanager realizes that he or she is going to be involved and engaged in appropriate ways, the need that is driving him or her to micromanage will be satisfied.

Project managers who work for micromanagers who do not understand the concept of delegation have an upward-mentoring challenge. Mentoring is usually viewed as more senior, more experienced people helping less experienced people develop the knowledge, skills, and attitudes they need to succeed in their jobs. A less known kind of mentoring is upward mentoring: when a subordinate mentors a supervisor. Upward mentoring is a concept that should be labeled "handle with care." Project managers who approach their supervisor and say "It is obvious you don't understand the concept of delegation so I am going to teach you" are not likely to get a good reception. Consequently, a more subtle approach is in order.

An effective way to approach upward mentoring is to begin by giving the micromanager the following message: "I appreciate your help, but know how busy you are. You can delegate this task to me and I will keep you informed of progress." Once the micromanager is able to see that the project manager is getting the work done, and keeping him informed, he will begin to understand how delegation works. However, patience is in order because it can take micromanagers' time to break old habits.

When a project manager senses that a supervisor is micromanaging her because of a lack of confidence, the challenge is straightforward and obvious: demonstrate competence. Concentrate on getting the work of the project done on time, within budget, and according to specifications. In other words, project managers in this situation should simply prove to the micromanager that they can do the job well. This will happen as a matter of course over the duration of the project. It might take completing an entire project or even more than one project before the micromanager becomes a believer, but with patience and persistence it will happen.

A project manager who works for a micromanager who is afraid that the project team will perform poorly and reflect badly on him is operating out of fear. Consequently, the best way to cure such an individual of micromanaging is to eliminate the fear factor. In the previous case, the micromanager had no confidence in the project manager. In this case the lack of confidence is in the team, but the solution is the same: demonstrate competence. Focus the team on getting the work of the project done on time, within budget, and according to specifications. When problem arise, develop effective solutions and implement them. Keep the micromanager fully informed at every step in the process. Do these things and the micromanager's fear will eventually be eliminated.

Project Management Scenario 18.2

My team is not handling this adversity very well

Janice Carter has been through tough times before at ABC, Inc. Consequently, the current hard times are just the latest she has had to deal with. Because the economy has been in a recession for an extended period of time, ABC is accepting almost any contract it can get. The company is getting enough work to keep it going, but taking on any and all projects is creating problems. Budgets and schedules are tight. Further, due to layoffs, personnel are scarce. One of the effects of the challenging economic times ABC is going through is that if employees resign or retire, they are not replaced. Carter has recently lost two members of her project team. Those who remain are expected to do their work in addition to that of the departed team members. She has never been expected to do so much with so little.

Carter is holding up well in view of the circumstances, but her team members are not. Her team members are beginning to show signs of stress, frustration, and fatigue. Carter has tried several strategies for bolstering the morale in her team, but without much success. She is beginning to worry. Finally, out of desperation Carter approached a more senior project manager—one who had mentored her earlier in her career. When asked how she was holding up, Carter responded "I'm doing fine but my team is not handling this adversity very well. I need some help. What can I do to give them hope that we can handle this situation until things turn around?"

Discussion Questions

In this scenario, Janice Carter is handling some tough times well but the members of her project team are not. Have you ever worked with people who did not handle adversity well? If so, describe the circumstances and how people responded to them. If you were Janice Carter's mentor, what advice would you give her concerning bolstering the morale of her team members?

SUMMARY

There is never a scarcity of adversity in organizations that are trying to compete in today's global environment, and adversity that affects the engineering and technology firm soon affects its project teams. For this reason, engineering and technology firms need project managers who understand that adversity comes with a gift in its hand. That gift is the opportunity to grow stronger by persevering and weathering the storm and getting better by learning from the experience so that one is better prepared next time.

Excellence in project teams is not just a matter of talent. There are plenty of talented people in organizations who never achieve excellence because when the work becomes difficult—when they are faced with adversity—they become overwhelmed and just give up. This is unfortunate because the best results are achieved by those who are willing to keep trying—those who refuse to give up and quit.

Perseverance means hanging in there a little longer when the natural inclination is to give up and quit. Persevering is more of a mental than a physical challenge. In fact, perseverance can be thought of as the mental equivalent of physical stamina. Strategies for persevering include: (1) thinking of others who have persevered and emulating their examples, (2) remembering that every time one fails at something she is better prepared to succeed the next time, and (3) focusing on the benefits of success rather than the consequences of failure.

Those who approach adversity as a one-time event will be unprepared when the next problem arises, as it surely will. Strategies for facing adversity include the following: (1) understand that adversity is a normal part of work and life, (2) refuse to get caught up in the here and now of the situation—look down the road past the difficulties, (3) focus on solutions rather than problems, (4) develop a course of action for getting past the difficulties and implement it, (5) prepare mentally and physically for the next round of adversity, and 6) stay positive and take adversity in stride.

It is important for project managers and their team members to understand and accept that life is not always fair. Getting hung up on the unfairness of life is a mistake. One of the best ways to respond to adversity is to find someone who is going through hard times too and help that person. Helping someone else who is facing difficulties will keep one from becoming self-focused and giving in to self-pity.

Micromanagers in an engineering and technology firm can cause difficulties for project managers. This is a common enough problem that project managers need to understand why some people feel compelled to micromanage as well as how to deal with micromanagers. People who micromanage do so because they: (1) think no one can do the job right but them, (2) struggle with letting go of work they used to do, (3) do not understand the concept of delegation, (4) have no confidence in the project manager, or (5) are insecure and fear the team will perform poorly and reflect badly on them. Project managers who report to a micromanager should be patient and refuse to give in to frustration. Rather, they should attempt to determine which of these reasons apply in their case and then do what is necessary to overcome that reason.

KEY TERMS AND CONCEPTS

Adversity	Focus on solutions
Do not give up	Stay positive
Never quit	Take adversity in stride
Perseverance strategies	Life can be unfair
Face adversity	Help team members
Adversity is normal	Micromanagers

REVIEW QUESTIONS

1. What is one of the most important factors that will help an individual become a champion?
2. Why do organizations need project managers who are willing to persevere?
3. Explain why talent alone will not produce excellence in a project team.
4. What is meant by the term "perseverance"? Give an example.
5. Explain three strategies for persevering in times of adversity.
6. What lesson does the life of President Franklin Delano Roosevelt teach about dealing with difficulties?
7. Explain what is meant by the phrase "adversity is a normal part of life."
8. List six strategies that will help project managers and their team members persevere when facing hard times.
9. Why is it important for project managers and their team members to understand that life is not always fair?
10. How can helping someone else who is facing difficulty help one to overcome adversity?
11. Explain the various reasons why some people become micromanagers.
12. How should project managers deal with superiors in their organizations who are micromanagers?

APPLICATION ACTIVITIES

The following activities may be completed by individual students or by students working in groups:

1. Assume that you are a new project manager. On your first project the team has run into several difficulties. Some of these difficulties are listed as follows: (1) contract revisions have shortened an already tight schedule by two months and (2) an important member of the project team has been injured in an automobile accident and will be out of work for six months. The team is not responding well to the adversity. Frustration is setting in, tempers are getting short, and complaints are mounting. Explain how you will go about leading the team to persevere through the hard times.

2. Assume you are an experienced project manager, but for the first time you have to report to a micromanager who just does not seem to be able to let go of the kind of work he used to do. His micromanaging is causing problems in your team. Team members are beginning to question who is in charge. Explain how you will go about reclaiming control of your team and project.

INDEX

Note: Page numbers followed by "*f*" indicate figures; those followed by "*t*" indicate tables